Processed Foods and the Consumer

*Published with assistance from
the Roger E. Joseph Memorial Fund
for greater understanding of
public affairs, a cause in which
Roger Joseph believed*

Processed Foods
and
the Consumer

*Additives, Labeling,
Standards,
and
Nutrition*

by Vernal S. Packard, Jr.

UNIVERSITY OF MINNESOTA PRESS □ MINNEAPOLIS

Library of Congress Catalog Card Number 75-32670

ISBN 0-8166-0784-2

Third Printing, 1978

Preface

This is a book about food—what's in food, why it's there, and how it came to be there. The volume is an overview of the food industry focusing on many of the major issues now being raised about food safety, ecology and the environment, and mankind's ability to feed itself. It deals with regulations governing food producers and processors; it relates supermarket foods to the latest nutritional recommendations of the Food and Nutrition Board, National Academy of Sciences/National Research Council; it provides a basis for understanding food labels and the use of label information in making purchases of food and vitamin supplements; and it brings together under one cover the health-related issues of food additives and nutrition.

If I were to point to one objective of this work, it would be to guide student and consumer alike through the maze of food ingredients, regulations, and standards in order to make as clear as present knowledge allows the critical issues confronting government and citizens in the development of a national food policy, a policy that must consider human health and nutrition—and food for the world's hungry. Like scattered bits of flint and steel, information pierces our sensitivities from a host of sources when an additive, a pesticide, or some disease is associated with the food we eat. There's a flash, an eyecatcher that holds our attention momentarily, then is gone; but usually a doubt lingers, a question is unanswered, there is a hesitance the next time we sit down to eat. Even for those who by profession find themselves teaching or doing research in the food sciences, there is seldom sufficient time to track down pertinent literature on any but a few of the questions that assail

them. This book may not provide all the answers, but it is an attempt to as-
semble the scattered details and to weave into some kind of logical sequence
what appears to be the scientific consensus at this time. Obviously new research may lead to conclusions and views different from those now seeming
to materialize. Likewise it is inherent in a work such as this that details will
change, additives be developed or discarded, food standards be revised, new
food names be conjured up to describe the products of the future. Yet it is
equally certain that questions we now seek to answer—about food's relation
to health, about the use of pesticides and drugs, about the relationship and
responsibility of consumer and government in dealing wisely with this, our
greatest of natural resources—will remain with us for the foreseeable future. If
there is value in viewing a problem more or less in its totality, and with an eye
to current research findings, may this book serve that purpose. Especially is it
dedicated to the beginning student, the homemaker, the extension specialist
and home agent, the supermarket home economist, and the institutional dietitian, to those persons who, each day, strive to understand food, to communicate what they learn about it, and to use wisely this resource so abundantly
given to our much-blessed nation.

V. S. P.

July 1, 1975

Contents

Processed Foods and the Consumer

I

Food Definitions and Standards

If a regulation prescribing a definition and standard of identity for a food has been promulgated under section 401 of the act and the name therein specified for the food is used in other regulations under section 401 or any other provision of the act, such name means the food which conforms to such definition and standard, except as otherwise specifically provided in such other regulation.

FDA definition

A standard of identity is much like a homemaker's recipe. It prescribes the kind and amount of ingredients and, in some instances, certain cooking procedures. Sometimes, as in home cookery, the standard allows the option of adding given ingredients; you can use them or not, as you prefer. If the recipe is followed, if no unspecified ingredients are included, if none are left out, if the processing steps are followed to the letter, the product can be called by its food name, i.e., breakfast cocoa, cherry pie, orange juice, milk, canned figs, pasteurized process cheese, artificially sweetened fruit jelly, and on and on. Many foods have standards of identity. Such standards offer reasonable assurance that a product is what you believe it to be, that each and every time you purchase a can or jar of it, a carton, pound, or ounce of it, the food, like batches of your favorite recipe, will always be the same. Legally, there can be no tinkering with the recipe (standard), no omission of an ingredient (at the prescribed level) because the processor happens to be out of it at the moment. And although one can take comfort in that fact, it is important to note that standards, if they guarantee flavor, consistency, and composition, also restrict innovation. A processor who deals in food with a standard of identity has little incentive to do research yielding new versions of that particular product—even if "recipe" changes would offer distinct advantages over the traditional item. Some segments of the food industry have put forth year after year the same product without in the least improving its flavor, texture, or nutritive value. Food standards are desirable; they ensure uniformity and freedom from deception. But there is, as usual, another side to the coin.

How a Standard of Identity Comes About

Before mayonnaise became readily available on the market shelf, many homemakers had favorite recipes for making this product. Some recipes called for two egg yolks, others three (or, if you liked it extra rich, maybe you would toss in four). Vinegar and lemon juice were added in particular proportions—not necessarily the same as your neighbor's who made his or hers too tart. You probably also used a vegetable oil of your choice depending upon cost and availability; you enhanced the recipe with spices, including just the right amount of paprika. Your family had a sweet tooth and in went an extra half cup of sugar, as you blended all the ingredients together to make your own special mayonnaise dressing.

Then one day, because all your friends told you how much they liked your mayonnaise and perhaps because of some enterprising streak in your nature, or simply because you needed the money, you may have decided to bottle it up and sell it. A good many products got started on the market in just this way, by pioneering business folks taking to the streets with something to sell.

As years went by, your business grew and you opened a plant which employed people to process and bottle mayonnaise; now you could sell your product in neighboring communities and perhaps even in different states. But lo and behold, others too had gotten the same idea and you found yourself bumping heads with a competitor whose product was not exactly the same as yours. (Doubtless to you the competitor's mayonnaise appeared inferior; it didn't have quite the right flavor and you were sure it had fewer egg yolks than yours.)

Still more people got into the business and around the country consumers in grocery stores began asking for mayonnaise, a product they knew from past experience. If the mayonnaise they purchased didn't come up to their expectations, they shopped around for a brand that was more appealing. In the meantime brands were proliferating. Some were the standard variety (or what consumers considered the standard to be), some were not. Certain mayonnaises took on innovative characteristics—a slightly different flavor, consistency, or color. There were buyers for both the standard and the innovative brands, which meant that the processor who altered his formulation—his recipe—had every right to believe his was a marketable "mayonnaise." Possibly he was using a salad oil produced nearby that was cheaper, though equally as good as another oil available only from a distance and at higher cost. Customers bought his mayonnaise; some considered it a premium product.

The time would come of course when there were so many different kinds of "mayonnaise" available that no one could be exactly sure what was being bought. One brand might contain a large percentage of egg yolks, another less. The cost of ingredients (and we should make no bones about it) would in all likelihood play an important role in the kind and amount of ingredients used. The almost inevitable possibility could not be discounted that someone would discover that a mayonnaise could be processed with fewer yolks and a fair sprinkling of turmeric or saffron which would impart a yellow-orange yolklike color to the product. Deceptive? Yes, if done strictly to deceive. But what if this mayonnaise were sold as a product low in egg yolk solids for those persons interested in a diet low in egg yolks (which are high in cholesterol). Well, that's not entirely deceptive. The seller has informed the buyer that the mayonnaise is low in egg yolks. But is this product *mayonnaise* as most persons know it to be? No, not at all. It is a mayonnaiselike product, perhaps valuable in its own right, but it is not mayonnaise.

Since your firm was an old, time-honored processor of mayonnaise, and since so many mayonnaise products were now on the market that consumers were becoming confused and possibly deceived, you decided to take action. You addressed a letter to the commissioner of food and drugs explaining the problem and setting forth examples and evidence that your interests and the interests of consumers would be better served if mayonnaise had a standard of identity, a uniform recipe of manufacture so that all mayonnaise sold in the country would be essentially the same. (Actually, any interested individual can so petition the commissioner, but ultimately he must be prepared to substantiate his petition, with evidence, at a public hearing.)

Seeing merit in the case (and obviously aware from a number of sources that a problem exists), the commissioner publishes a proposal in the Federal Register. Keep in mind that it is, at this stage, only a proposal. Other interested persons—in this instance mayonnaise processors *and* consumers—are then given an opportunity to voice their opinions, pro or con. After taking all comments into consideration, the commissioner, if he thinks the facts warrant it, will publish an order in the Federal Register stating his intent to establish a standard of identity for mayonnaise. The recipe (ingredients and processing requirements) is spelled out, and further opportunity is provided for comment. A manufacturer can plead his case for more or fewer eggs, for certain kinds of salad oil, spices, and so on. Consumers, too, can enter the debate, and eventually a public hearing is held. Ultimately the commissioner of food and drugs decides the merits of the issue. If evidence seems to weigh in favor of a standard of identity, a final order is placed in the Federal Register, along with an effective date, the date upon which the order becomes law and all processors shipping interstate are expected to comply. Later a confirming notice is published, at which time, legally, the product is considered standardized; it has a standard of identity.

Under similar sets of circumstances, many foods have been assigned a standard of identity over the years. Whether you buy mayonnaise in Maine or Mississippi, in a supermarket or from the corner grocer, the product, if it is shipped interstate, is essentially the same. In practice this means a food meets legal requirements (conforms to the standard) if (1) no ingredient is used over and above the ones listed in the standard, (2) no ingredient is left out, and (3) quantities (proportions) of ingredients (where specified) conform. In reality, then, the recipe is fixed. Mayonnaise is mayonnaise not by whim or fancy or taste, but by legal definition.

To Whip Up a Batch of Mayonnaise

A standard of identity fixes the formulation of a food product. Only those ingredients set forth in the standard may be utilized and the only excepted ingredients, if analysis were made to determine the composition of the food, are incidental additives (minor contaminants) arising from allowable use in or on one or more of the ingredients. Even then the incidental additive must be granted clearance from label declaration. In effect, then, the processor of a food is responsible for its content down to and including minuscule amounts of chemicals that may enter his product from one or more ingredients he purchases from suppliers. (See Chapter III for a discussion of incidental additives.)

An intentional additive, an ingredient added for some specific functional purpose—to smooth, to thicken, to clarify, or to leaven—can be part of the original standard of identity or may be added later, by amendment. All such additives must be cleared according to a set of preestablished rules which prescribe methods of evaluating usefulness, amounts required, and safety. These aspects will be dealt with later, but the point here is that a food standard is all-inclusive. Literally all ingredients are recognized and accounted for, and the processor is bound, under threat of prosecution, to these ingredients solely. For mayonnaise, the standard of identity (with explanatory notations) is shown in Figure 1.

Figure 1. Standards of Identity for Mayonnaise (from *Code of Federal Regulations*, 1974, Title 21, pts. 10-129, pp. 125-126) and Explanatory Notations

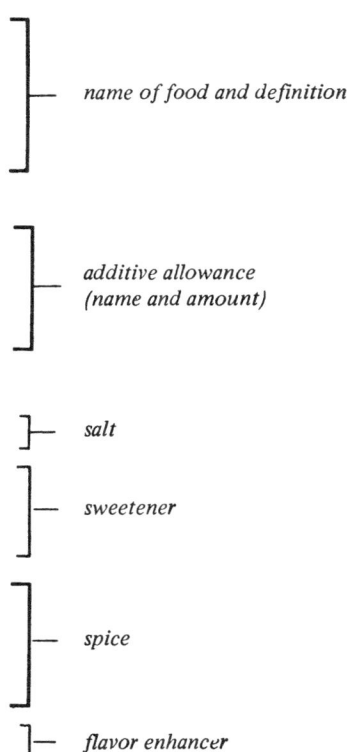

25.1 Mayonnaise, mayonnaise dressing; identity: label statement of optional ingredients.

(a) Mayonnaise, mayonnaise dressing, is the emulsified semisolid food prepared from edible vegetable oil and one or both of the acidifying ingredients specified in paragraph (b) of this section, and one or more of the egg-yolk-containing ingredients specified in paragraph (c) of this section. For the purposes of this section, the term "edible vegetable oil" includes salad oil that may contain not more than 0.125 percent by weight of oxystearin to inhibit crystallization as provided in the food additive regulation in 121.1016 of this chapter. Mayonnaise may be seasoned or flavored with one or more of the following ingredients:

(1) Salt.

(2) Sugar, dextrose, corn sirup, invert sugar sirup, nondiastatic maltose sirup, glucose sirup, honey. The foregoing sweetening ingredients may be used in sirup or dried form.

(3) mustard, paprika, other spice, or any spice oil or spice extract, except that no turmeric or saffron is used and no spice oil or spice extract is used which imparts to the mayonnaise a color simulating the color imparted by egg yolk.

(4) Monosodium glutamate.

(5) Any suitable, harmless food sea-

name of food and definition

additive allowance (name and amount)

salt

sweetener

spice

flavor enhancer

soning or flavoring (other than imitations), provided it does not impart to the mayonnaise a color simulating the color imparted by egg yolk.

Mayonnaise may contain one or both of the optional ingredients specified in paragraph (d) of this section, subject to the conditions prescribed in that paragraph. Mayonnaise may be mixed and packed in an atmosphere in which air is replaced in whole or in part by carbon dioxide or nitrogen. Mayonnaise contains not less than 65 percent by weight of vegetable oil.

(b) The acidifying ingredients referred to in paragraph (a) of this section are:

(1) Any vinegar or any vinegar diluted with water to an acidity, calculated as acetic acid, of not less than 2½ percent by weight, or any such vinegar or diluted vinegar mixed with the additional optional acidifying ingredient citric acid, but in any such mixture the weight of citric acid is not greater than 25 percent of the weight of the acids of the vinegar or diluted vinegar calculated as acetic acid. For the purpose of this paragraph, any blend of two or more vinegars is considered to be a vinegar.

(2) Lemon juice or lime juice or both or any such juice in frozen, canned, concentrated, or dried form, or any one or more of these diluted with water to an acidity, calculated as citric acid, of not less than 2½ percent by weight.

(c) The egg yolk-containing ingredients referred to in paragraph (a) of this section are: Liquid egg yolks, frozen egg yolks, dried egg yolks, liquid whole eggs, frozen whole eggs, dried whole eggs, or any one or more of the foregoing with liquid egg white or frozen egg white.

(d) Mayonnaise may contain calcium disodium EDTA (calcium disodium ethylenediaminetetraacetate) or disodium EDTA (disodium ethylenediaminetetraacetate), singly or in combination. The

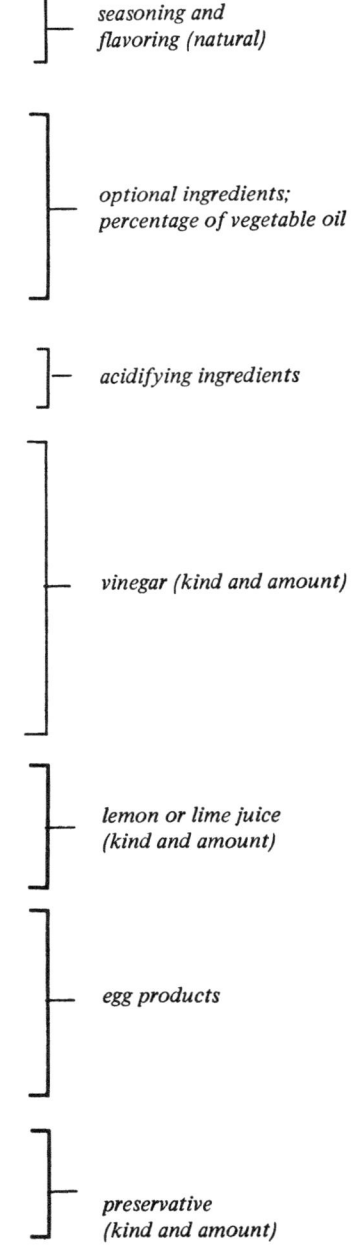

seasoning and flavoring (natural)

optional ingredients; percentage of vegetable oil

acidifying ingredients

vinegar (kind and amount)

lemon or lime juice (kind and amount)

egg products

preservative (kind and amount)

quantity of such added ingredient or combination does not exceed 75 parts per million by weight of the finished food.

(e) (1) When the additional optional acidifying ingredient as provided in paragraph (b) (1) of this section is used, the label shall bear the statement "citric acid added" or "with added citric acid."

(2) If mayonnaise contains calcium disodium EDTA or disodium EDTA or both, the label shall bear the statement "- - - - - - - - added to protect flavor" or "- - - - - - - - added as a preservative," the blank being filled in with the words "calcium disodium EDTA" or "disodium EDTA" or both, as appropriate.

(3) Wherever the name "mayonnaise" or "mayonnaise dressing" appears on the label so conspicuously as to be easily seen under customary conditions of purchase, the statements specified in this paragraph, showing the optional ingredients present, shall immediately and conspicuously precede or follow such name, without intervening written, printed, or graphic matter.

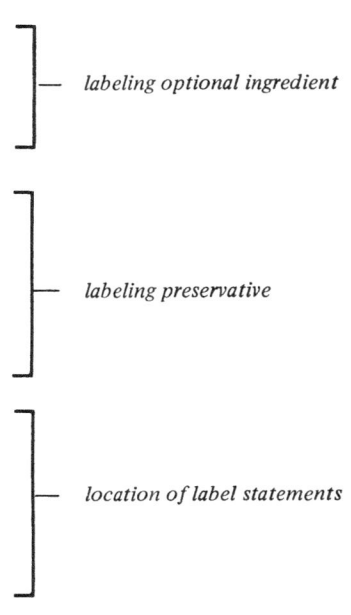

— *labeling optional ingredient*

— *labeling preservative*

— *location of label statements*

To Bake a Cherry Pie

Like mayonnaise, cherry pie has a standard of identity that spells out kinds and amounts of ingredients and leaves little to the culinary imagination. So it is with all foods for which standards have been promulgated. But since pie baking is not yet a lost art, the standard of identity for cherry pie is included here as evidence of government controls set forth to protect the consumer. A cherry pie—that sweet springtime dessert—under government regulations bakes out in dry legalistic terms as shown in Figure 2.

Figure 2. Standard of Identity for Frozen Cherry Pie (from *Code of Federal Regulations*, 1974, Title 21, pts. 10-129, pp. 192-193) and Explanatory Notations

28.1 Frozen cherry pie; identity; label statement of optional ingredients.

(a) Frozen cherry pie (excluding baked and then frozen) is the food prepared by incorporating in a filling contained in a pastry shell mature, pitted,

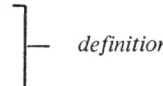

— *definition*

stemmed cherries that are fresh, frozen, and/or canned. The top of the pie may be open or it may be wholly or partly covered with pastry or other suitable topping. Filling, pastry, and topping components of the food consist of optional ingredients as prescribed by paragraph (b) of this section. The finished food is frozen.

(b) The optional ingredients referred to in paragraph (a) of this section consist of suitable substances that are not food additives as defined in section 201 (s) of the Federal Food, Drug, and Cosmetic Act or color additives as defined in section 201(t) of the act; or if they are food additives or color additives as so defined, they are used in conformity with regulations established pursuant to section 409 or 706 of the act. Ingredients that perform a useful function in the formulation of the filling, pastry, and topping components, when used in amounts reasonably required to accomplish their intended effect, are regarded as suitable except that artificial sweeteners are not suitable ingredients of frozen cherry pie.

(c) The name of the food for which a definition and standard of identity is established by this section is frozen cherry pie; however, if the maximum diameter of the food (measured across opposite outside edges of the pastry shell) is not more than 4 inches, the food alternatively may be designated by the name frozen cherry tart. The word "frozen" may be omitted from the name on the label if such omission is not misleading.

(d) (1) Each of the optional ingredients used shall be declared on the label as required by the applicable sections of part 1 of this chapter.

(2) The label shall not bear any misleading pictorial representation of the cherries in the pie.

28.2   Frozen cherry pie; quality; label statement of substandard quality.

(a) The standard of quality for frozen

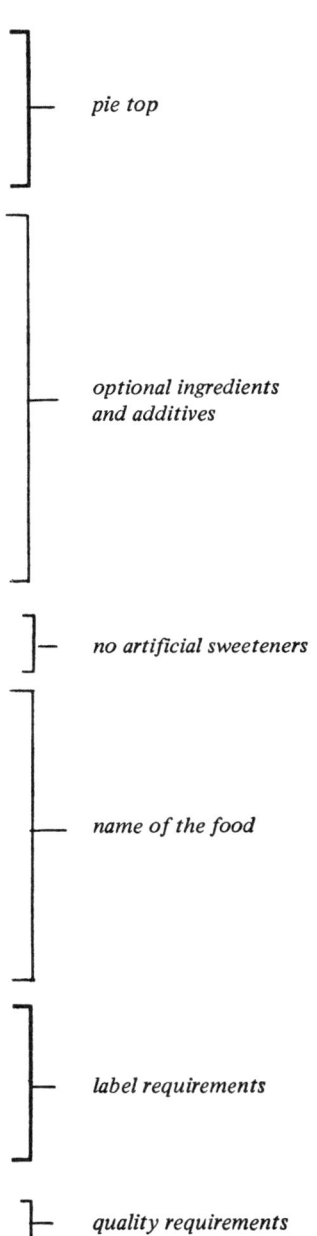

pie top

optional ingredients and additives

no artificial sweeteners

name of the food

label requirements

quality requirements

cherry pie is as follows:

(1) The fruit content of the pie is such that the weight of the washed and drained cherry content is not less than 25 percent of the weight of the pie when determined by the procedure prescribed by paragraph (b) of this section.

amount of cherries

(2) Not more than 15 percent by count of the cherries in the pie are blemished with scab, hail injury, discoloration, scar tissue, or other abnormality. A cherry showing skin discoloration (other than scald), having an aggregate area exceeding that of a circle nine thirty-seconds of an inch in diameter is considered to be blemished. A cherry showing discoloration of any area extending into the fruit tissue is also considered to be blemished.

blemish limitations, definition

(b) Compliance with the requirement for the weight of the washed and drained cherry content of the pie, as prescribed by paragraph (a) (1) of this section, is determined by the following procedure:

(1) Select a random sample from a lot:

(i) At least 24 containers if they bear a weight declaration of 16 ounces or less.

(ii) Enough containers to provide a total quantity of declared weight of at least 24 pounds if they bear a weight declaration of more than 16 ounces.

(2) Determine net weight of each frozen pie.

(3) Temper the pie until the top crust can be removed.

(4) Remove the filling and cherries from the pie and transfer to the surface of a previously weighed 12-inch diameter U.S. No. 8 sieve (0.094-inch openings) stacked on a U.S. No. 20 sieve (0.033-inch openings).

procedure for determining compliance

(5) Distribute evenly over the surface and wash with a gentle spray of water at 70°-75° F. to free the cherries and cherry fragments from the adhering material.

(6) Remove the U.S. No. 8 sieve and examine the U.S. No. 20 sieve and trans-

fer all cherry fragments to the U.S. No. 8 sieve.

(7) Drain the cherry contents on the No. 8 sieve for 2 minutes in an inclined position (15°-30° slope). Weigh the U.S. No. 8 sieve and the washed and drained cherries to the nearest 0.01 ounce.

(8) The weight of the washed and drained cherries is the weight of the sieve and the cherry material less the weight of the sieve. Calculate the percent of the cherry content of each pie with the following formula, and then calculate the average percent of the entire random sample:

Percent of the cherry content of the pie =
$$\frac{\text{Weight of washed and drained cherries}}{\text{Net weight of pie}} \times 100$$

must equal 25%

(c) If the quality of the frozen cherry pie falls below the standard of quality prescribed by paragraph (a) of this section, the label shall bear the general statement of substandard quality specified in 10.7(a) of this chapter, in the manner and form specified therein; but in lieu of the words prescribed for the second line inside the rectangle, the label may bear the alternative statement "Below standard in quality——   ," the blank being filled in with the following words, as applicable: "too few cherries," or "blemished cherries." Such alternative statement shall immediately and conspicuously precede or follow, without intervening written, printed, or graphic matter, the name of the food as prescribed by 23.1.

label statement,
substandard quality

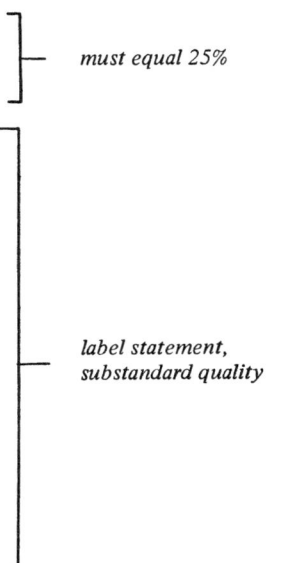

Some Food Categories Covered by Standards of Identity

Food standards cover a wide variety of foods. Most basic foodstuffs come under their protective umbrellas. The standards change from time to time, amendments bring them up to date in accordance with the latest technologies and ingredient sources, new subcategories are added to account for innovations (new recipes of old standards), an additive may be allowed to enhance flavor or texture or to prevent spoilage, a whole new category may be added, but the basic food standards, like an old house which has been renovated and

and remodeled, remain unaltered. Table 1 is provided to acquaint the non-technical reader with categories of foods for which standards are prescribed (and to offer some assurance that consumer interests are being served). Within each category are a number of different, though related food products. For example, under canned vegetables standards are given for peas, green beans, wax beans, corn, mushrooms, and other vegetables; under bakery products are included white bread and rolls, enriched bread, milk bread, raisin bread, and whole wheat bread (and bread products). Market basket foods, for the most part, are what they purport to be within very strict limits of ingredient usage and process control. Meat and poultry products, though under the jurisdiction of the United States Department of Agriculture (USDA), not the FDA (Food and Drug Administration), are likewise rigidly controlled. In Appendix I may be found a listing of specific standardized foods with the more significant requirements isolated and identified. But note that in all categories and for all standards, basic coverage calls for (1) kinds and amounts of ingredients (compositional requirements), (2) quality standards, and (3) standards of fill of containers. The consumer is assured, in a sense, commonality of recipe, quality, and amount of product in a given package. Where standards permit fortification, fixed levels of vitamins and/or minerals—the nutritional makeup—are also guaranteed. Conversely, by preventing (or not providing for) fortification, food standards minimize risk of excessive nutrient intake. Typical of this latter consideration were questions raised during debate over the enrichment of flour and bread. For a time, an increase in the content of iron was ordered for these products. Then, following objections by several medical doctors, the order was stayed and the levels of iron fortification reverted to the original lower standards. The debate continues, but the objections still stand: higher dietary levels of iron would (1) create a health hazard for persons suffering from iron storage disorders, (2) mask diagnoses signaled by anemia (intestinal cancer, for one), (3) possibly contribute to the development of Parkinson's disease in individuals with high stores of brain iron, and (4) possibly cause elevated hematocrit and hemoglobin, which, by one study, has been found associated with early mortality. On the other side of the ledger, iron remains to this day a nutrient as likely as any to be in short supply in many diets; some fortification would appear justified. However, there is more to nutrient utilization by the body than the mere presence of a nutrient (as a fortification substance) in food. Biological availability, especially as related to iron, is of utmost significance. This aspect is discussed in a later chapter.

At the same time that iron fortification standards were reduced to original

Table 1. Major Categories of Foods Covered by Standards of Identity

Cocoa and cocoa products	Fruit butters, jellies, preserves, and
Cereal flours and related products	related products
Macaroni and noodle products	Frozen fruits
Bakery products	Nonalcoholic carbonated beverages
Milk and cream	Shellfish
Cheeses and cheese products	Fish
Frozen desserts	Eggs and egg products
Food flavorings	Margarine
Dressings for food (salad dressings,	Nut products
mayonnaise, french dressing)	Nutritive sweeteners
Table syrups	Frozen vegetables
Canned fruits and fruit juices	Canned vegetables
Fruit pies	Tomato products

Source: Adapted in large part from *Code of Federal Regulations*. 1974, Title 21, pts. 10-129, pp. 12-254.

levels, provisions for vitamin D fortification were deleted. Again evidence suggests that excessive intake of this common "sunshine" vitamin can result in serious health hazards, but more about this later. For now, suffice it to say that food standards have significant implications for human health.

What Food Standards Mean to the Consumer

You may search food labels from top to bottom and no word will you find conveying the message that the food contained therein is regulated under a standard of identity (see Table 1 and Appendix I). Significant requirements of consumer interest have been extracted for the appendix information.

But it is not enough simply to list these products; some comments should also be offered. First, concerning the law. At present a concerted effort is underway to amend the law and make it mandatory to declare all ingredients in all foods, both standardized and nonstandardized. And there is some justification for such a move—*if* we train ourselves to read the labels. Few of us do at present, but for those with specific nutritional deficiencies or sensitivities, ingredient labeling can be most important. For allergy victims this is especially true. Therefore it is essential—very much so—to understand that *no statutory authority exists to require declaration of mandatory ingredients in standardized foods.* If the food standard says, in essence, thou shalt have egg yolk solids in mayonnaise, the processor, while he is legally obliged to provide those solids, is not under legal obligation to declare them on the label. On the other hand, if a food standard allows *optional ingredients*, regulatory authority (though not always imposed) exists to demand their declaration on the label when such ingredients are used. Some standardized foods consist entirely of optional ingredients; full ingredient labeling can be required. For those processors who voluntarily list all ingredients, both optional *and* mandatory, the format for listing them is standardized.

The "Standards" in Standards of Identity

So much for the law. Now, in a general way, let us consider those classes of foods for which standards of identity have been adopted and some of the major requirements such standards impose.

Cocoa products. That consumers will not be deceived, there are indeed milk solids, to a level of at least 12 percent, in milk chocolate. Skim milk chocolate has in it less milkfat than milk chocolate. Breakfast cocoa is a high fat product with not less than 22 percent *cacao* fat (not milkfat). Medium fat cocoa has less fat than breakfast cocoa—between 10 and 22 percent. Lowfat cocoa, while lower in fat content than either medium or breakfast cocoa, may still be a product, comparatively speaking, high in fat. Lowfat milk, for example, contains no more than 2 percent fat. Lowfat cocoa may, under the law, contain nearly 10 percent fat.

Wheat flour and related products. As previously mentioned, scientific concern has been expressed concerning the overfortification of enriched flour products. Nonetheless, if flour or bread is labeled *enriched*, you can be sure it is supplied with certain nutrients to levels higher than those ordinarily found in unenriched products. Durum flour must be made from durum wheat, self-rising flour must rise of itself. If it is *enriched* self-rising flour, then indeed it is enriched.

In whole wheat flour, proportions of natural wheat constituents must remain unaltered; no nutrient addition is allowed, none may be removed.

Most flours are limited to 15 percent moisture. In all products—even dried products—there is always a certain amount of water, but food standards. where existent, limit this cheapest of food ingredients to rather precise amounts.

Cornmeal, corn flour, and grits are sold under standards similar to those for wheat products. Moisture levels are restricted, as are additions of fiber, fat, and nutrients. Yellow grits, unlike plain grits (white grits), are processed from yellow corn.

Rice feeds much of the world. Here in the United States it is a packaged food and its label tells whether or not it is coated with talc and glucose. If it is, it is termed *coated rice*. Enriched rice is enriched—with such nutrients as thiamine, riboflavin, niacin, iron, and possibly vitamin D and calcium. If this is the rice product you want, look for the word *enriched* on the label.

Macaroni and noodle products. Like rice, macaroni and noodle products may or may not be enriched. The label spells the difference. Furthermore, milk macaroni has milk in it—the sole moistening ingredient. Whole wheat macaroni is made from whole wheat flour; vegetable macaroni contains tomato, artichoke, beets, carrots, parsley, or spinach to a level of at least 3 percent. If soy flour is added, the label will declare "soy" noodles.

Table 2. Milkfat Levels in Various Dairy Products

Product or Product Category	Milkfat Level (Percentage by Weight)
Fluid milk products	
Skim milk	Less than 0.5
Lowfat milk	0.5-2.0
Whole milk	3.25
Fluid cream products	
Half and half	10.5-18.0
Light cream	18.0-30.0
Light whipping cream	30.0-36.0
Heavy cream	36.0 (or higher)
Condensed and dried products	
Nonfat dry milk	Not over 1.5 (unless label specifies more)
Evaporated milk	7.5
Sweetened condensed milk	8.5
Ice cream and frozen desserts	
Ices	None
Fruit and nonfruit sherbets	1.0-2.0
Ice milk	2.0-7.0
Ice cream and frozen custard	10.0
Natural and processed cheeses	
Dry curd cottage cheese	Less than 0.5
Lowfat cottage cheese	0.5-2.0
Cottage cheese	4.0 (or slightly higher)
Neufchatel	20.0-33.0
Pasteurized processed cheese spread	20.0
Pasteurized processed cheese food	23.0
Hard grating cheeses	32.0
Part skim mozzarella	30-45
Most other cheeses[a]	40.0-50.0

Source: Adapted in part from Federal Register. 1974. 39(235): 42351.
[a] See Appendix I for a complete listing.

Bread products. By now it should be clear that bread is either enriched or unenriched. Raisin bread has raisins in it (not less than 50 parts to 100 parts of flour), milk bread has milk added as the sole wetting agent, whole wheat bread is made from whole wheat—standards of identity require it.

Milk and cream products. The great majority of dairy products are processed under standards of identity. Fat content is set, as are skim milk solids content and fortification standards. In ice cream (and similar frozen desserts), where air may be whipped into the product diluting out food ingredients, guarantees of the food solids content and weight per gallon of the finished product are built into the standard. Preservatives are not allowed in fluid milk and cream products, nor in ice cream or evaporated or dried milk. Provision is made for their use in some cheese products to retard growth of mold.

A number of surveys recently spotlighted consumers' confusion over milkfat (butterfat) levels in various dairy products. To help clarify the point, Table 2 lists the milkfat content of the majority of dairy products currently available in the marketplace. Many of the standards covering these items went into effect January 1, 1975.

Read the label carefully. Quite often standards require a listing of the percentage of fat on the food label. Vitamin fortification can be another point of confusion. Usually the addition of vitamins A and D is desirable in lowfat foods; both these vitamins are associated with fats and oils and are lost when fats and oils are removed. Lowfat products, then, are usually low in these two vitamins. Although vitamin A must be added to lowfat milk, its addition is optional in whole milk and evaporated milk. Vitamin D, though added to most whole milk, is not a required nutrient additive in this product or in skim milk and lowfat milk. Ordinarily it will be present, but reading the label is the only way to know this for certain.

Protein of milk is a naturally complete, high-quality protein of significant nutritional value. It is found in the skim milk portion of milk and is regulated to the extent that all fluid milk products (milk, lowfat milk, and skim milk) are required under federal standards to contain a minimum of 8.25 percent skim milk solids. These solids average between 3.0 and 3.5 percent protein. To increase the protein content of milk, skim milk solids are often added to fluid products in amounts ranging from 0.5 to 1.75 percent. Larger amounts are self-limiting because they cause the product to taste salty.

The amount of skim milk solids added to milk is indicated on the label. However, present regulations (effective January 1, 1975) allow the use of a specific name for milk containing added solids to a level of 10.0 percent total milk solids. This high-protein milk has about 1.75 percent added solids and is labeled "protein fortified." Only the 10 percent solids product will bear this name.

Cheeses line the market shelves in almost unending varieties. New cheeses are introduced with splashy regularity. There are natural and processed cheeses, cheese foods and cheese spreads, full fat, skim milk, and part skim cheeses. Most cheeses are high in fat content (40-50 percent on a dry weight basis), although cheese foods and cheese spreads contain about half the fat content of other processed cheeses. (Cottage cheeses, by comparison, are very low in milkfat at 0.5 to 4.0 percent.) If it's a lowfat processed cheese food or processed cheese spread you desire, you'll have to observe the label carefully, looking specifically for the wording cheese *food* or cheese *spread*. In energy value you'll be getting proportionately less than you would if you bought conventional processed cheese, so you will want to take this into consideration in the price you pay.

Before 1975, cottage cheese with a cream dressing was called creamed cottage cheese. Now the same product, at about 4.0 percent fat, is labeled simply cottage cheese. A similar food, only lower in the percentage of fat, will be labeled lowfat cottage cheese. This food contains between 0.5 and 2.0 percent fat. The label will designate the amount to the nearest 0.5 percent. Still a third product in this line of foods is even lower in fat—a skim milk cottage cheese of 0.5 percent (or less) fat. This cheese is called cottage cheese dry curd, or dry curd cottage cheese.

In certain soured dairy products, like cottage cheese and sour cream, there are two methods of producing acid (souring). One is to add bacterial cultures which ferment lactose (milk sugar) to lactic acid, the other is to add acid directly. Generally the label will indicate, by appropriate reference to direct acidification, when the latter method is employed. Otherwise, it can be assumed that bacterial fermentation has taken place. If there is reason for noting the process used, it is that bacterial cultures, under good process control, produce a mixture of delicate flavors and fragrances that cannot be duplicated by acid addition.

Food flavoring. What you pay for and what makes vanilla extract costly is the amount of flavoring present *and* its alcohol content. Therefore identity standards establish minimum levels for both these ingredients. When vanillin is present, the maximum level is set at one ounce for each unit of vanilla.

Dressings for food. Mayonnaise, french dressing, and salad dressings are all standardized products. Since the major ingredient is vegetable oil, its level is established for each product. In all cases a preservative, EDTA, is allowed to retard rancidity (oxidation). French dressing is permitted to have "safe and suitable" color additives.

Canned fruit and fruit juices. Standards for fruit and fruit juices specify not only varieties of fruits, the size and shape of the fruit (whether whole, cut, diced, or crushed), and optional ingredients, but also the packing media, kind and amount of blemishes, and standards of fill.

Table 3 indicates the optional ingredients (additives) allowed in some common canned fruit products. Table 4 gives permissible sucrose concentrations for packing syrups of various strengths.

Table 3. Optional Ingredients Permitted in Some Common Fruit Products

Canned Fruit Product	Optional Ingredients Permitted
Peaches, apricots, fruit cocktail	Natural and artificial flavors, spice, vinegar, lemon juice or organic acid, ascorbic acid[a]
Prunes	Natural and artificial flavors, spice, vinegar, lemon juice or organic acid, unpeeled pieces of citrus fruits
Pears, plums	Natural and artificial flavors, spice, vinegar, lemon juice or organic acid, salt
Grapes (seedless), cherries	Natural and artificial flavors, spice, vinegar, lemon juice or organic acid
Berries	Natural and artificial flavors
Figs	Natural and artificial flavors, spice, vinegar
Plums	Natural and artificial flavors, spice, vinegar, lemon juice or organic acid, artificial colors

Source: Federal Register. 1975. 40(27):5762-5775.
[a] In amount no greater than needed to preserve color.

Have you ever wondered what proportions of fruit are present in fruit cocktail? Well, standards require the following: peaches, 30-50 percent; pears, 25-45 percent; whole seedless grapes, 6-20 percent; pineapple, 6-16 percent; cherry ingredients, 2-6 percent. If the product is called fruit cocktail, the fruits above will be present within the ranges indicated.

Canned fruit poses no real difficulty in understanding clearly what it is you are buying. The same can be said for canned fruit juices—pure fruit juices, that is. If you buy orange juice, you will get just that. But don't be misled. Unless the container, be it frozen concentrate or single strength canned juice, is labeled orange juice, grapefruit juice, or whatever—unless those words specifically identify the beverage—then you will not be getting pure fruit juice. The product you buy may look and taste like orange juice, but it will not be 100 percent pure juice of the fruit, if that is indeed what you are looking for.

Simulated or diluted fruit juice products are usually termed *fruit drink* or *powdered mix* (in the case of powders) or are called by some trade name such

Table 4. Percentage by Weight of Sucrose[a] Required
in Various Canned Fruit Packing Syrups

Canned Fruit Product	Extra Light[b]	Light[b]	Heavy[b]	Extra Heavy[c]
Peaches.	10-14	14-18	18-22	22-35
Apricots	10-16	16-21	21-25	25-40
Prunes	20	20-24	24-30	30-45
Pears	< 14	14-18	18-22	22-25
Grapes (seedless)	< 14	14-18	18-22	22-35
Cherries				
Sweet.	< 16	16-20	20-25	25-35
Red tart	< 18	18-22	22-28	28-45
Berries				
Raspberries	11-15	15-20	20-27	27-35
Blackberries	< 14	14-19	19-24	24-35
Blueberries.	< 15	15-20	20-25	25-35
Boysenberries	< 14	14-19	19-24	24-35
Dewberries.	< 14	14-19	19-24	24-35
Gooseberries.	< 14	14-20	20-26	26-35
Huckleberries	< 15	15-20	20-25	25-35
Loganberries.	< 14	14-19	19-24	24-35
Strawberries	< 14	14-19	19-27	27-35
Youngberries	< 14	14-19	19-24	24-35
Fruit cocktail	10-14	14-18	18-22	22-35
Figs.	11-16	16-21	21-26	26-35
Plums.	11-15	15-19	19-25	25-35

Source: Adapted from Federal Register. 1975. 40(27):5762-5775.

[a] Sucrose concentration is determined by Brix hydrometer.

[b] Upper levels in this category must be less than the value shown, i.e., for peaches, extra light syrup: density of the syrup must be at least 10 percent or more, but less than 14 percent.

[c] Upper levels in this category may not be exceeded, i.e., for peaches, extra heavy syrup: density of the syrup must be at least 22 percent, but not more than 35 percent.

as Tang and Hi-C. A lot, a little, or no actual fruit juice may be present, and standards are currently under review. Still, you need not be fooled. Real juice amounts are declared on the label directly associated with the food name. The label will tell you how much (if any) fruit juice is actually present.

Fruit pies. Frozen cherry pies are processed according to standard quantity and quality of cherries. No artificial sweeteners are allowed. Not more than 15 percent (by actual count) of cherries may be blemished. This is a relatively new category of food standards, and at present only cherry pies are standardized.

Fruit butters, jellies, and preserves. Fruit butters must contain 43 percent soluble solids in the finished product; preservatives are permitted. Jellies and jams are required to have 65 percent soluble solids and 47 pounds of fruit for each 55 pounds of sweetener, artificially sweetened jellies and jams 55 percent fruit juice. The amount of fruit in fruit jellies and preserves is thereby standardized, your assurance of minimum amounts of specific ingredients when you purchase these goods.

Table syrups. Table syrups must consist of not less than 65 percent soluble sweetener solids by weight. Syrups may be prepared with or without added water and may contain a variety of safe and suitable ingredients, but not added vitamins, minerals, protein, or artificial sweetener. Among ingredients permitted are honey, butter (not less than 2 percent), emulsifiers and stabilizers, natural and artificial flavoring, color additives, salt, chemical preservatives, viscosity adjusting agents, acid adjusting agents and buffers, and defoaming agents. When honey and maple syrup are present, either singly or in combination, they must by weight consist of at least 10 percent of the finished food.

Unlike the category of products classified generally as table syrups, maple syrup must contain by weight 66 percent of soluble solids derived solely from maple sap. Salt, preservatives, and defoaming agents are allowed. Standards for cane syrup require 74 percent soluble solids from sugarcane juice. Optional ingredients are the same as those for maple syrup. As distinguished from maple and cane syrup, sorghum syrup may contain enzymes, anticrystallizing agents, and antisolidifying agents over and above the additives allowed for these other products. Soluble solids content is set at 75 percent.

Nonalcoholic carbonated beverages. This is a class of products (soft drinks), made from carbonated water. Included are soda water, colas, pepper products, ginger ales, root beers, etc. Standards limit the amount of carbon dioxide. Such products may contain alcohol, but are restricted to 0.5 percent by weight, and then only to alcohol contributed by flavoring materials. The colas and pepper products contain caffein, but to a level no greater than 0.02 percent by weight. Safe and suitable optional ingredients are allowed, except for vitamins, minerals, protein, and artificial sweeteners. Optional ingredients (coloring, flavoring, preservatives, etc.) must be declared on the label.

Frozen fruits. A new addition to standardized products, frozen strawberries, may shortly be standardized by genus (*Fragaria*), by pack style (whole, halves, or slices), and by size (small, medium, and large) of berry. Extraneous matter, stems, leaves, and so forth, will be regulated by quantity, as will the amount of blemished fruit.

Shellfish. When you purchase canned shrimp or oysters, you are interested in getting a specific amount and size in a given container. Both factors are regulated by identity standards. Each container must be filled with the product to a given percentage of its water-holding capacity. The size of oysters is specified by the actual numbers of oysters per gallon, larger for smaller oysters and vice versa. Breaded shrimp (regular or lightly breaded) is also controlled by the amount of shrimp in the breaded mixture.

Fish. Two common canned fish foods, tuna and salmon, are limited to species of fish which may be used under these food names. Tuna also has restrictions on parts of the fish that may be included. Tuna styles include solid or solid pack, chunks, flakes, and grated. Color designations (part of the food name) are white, light, dark, and blended. Packing media, whether oil or water, are restricted, with appropriate regard to designating such information on the label. For seasoning, tuna processors may use salt, monosodium glutamate, hydrolyzed protein, spices, garlic, and lemon flavoring.

Salmon has similar restrictions. Forms of canned salmon may be skinless (with backbone removed), or regular, minced, tips, or tidbits. Optional ingredients are salt and edible salmon oil.

Eggs and egg products. The emphasis in standards for egg products falls on the need to render them free from salmonella, which is a bacterium that can cause food poisoning. Egg products may be pasteurized, but no chemical preservatives are allowed. Liquid and dried yolks are further required to meet specific standards of yolk solids (percentage of yolk solids).

Oleomargarine. This butter substitute must, like its forebear, contain not less than 80 percent fat. Although preservatives are not allowed in butter, they are in margarine. Margarine is vitamin fortified, butter is not.

Nut products. In peanut butter consumers are assured of a fat content of no more than 55 percent. That's still a very high fat product, however. No artificial flavors, color additives, or vitamins are permitted.

Standards also go so far as to designate the minimum number of nut varieties (four) that must be present in a can or package of mixed nuts. Amounts of each are restricted within the broad range of 2-80 percent. If a single nut species is present to between 50-80 percent of the total, the percentage is declared on the label.

Nutritive sweeteners. Certain sweetening foods with inherent nutritional value are controlled in type of sugar and solids content. Dextrose, for example, must contain not less than 90 percent total sugar solids, glucose syrup not less than 70 percent total solids; the amount of glucose is specified in both instances.

Frozen and canned vegetables. Standards for frozen peas assure no more than certain specified percentages of blind peas, blemished or seriously blemished peas, and extraneous vegetable matter—pods, stems, leaves, and so on. This is a relatively new category of standardized food and no doubt will be followed by standards for certain other types of vegetables. Like standards for canned fruit, those for canned vegetables set requirements for quality and fill of container. Length and style of cut in green and wax beans are also regulated, as is the style of corn (whole kernel, cream). Because texture—firmness, crispness, crunchiness—is an important criterion of quality, texture is controlled, in some cases by allowances for additives that serve as firming agents. For canned mushrooms, weight of mushrooms per container is regulated, and the addition of vitamin C is optional. Practically all other vegetables are standardized by specific plant part allowed and also form (style) and optional ingredients. Salt and monosodium glutamate are two common optional ingredients. Artificial flavoring is not permitted.

Tomato products. Standards of identity for tomato products declare tomato juice to be the unconcentrated liquid strained from red or reddish varieties of tomatoes. Vitamin C (ascorbic acid) may be added to a level of 10 milligrams per fluid ounce. If ascorbic acid is added the label states, "Enriched with vitamin C."

Catsup (ketchup) standards allow salt, vinegar, spices and/or flavoring, onions and/or garlic, and sweetener to be added to the tomato juice concentrate. The puree of tomato has to be concentrated to between 8 and 24 percent soluble tomato solids, the paste to more than 24 percent soluble solids.

Meat and poultry. Although standards for meat and poultry are regulated by the USDA, the percentages of meat (or poultry) in a given food are established just as percentages of fat are fixed for milk and cheese products. Consumers do indeed (and should) get the same amount of meat in meatballs in sauce, ham in ham spread, hot dogs in beans and frankfurters every time the product is purchased, no matter who the processor is. A large number of similar everyday grocery items are regulated and serve to safeguard consumer interests (while never perfectly, at least within reasonable bounds of personal expectations).

Appendix I lists a number of meat and poultry products which are bound by standards of composition. These are federal (USDA) standards and apply to products shipped interstate. States may and do promulgate regulations to control the composition of meat products sold within state boundaries. Hamburger, a common meat item, under federal regulations may contain no more than 30 percent fat. Some state requirements further subdivide this standard to account for relative leanness. Lean ground beef may be limited to 22 percent fat, extra lean ground beef to 15 percent fat. Of added importance, meat products labeled hamburger, hamburg, burger, ground beef, or chopped beef are not permitted to contain vegetable protein extenders. When extenders are present (as they may be under specific regulations covering their use) meat or poultry/vegetable protein blends will carry information on the label to that effect. Either the name of the food or the list of its ingredients will state their presence (see Chapter II). For the present it is perhaps sufficient to indicate that food names should distinguish between pure meat and blended meat and blended meat products. Further, a movement now underway would drastically reduce the number of fanciful names applied to retail cuts. The National Live Stock and Meat Board has proposed a labeling program, expected to be adopted nationally, that would name meat cuts by (1) species, (2) primal area, and (3) a recommended retail name. The main elements in this naming system would include the following:

Species: beef, pork, lamb, veal

Primal cut

    beef: chuck, rib, loin, round, shank, plate, brisket, flank

    pork: shoulder, loin, leg

    veal: shoulder, rib, loin, leg, breast, shank

    lamb: shoulder, breast, shank, rib, loin, leg

Standardized name: Under this system, fanciful names—Family, His and Hers, Delmonico, Minute, Pan, Penthouse, Saratoga, Boston, New York, etc. —would still be used, but as a separate sticker applied to the retail cut apart from the basic food name. Fanciful names are also being proposed for use in meat/plant protein blends, but with appropriate additional name components (or ingredient listings) to make known the presence of nonmeat protein. Even so, it is readily apparent that food buying will continue to call for careful reading and evaluation of all label information.

Safe and Suitable Ingredients

Throughout the many volumes of food regulations will be found frequent reference to allowances for "safe and suitable" ingredients. Since this is an important, commonplace provision, an official definition has been established for the term. It means, for legal purposes, that the ingredient (1) performs an appropriate function in the food in which it is used; (2) is utilized at a level no higher than necessary to achieve its intended purpose in that food; (3) is not a food additive as defined by the federal Food, Drug and Cosmetic Act, unless used in conformity with appropriate regulations covering these additives.

Nonstandardized Foods

Noting the strict requirements for standardized food, one might assume that processors of nonstandardized foods are entirely free to formulate and alter processes at will in the manufacture of all other food products. This is not so, however. But even if it were true, one could be assured that most basic foods, those esteemed and recommended by nutritionists, are covered by suitable standards. Moreover, all other foods must comply to certain specific standards of quality, in some cases composition and in other instances nutrient makeup.

As will be seen in Chapter II, a common or usual name regulation applied to nonstandardized foods can serve the same purpose as a standard of identity. As for quality, any food containing food poisoning microbes or toxins is considered adulterated and subject to seizure, and the processor to prosecution, even though no special reference to microbial adulteration is made either in a specific standard of identity or in other regulations applying to certain nonstandardized foods. Composition may be controlled to some extent by promulgation of a common or usual name, and nutritive value of certain foods is now regulated by nutritional quality guidelines.

Cream Pies and Food Grade Gelatin

Referring solely to quality, recent regulations established bacterial limits on several nonstandardized foods known to be favorable carriers of heavy bacterial loads. These products are frozen ready-to-eat banana, coconut, chocolate, or lemon cream pies and food grade gelatin.

For the pies, bacterial limits are set at 50,000 per gram aerobic plate count and 10 or fewer coliform bacteria. Aerobic counts measure total numbers of bacteria that will grow in the presence of oxygen (air). Coliform bacteria, while not necessarily disease germs themselves, are known to reside in those habitats where disease-producing bacteria thrive. Thus the presence of coliform is indicative of food contamination from sources known to harbor disease germs.

Food grade gelatin is now restricted to an aerobic microbe count of 3000 or less per gram, a coliform count of 10 or less per gram. It must bear no foreign odor, be clear and light in color.

When either cream pies or food grade gelatins fail to meet the requirements indicated above, the processor must label the product substandard and state the specific quality factor that has been violated. In practice, of course, few processors would be willing to market obviously substandard foods; regulatory restrictions would be expected to be observed almost without exception.

II

Food Names

The common or usual name of a food, which may be a coined term, shall accurately identify or describe, in as simple and direct terms as possible, the basic nature of the food or its characterizing properties or ingredients.

FDA definition

By definition a common or usual name is any food name, possibly coined, that accurately and directly identifies or describes the basic nature of a food or its characterizing properties or ingredients. Such a name cannot be misleading nor is it allowed to be confusingly similar to the name of another food "not reasonably encompassed within the same name." Once established, a common or usual name remains uniform for all identical or similar products.

Because a food name describes a food, to some extent a name establishes the composition of that food. But further than that, common or usual names of nonstandardized foods must specifically include the percentage(s) of any characterizing ingredient(s) when the proportion of any such ingredient "has a material bearing on price or consumer acceptance," when an "erroneous impression" of ingredient content (amount) might otherwise be created, or "when the user has to add characterizing ingredient(s)." Though composition is not regulated specifically, label information must at least inform the buyer of the amount of significant ingredients in the food. Conversely, absence of ingredients and/or the need to add characterizing ingredients must be indicated as a part of the common or usual name. In essence, then, a common or usual name regulation, as currently applied, can serve about the same function as a standard of identity.

When Ingredients Are Listed by Percentage

In the future you will probably find ingredient percentage(s), when listed, closely associated with the food name. The kind of information to expect is a listing of the percentage of each characterizing ingredient based on its quantity in the finished food (weight/weight, or volume/volume) in *descending order by predominance*. Other information regarding absence of ingredients or need to add characterizing ingredients will follow percentage ingredient declaration.

How Common or Usual Names Are Established

Common or usual names may come about through usage alone. If a product, sold under a specific name, becomes familiar to consumers, that name becomes the common or usual name for that product. So it was that "mayonnaise" became named, and "catsup" (or ketchup), "half and half," and "fish sticks made from minced fish." But the names become established in law only when they are adopted through some specific regulation. Names may be legalized, then, through a standard of identity, through a regulation in subpart B of part 102 or part 100 or other regulations in the food laws.

Any interested person may submit a petition to establish a common or usual name. If a proposal is published in the Federal Register, at least sixty days will be allowed for comment. During that time counterproposals may be submitted or a request may be made that a public hearing be held. When good cause in shown, a hearing is granted.

When you ask by name for a food that has been legally christened with a common or usual name, you are reasonably assured that the product is what you think it is. Some examples of recently promulgated common or usual names follow.

Seafood Names

Shrimp and crabmeat cocktail are two common foods. If shrimp is the sole seafood ingredient, the common or usual name regulation permits the food to be called shrimp cocktail (or crabmeat cocktail if crabmeat is the sole seafood ingredient). At the same time, the processor must declare, associated directly with the food name itself, the percentage of shrimp or crabmeat as the case may be. Where more than one seafood is present, the percentage by weight of each seafood is declared on the label.

Greenland turbot. "Greenland turbot" is the common or usual name allowed to describe the food fish *Reinhardtuis hyppoglossoides*, a species of Pleuronectidae right-eye flounders. The term *halibut* may be applied only to Atlantic halibut (*Hippoglossus hippoglossus*) or Pacific halibut (*Hippoglossus stenolepsis*).

Crabmeat. As with Greenland turbot, only certain species of crabmeat may be called by specific given names. Names and crab species follow:

Scientific Name	*Common or Usual Name*
Paralithodes camtschatica and *Paralithodes platypus* . . .	King crabmeat
Paralithodes brevipes	King crabmeat or Hanasaki crabmeat
Erimacrus isenbeckii	Korean variety of crabmeat or Kegani crabmeat
Chionoecetes oplio, Chionoecetes tanneri, Chionoecetes bairdii, and *Chionoecetes angulatus*	Snow crabmeat

Bonito. To be called Bonito or Bonito fish, the food fish must be of the species *Sardi chilensis* or *Sardi velox*.

Some Noncarbonated Beverages

There are a number of beverages on the market that imitate fruit and vegetable juice products but may contain no actual fruit or vegetable juice. Therefore, to promote consumer understanding, the common or usual name of these beverages must include (as a part of the name itself, the information associated directly with the name) the statement "contains no _____ juice." The blank is filled in with the specific juice or juices that are not present.

For fruit or vegetable juice beverages (other than orange juice) that contain some, but less than 100 percent, real fruit or vegetable juice, the food name will include a statement of the amount of fruit or vegetable juice present in increments of 5 percent. In no case may the juice content be less than the percentage indicated on the label, the incremental requirements notwithstanding. At present, orange juice products are not included, pending adoption of specific requirements that will apply only to orange juice products.

Mixers for "Main Dishes"

A large number of packaged foods, some of which are recent additions to the marketplace, are intended solely for the preparation of main dishes or dinners. These foods are not dinners in themselves, but become the mix or the ingredients from which a dinner may be prepared. These are the Hamburger Helpers, the casserole mixes, the noodles and tomato sauce used in hot dishes. For these products the common or usual name regulation requires that each important ingredient be named (by its common or usual name) in descending order by weight.

Heat and Serve Dinners

More will be said about frozen heat and serve dinners when guidelines for nutritional quality are discussed. These guidelines set nutrient standards for certain classes of food and allow labels to declare that the United States government nutrient standards have been met (where such is actually the case). The common or usual name of these foods, which are marketed as complete dinners, includes a listing, in order of decreasing predominance by weight, of the three or more components (dishes) specified in the guidelines. For example, a frozen ham dinner is labeled as such and includes, directly associated with the name, the components ham, mashed potatoes, and peas (and possibly tomato soup and apple tart dessert). This information can be found on the label along with the statement that government nutrient guideline standards have been met. This latter statement is important, for it assures appropriate nutrient content, levels of nutrient, and nutrient mix.

Mixtures Containing Olive Oil

A proposal made June 14, 1974, allows the use of some descriptive common or usual name for mixtures of fat or oil with olive oil followed by a listing of all fats and oils in the products in order of descending predominance. This format would be used for products containing some, but less than 100 percent, olive oil.

Potato Chips Made from Dehydrated Potatoes

As is no doubt apparent, both consumers *and* processors have a vested interest in food names and food naming. Processors know that consumers seek out foods by names with which they have become familiar. There is built-in advertising value in that fact, or, conversely, there is less need to spend funds for advertising when a food name is well known; people automatically look for it or ask for it by name. Potato chips are a good example of the subtleties and special-interest implications that arise when a given food name is assigned to a product.

By now most consumers are familiar with the stack-pack brands of potato chips, all chips precisely the same size and shape and neatly stacked in a protective carton. These potato chips are not the same as conventional potato chips in that they are molded and formed from dried potatoes, not fresh potato slices. Ultimately, competing firms asked, Should manufacturers of the fried chiplike food be allowed to call their product potato chips? It's a simple question replete with a host of significant factors related to industrial competition in the marketplace. Eventually the issue was resolved by establishing a common or usual name which allowed use of the name "potato chip," but with the additional explanatory statement: made from dehydrated potatoes.

Restructured Foods

Advances in food technology which make it possible to form identically shaped individual potato chips presage the potential for combining fragments and pieces of other foods to form finished products of uniform size and shape. The technology is possible and, for some foods, quite routine. The development, brought on by more than aesthetic considerations, though this quality factor is important, hinged on the need to utilize small units of food and reduce waste of shattered or broken foods when distinctive shape and appearance characterize the final product. Uniformity can also be advantageous in filling operations where net weight must be geared to package design and efficient use of package materials. Necessity, then, accounts for the presence on market shelves of "onion rings made from diced onion" (the legal name given this food), "fish sticks made from minced fish," "fried clams made from minced clams," and breaded shrimp products (cutlets, shrimp sticks, etc.) "made from minced shrimp."

An earlier comment suggested that common or usual name regulations have the effect, in given instances, of fixing the composition of food. The products above are good examples, the names as indicated being permitted for use only on products in which the composite food contains "not less than the quantity present in, or similar to," whole units. Thus is there legal assurance that consumers will receive their full measure in their purchases of these foods.

Filled Milk Products

Someday, probably in the near future, a whole new array of dairy products will very likely find their way into the refrigerated display cases of supermarkets. A court case and a resultant FDA proposal portend things to come.

On November 9, 1972, the United States District Court for the Southern District of Illinois handed down a decision in *Milnot Co. v. Richardson* holding as unconstitutional the Filled Milk Act of 1923. This act prohibited shipment of filled milk products across state lines. Filled milk or filled dairy products are legally defined as a combination of milk (or milk product) with fat or oil other than milkfat. Milkfat is skimmed from the milk and replaced with another fat, either animal or vegetable. The potential value of such products has to do with the relationship, as yet unresolved, between dietary fat and heart disease. Conceivably, and within the realm of technical capability, the relatively hard (saturated) fat found in milk could be replaced with soft (unsaturated) vegetable fat. The implications are readily apparent and signal the availability of filled milk, filled cream (half and half, light and heavy creams), filled evaporated and dried products.

Yet to be finalized are proposals that would require filled milk products to be labeled imitation unless they meet certain nutrient standards—those standards of fortification and formulation that would result in a product of equivalent nutrient value to the product being imitated. Basic requirements of the proposals for nutritional equivalency are shown in Table 5. Under these standards the end product would be considered of equivalent nutrient value to its conventional counterpart. Nutritional equivalency is now required of food substitutes marketed under anything but an imitation label.

Formulated Meal Replacements

Since the late 1950s, a large market has developed for food products in liquid, cooky, or some other form that could serve as a complete meal. These products are marketed as having all the nutritional requirements for a standard meal (breakfast, lunch, or dinner). Henceforth, the nutritional makeup will be standardized through nutritional quality guidelines set by the FDA. The name for these foods, as proposed, is "formulated meal replacement" (or base), assuming that the product requires no supplementation with other foods. When a meal replacement requires the addition of other components (milk, for example), the common or usual name must include a statement indicating that "you must add _____ ."

Table 5. Proposed Nutritional Equivalency for Filled Milk Products[a]

Nutrient	Milk Products[b]	Cream Products[c]
Vitamin A	2000 I. U.	40 I. U.
Vitamin D	400 I. U.	
Unsaturated fatty acid content[d]	4% of fat	4% of fat
Vitamin E	1.0 I. U.	1.0 I. U.

[a] These standards were proposed by the FDA August 2, 1973. Federal Register. 1973. 39(148):20748.

[b] Milk products include filled milk, filled lowfat milk, filled skim milk, and the condensed, evaporated, concentrated, or dried form of these products.

[c] Cream products include half and half, light cream, light whipping cream, heavy cream, or dried form of these products.

[d] Cis, cis-methylene interrupted polyunsaturated fatty acids (as glycerides).

Protein Extenders and Replacers

Cars and the automobile industry would seem an unlikely source of ideas leading ultimately to plant protein products that look and taste like meat. But history records that plant protein extenders and replacers, those proteins concentrated and purified and made into meat analogues (or poultry, fish, and cheeselike foods), had their origin in products first conceived in research laboratories of the Ford Motor Company. It was here that researchers obtained experience in working with soy protein, not for use as food but for conversion to enamels and plastics. Later this knowledge was put to work in devising the technology for spinning and extruding plant protein into spongy fibers and units with the texture of meat. With the addition of flavoring, binders, and nutrient isolates a shoulder of ham or a slab of bacon could be made, with no loss in nutritional value. Pigless ham! Steerless beef! Chickenless poultry and eggs! They are reality today. And like so many inventions of man's restless energy, the work, though done during that period in United States history which may well be the highest meat-consuming era per capita of all time, now, in a protein-short world, is of incalculable significance, as though planned that way.

Plant protein extenders and replacers will be commonplace food ingredients in the foreseeable future. Their presence heralds an era of new products and new versions of standard products. Now is the time to become familiar with them, to give them a name, and to understand their basic nutritional nature.

For the present, plant protein products—sometimes called textured vegetable protein (TVP) because of a texturizing step in their manufacture—are being regulated under an application of common or usual name regulations.

(An FDA proposal suggesting common or usual names of plant/animal protein products along with requirements for nutritional equivalency appeared in the Federal Register, June 14, 1974, p. 20892.) A standard of identity, once proposed for these protein sources, was discarded in favor of the former mainly because of the rapidly expanding development of the art. As mentioned earlier, rigid identity standards have the effect of stanching research, particularly that involved in the creation of new products. Moreover, a common or usual name can serve essentially the same function as an identity standard, including that of regulating composition.

Three classes of textured plant proteins are proposed: (1) slightly modified plant products (which maintain a fair degree of original plant integrity) of 65 percent or less protein; (2) concentrates of between 60 and 90 percent protein; and (3) isolates of more than 90 percent protein. The key words are *concentrate* and *isolate*. By these names will the protein-rich textured vegetable products be known. Important to consumers is the fact that protein content influences both flavor and texture.

Slightly modified plant protein products. Protein foods in category (1) include vegetable and cereal flours and granular products. The foods are defatted, dried, and altered physically by grinding or extrusion. On package labels these foods will be recognizable perhaps more from what is lacking in their name than what is present. They may *not* be termed protein. Rather they are named (described) by plant source(s) and a word which describes their physical form, i.e., soy flour, soy granules, corn granules, wheat flour, etc. In all cases, including examples to follow, the words *textured* or *texturized* may be applied if appropriate.

Whatever the source, the finished product must be derived from "safe and suitable" edible plant protein substances. Soybean and wheat are common sources. The future will see greater use of corn, barley, and rice, mustard seed and rapeseed. Peanuts, too, are possible protein sources as are cottonseed and safflower and sunflower seeds. Any plant harboring edible extractable protein is potentially usable for this application of food technology.

Plant protein concentrates. Concentrate is the term suggested to designate products consisting of 65 to 90 percent protein. These plant protein foods are literally concentrated protein. When the protein is derived from more than two plant sources, the name applied to them may be "vegetable," "cereal," or "plant" as in *plant protein concentrate bits.* Such products are prepared by a process of extraction and washing which removes a considerable amount of fiber, carbohydrate (sugar), and minerals.

Plant protein isolates. The term suggested for the most concentrated plant protein products is *isolate.* Isolates contain more than 90 percent protein and may be highly purified protein sources. The common or usual name may or may not identify the source since the nonprotein materials that may have characterized the source have to a great extent been eliminated. Therefore, these products may be identified simply as *plant* (or vegetable or cereal) *protein isolate.*

Mixtures of plant protein concentrates and isolates. Where mixtures of concentrates and isolates are prepared, the common or usual names must be listed according to descending order of predominance in the mixture. If the mixture includes three or more such products, the designation may be simplified to plant (or cereal or vegetable) protein product, along with a suitable description of the physical form, i.e., bits, granules, etc.

Mixtures of modified plant products, plant protein concentrates, and plant protein isolates. Conceivably mixtures of plant protein products could consist of slightly modified products of less than 65 percent protein, along with concentrates and isolates. These forms of vegetable protein must show predominance by weight in decreasing order. Or the designation may state: plant (or vegetable or cereal) protein product and _____ (or _____ and plant protein product), whichever reflects predominance in descending order. The blank in this case will carry the common or usual name of the other component(s). An example might be vegetable flours and plant protein product.

When flavors are present. When plant protein products are flavored to resemble meat, seafood, poultry, eggs, or cheese, a flavor designation will accompany the product name, i.e., corn flour, ham flavored, or artificially ham flavored soy protein isolate.

Mixtures of animal and plant protein products. Presence of animal protein —milk, fish, egg, etc.—in plant protein products will reflect this fact in the product name. Predominance will also be indicated in decreasing order of listing. Names might include *plant and fish protein concentrate* or *animal and plant protein concentrate.* A careful examination of the label will automatically indicate which protein source predominates. A summary of common or usual name designations for plant protein products is shown in Table 6.

Plant protein as imitation food. Because plant protein can be processed to simulate meat, poultry, eggs, and cheese, such foods are marketable as substitutes for these conventional foods. As extenders or as complete replacement foods, they are then considered, for legal purposes, imitations. Accordingly, textured vegetable protein products must be labeled imitation *unless* they have nutrient value equivalent to the foods they imitate (under regulations promulgated in conjunction with nutritional labeling). For most foods, nutritional equivalency is defined as the presence in the food substitute of those nutrients at those levels found in the conventional food. This is an important distinction in that substitutes nutritionally the equal of their traditional counterparts need not be labeled "imitation."

Table 6. Common or Usual Name of Plant Protein Products

Product Description	Common or Usual Name[a]
(1) Less than 65% protein	Source name + physical form, i.e., soy flour, soy granules
(2) 65%-90% protein	Source(s) name(s) + protein concentrate + physical form, i.e., soy protein concentrate bits. If more than 2 sources, the term *vegetable* or *cereal* or *plant* may be used, i.e., vegetable protein concentrate
(3) Over 90% protein	Source(s) name(s) + protein isolate + physical form, i.e., soy protein isolate granules, or vegetable protein isolate
(4) Mixture of (2), (3)	Common or usual names of each component in descending order of predominance + physical form, i.e., soy protein isolate and peanut concentrate granules. If mixture consists of 3 or more products, the word *plant* or *vegetable* or *cereal* may be used + protein product, i.e., vegetable protein product
(5) Mixture of (1), (2), (3)	Common or usual name of the components in descending order of predominance; if the product consists of a mixture of three or more of these products: "Plant protein product and _____" or "_____ and plant protein product," whichever accurately reflects predominance; the blank will be filled in with the common or usual name of the component(s) in (1), or, if more than source is involved, the word *vegetable* or *plant* or *cereal* may be used; physical form also indicated, i.e., soy flour and plant protein product granules
Any product in (1)-(5) flavored to resemble meat, seafood, poultry, eggs, or cheese	Appropriate common or usual names of products (1)-(5) + flavor declaration, i.e., soy flour, shrimp flavored, or artificially ham flavored soy protein isolate
Any product in (1)-(5) which also contains animal protein (milk, fish, egg, etc.)	Common or usual name source in descending order by predominance, i.e., plant and fish protein concentrate or animal and plant protein concentrate

Source: Adapted from Federal Register. 1974. 39(116):20892-20895.

[a] The term *textured* or *texturized* may also be included when appropriate.

Common or usual name standards for vegetable protein products specify conditions which must be met to satisfy requirements of nutritional equivalency. These conditions include (1) amount of protein, (2) protein quality, and (3) nutrient content.

Nutritional equivalency rules, as originally proposed for plant protein products, considered the need for setting separate standards of protein, vitamin, and mineral content for each type of food simulated. Ultimately such specificity was abandoned as unnecessary (from a nutritional standpoint) and impractical. Uniform nutrient content has been proposed, with an exception made for micronutrient (vitamin and mineral) levels in cheese products. Protein content, when used to replace meat, seafood, poultry, eggs, or cheese, would be set at 18 percent.

The quality of protein varies, among protein sources and within the same source. For labeling purposes, protein quality is measured by the "protein efficiency ratio" (PER) method (see Chapter XI). PER is a measure of the growth rate of rats, with the test protein compared to a standard protein called casein (the major milk protein). Since the overall quality of protein depends on the proportion of plant protein in the finished product, PER standards must appropriately reflect this difference. At low levels of plant protein (less than 30 percent), PER is proposed at 80 percent that of casein, at high levels (over 30 percent), 108 percent that of casein. In amounts less than 30 percent plant protein, the plant protein is considered an extender and the overall protein quality is assumed to be fixed primarily by the animal product which predominates; thus the provision for a lower PER standard.

Except for cheese, which is significantly different in vitamin and mineral content (particularly vitamin A, calcium, and phosphorous) from other animal or poultry products, nutritional equivalency allows for uniform fortification of vitamins and minerals. These and other considerations of nutritional equivalency are summarized in Table 7.

Vegetable protein extenders and replacers in meat products. In 1972, consumption of soy protein meat substitutes was estimated at 1 percent of all meat consumed. By 1980 it is anticipated that this figure will jump to between 10 and 20 percent. Most plant/animal blends will come in the form of ground meat and formulated meat products—sausage and wieners—and a good share of these products will ultimately be utilized in a variety of meat dishes, i.e., stews, meatballs, hashes, chili, and so forth. It is well to note that beef consumption in the United States averages about 116 pounds per capita, of which some 22 percent (26 pounds) can be classified as ground beef. Much of

Table 7. Plant Protein Nutrient Requirements for Achieving
Nutritional Equivalency to Various Animal Products

Nutrient or Nutritional Factor	Meat, Seafood, Poultry, Eggs[a]	Cheese[a]
Amount of protein (on hydrated basis)	18%	18%
PER of protein[b] when present at less than 30%.	80	80
PER of protein[b] when present at more than 30%.	108	108
Vitamin A	15　　I. U.[c]	54 I. U.[c]
Thiamine.	.014 mg	
Riboflavin	.01　mg	
Niacin	.30　mg	
Panthotenic acid	.040 mg	
Vitamin B_6	.02　mg	
Vitamin B_{12}	.09　μg	
Iron.	.13　mg	
Calcium	10　　mg	26 mg
Phosphorus	10　　mg	26 mg
Folic acid	.40　μg	
Magnesium.	1.14　mg	
Zinc.	.23　mg	

Source: Adapted from Federal Register. 1974. 39(116):20892-20895.

[a] The plant protein would be used as extenders or replacers for these products.
[b] As percentage of PER of casein.
[c] The following figures are per gram of protein.

the reason for this relatively high intake of ground beef is that it can be used in a variety of meat dishes. Almost without realizing it, Americans eat hamburger two to three times a week. And without resorting to a crystal ball, it seems safe to assume that vegetable/animal protein blends will in the future be largely utilized in food combinations that contain meat (or poultry or fish) as an ingredient. Naming these foods to inform the consumer adequately of their nature is a complex task. Proposals have been published toward this end, and though they may eventually be altered, the essential elements indicate things to come in food labeling. The proposals, therefore, will suffice to illustrate trends, if not substantive labeling requirements.

Particle size. Vegetable protein particles that may suitably serve as meat replacers must be of similar size to the meat particles they replace. For regulatory purposes this size is determined to be that of particles which will not pass through a sixteen-mesh screen. These are termed large particles; all others become small particles and are subject to different regulations. The reasoning is that small particles can be part of the product without misleading consumers to believe that meat is present to a greater extent than is actually the case. Small particles, therefore, are treated more or less as just another ingredient, large particles as actual meat replacer.

Labels of meat food products. For meat products in casing or other food form the label will indicate the common or usual name of the plant product in the food name, i.e., chili with textured vegetable protein. Ingredients of the vegetable protein will be listed with the other ingredients, for example, soy flour, salt, and caramel coloring. When artificial coloring is used, the label must bear a statement to that effect.

Amount of nonmeat replacer allowed. Precise limitations on the addition of textured vegetable protein are mandatory under law. The amount allowed generally parallels the amount of plant-base or dairy-base foods permitted under present standards for the specific meat food product. Whereas chili may contain up to 8 percent dairy and/or cereal food, this amount (8 percent) could be replaced in part or in total by textured vegetable products.

For some meat foods a small amount of textured vegetable product in large particles is permitted without the processor being obliged to declare its presence in the food name. Likewise the presence of textured vegetable product in small particles need not be indicated on the food name per se. But in both cases the formulation of meat foods with these products would require declaration in the statement of ingredients. Percentages of large particle vegetable protein replacer which is permitted but need not be declared with the food name are given in Table 8.

Imitation sausage. Sausagelike products that do not comply with conventional sausage standards are expected to bear the word *imitation* unless, clearly associated with a fanciful name, a descriptive phrase indicates the general nature of the product. It is possible, then, that future products of this type may bear a fanciful name (Lushus Pup, Tasty Treat, or some such coined name). But, accompanying the food name will be a phrase describing the basic character of the product, i.e., "meat and soy protein concentrate stick," "pork and soy flour chunks," or "beef and nonfat dry milk bits." Closely associated with the descriptive phrase will appear a "conspicuous ingredient statement" providing the percentage of each ingredient. Additionally, protein quantity and quality must match the product being imitated; vitamin and mineral content must range within 90 and 150 percent of that found in the traditional food.

Meat patties and meat patty mixes. Glance quickly at the following two phrases: "meat patty" and "patty with meat." Do you know what you read? Do the two phrases say something different to your eyes? Conspicuously so? Well, in the future they must or you'll be going home with something other

Table 8. Maximum Percentage of Large Particle Textured Vegetable Product Allowed in Various Meat Food Products and Exempt from Label Declaration in the Food Name[a]

Product Type	Examples	Percentage of Large Particle Textured Vegetable Product
High meat product (not including sausage)	Meat loaves for baking or meat loaf mixes for baking, meatballs or meatball mixes, pizza with meat toppings or pizza with meat topping mixes, salisbury steaks	5
Meat with gravy (or sauce)	Sliced beef and gravy	4
Sauce (or gravy) with meat	Chili, sauce or gravy with sliced meat, sauce or gravy with ground beef	3
Meat salads or hashes	Meat salad, meat salad spread, roast beef hash	3
Sauce (or gravy), meat and vegetables	Meat pie, chili with beans, meat stew	2
Starch (or beans) with meat in sauce	Spaghetti with meat in sauce, macaroni with meat in sauce, chili macaroni, beans with ham (or bacon) in sauce	1
Meat sauce	Spaghetti sauce with meat, chili sauce with meat, hot dog sauce with meat	½

[a] These percentages are USDA proposals published in the Federal Register, May 4, 1973. The exemption from label declaration applies solely to information applied in association with the food name. Under the proposal, these products would still have to be declared in the statement of ingredients.

than what you expect. The USDA has made certain proposals to regulate the naming (and composition) of these foods. According to the USDA, meat patty (or meat patty mix) is a product made from meat. No water, binder, extender, or other nonmeat substances, poultry products, or meat by-products are permitted. The fat content (which may include added fat) cannot exceed 30 percent. If a species name is used—veal, lamb, etc.—only that meat species can be used. "Patty with meat" products, on the other hand, consist of the following: (a) a minimum of 60 percent meat; (b) a maximum fat content of 30 percent; (c) a minimum of 13.5 percent protein; (d) a protein efficiency ratio (PER) of not less than 90 percent of the PER of "meat patties."

Though similar, "meat patties" and "patties with meat" are not identical. And while there is no need to be alarmed over nutrient differences, at least for most of us, it is apparent that a careful reading of the label will be necessary to appreciate fully the nature and cost of these food products.

The Statement of Ingredients

There seems little doubt that, voluntarily or by legal mandate, all food labels will soon list *all* ingredients. As indicated earlier, statutory authority does not currently exist for requiring a complete list of ingredients in every instance. But considerable consumer pressure, with some justification, appears to have set the gears in motion. Especially where food allergies may be involved, a listing of all potential allergens is a must. Since allergic tendencies are almost as variable as the foods we eat, a comprehensive listing of ingredients appears to be the only suitable procedure to follow. On the label, the ingredients themselves will be declared by their common or usual names.

A look at a sample statement of ingredients might serve as a guide to concerned shoppers. Following is an example of an ingredient statement from a cake mix. First, note how the information appears on the label.

Cake Mix

Ingredients: Sugar, enriched flour (bleached), shortening (with freshness preserved by BHA and BHT), nonfat dry milk, leavening, wheat starch, propylene glycol mono esters and mono and diglycerides, salt, artificial flavors, guar gum, soy lecithin

Identified more specifically, the ingredients are as follows:

1. Sugar—sweetener (since ingredients are listed in descending order of predominance by weight, sugar is the ingredient found in greatest amount).

2. Enriched (bleached) flour—flour fortified with vitamins and minerals as provided in the standard of identity for enriched flour.

3. Shortening—a fat or oil, kept from going rancid during storage by two antioxidants, BHA and BHT.

4. BHA and BHT—antioxidants necessary to prevent rancid flavors from developing during storage.

5. Nonfat dry milk—dried, skim milk powder.

6. Leavening—a product (such as baking soda) required to cause the cake to rise.

7. Wheat starch—a starch made from wheat; note that the source (wheat) is indicated for those who might react adversely to wheat products.

8. Propylene glycol mono esters; mono and diglycerides; guar gum; soy lecithin—thickeners and binders (emulsifiers and stabilizers).

9. Salt—take note if on a low-sodium diet.

10. Artificial flavors—flavoring is present and it is *artificial* flavoring.

Another ingredient statement appears below, with ingredients identified. Can you tell what the product is?

1. Partially hardened soybean oil—the major ingredient.

2. Nonfat dry milk—another major ingredient.

3. Salt.

4. Sodium benzoate and citric acid—preservatives.

5. Lecithin—a naturally occurring emulsifier.

6. Artificial coloring—coloring, artificial.

7. Artificial flavoring—flavoring, artificial.

8. Vitamin A palmitate—the chemical form of vitamin A, used to fortify the product.

The ingredients above make up a common table spread, one of many different kinds collectively referred to as margarine.

A statement of ingredients will probably change somewhat as time goes by. There is need, for those who suffer from food allergies, to require appropriate changes in the list of ingredients whenever pertinent alterations in formulation are made. Such alterations can take place rapidly depending upon the cost of ingredients. One change to expect, if recent government proposals become law, is a percentage figure following all characterizing ingredients (major ingredients bearing on consumer price and acceptance) in the food. You'll be informed, both on the front of the package and on the ingredients panel, how much of each food component is actually present.

Naming Ingredients by Class Names

To simplify ingredient listing, a naming system has been developed which allows the grouping of certain foods in a single food name class. This technique was first introduced for some standardized foods and has now been expanded to include nonstandardized food. A label may indicate the presence of "milk," for example, without indication whether the ingredient is fluid, concentrated, or dried milk. The phrase "egg whites" could be used to denote the presence of dried, frozen, or liquid whites of eggs. Because the physical state of these foods in no way alters their potential as allergens, no serious consumer considerations go unstated. It is appropriate to assume that no such allowance will be made when allergenic properties are involved.

Table 9 lists, by name and food species, the various food classes identified thus far (in proposal form). In addition to the classes in Table 9, fat and/or oil will be declared according to the source from which it is derived (beef fat, cottonseed oil, etc.) and flour products will specify the kind of flour—bromated, enriched, durum, whole wheat, and so on. The necessary information for making informed purchase decisions will thereby generally be available.

Table 9. Class Names for Various Groups of Foods

Name	Products
Skim milk	Skim milk, concentrated skim milk, nonfat dry milk
Milk.	Milk, concentrated milk, dried milk
Cultured _____	Bacterial cultures of the product, the blank being filled in with the appropriate substrate, i.e., skim milk, buttermilk, etc.
Buttermilk.	Sweet cream buttermilk, concentrated sweet cream buttermilk, dried sweet cream buttermilk
Whey	Cheese whey, concentrated cheese whey, dried cheese whey
Cream	Fluid cream, dried cream, plastic cream
Butter	Butter, butter oil, anhydrous butterfat
Eggs.	Liquid, dried, or frozen whole eggs
Egg whites	Liquid, dried, or frozen egg whites
Egg yolks.	Liquid, dried, or frozen egg yolks
Sugar	Sucrose and invert sugar
Corn sweeteners.	Sweeteners derived from corn
Vinegar.	Cider vinegar, apple vinegar, wine vinegar (grape vinegar), malt vinegar, sugar vinegar, glucose vinegar, spirit vinegar

Summary

A food name, like the name of a person, carries with it a connotation of character and consistency. A food name that is established by regulation may set the standard by which the food must be formulated. Thus the food name becomes the standard of identity. The essential features of a food name demand that it be informative (accurately depicting the contents of a food package) and that it not be misleading.

But even though a regulation should serve to prevent deception, it should not impede or deter the development of new products and new technology. Toward this end, recent proposals and regulations for food naming have attempted to provide informative labeling while leaving the door open for imaginative processes and formulations which utilize plant sources heretofore limited in their usefulness as human food. A crossroad has been reached. The abundance of the plant kingdom promises now to be a source of many new protein products.

III

Food Additives

A food additive is any substance the intended use of which results or may reasonably be expected to result, directly or indirectly, in its becoming a component or otherwise affecting the characteristics of any food (including any substance intended for use in producing, manufacturing, packing, processing, preparing, treating, packaging, transporting, or holding food; and including any source of radiation intended for any such use), if such substance is not generally recognized, among experts qualified by scientific training and experience to evaluate its safety, as having been adequately shown through scientific procedures (or, in the case of a substance used in food prior to January 1, 1958, through either scientific procedures or experience based on common use in food) to be safe under the conditions of its intended use. . . .

FDA definition

In the home sodium bicarbonate (baking soda) is added to bread dough to make it rise. The commercial baker uses sodium bicarbonate, and it is a food additive. You shake salt over a buttery ear of corn fresh from the farm. The dairy plant operator adds salt to butter, and it is a food additive. Turkey dressing—a recipe for which you are particularly proud—calls for a gourmet's mix of spices and seasoning. At Thanksgiving you dig out your recipe and mix the ingredients with anticipation and relish. Those same spices added to bread cubes by a "commercial" chef are food additives. With your new strawberry bed in mind, you decide to put up some strawberry preserves, thick, fruity preserves the way mother or grandmother used to make them. You like your jam especially thick and so mix in ample amounts of thickener—pectin. You guessed it—another food additive for the processor who "puts up" preserves by the hundreds of thousands of jars each year.

Then there are certain foods parents like to see-to-it that their kids eat because these foods are "good for them," say, carrot sticks, milk, good red meat. The same vitamins and minerals found naturally present in such foods, when extracted or synthesized by the food processor and added back to these or other foods to standardize or supplement them (because nature doesn't always standardize nutrient levels very well), are food additives. Vitamins in vitamin tablets are food additives. And so are the ingredients that carry the vitamins—the preservatives that maintain potency, and the sugar coating.

And now let's reach a little—into a sudsy froth of dishwater to clean up the dinner dishes. You wash them, set them in a rack, and then rinse them with a hot spray of water. You spray carefully but, piled together as they are, a few bubbles of detergent remain on some of the plates and silverware. A few bubbles won't hurt anybody, and so after they dry, the dinner dishes are stacked away with traces of dried detergent spotting their faces. When a food processor cleans his dirty dishes—food processing equipment—a similar detergent remaining after rinsing (in the same minute amounts) becomes a food additive, an *incidental* food additive.

Or take flies—you hate them—and mosquitoes, too. So whenever one gets into the house, or into the kitchen, you go after it with a can of household spray. That same spray (or others similar to it) used in a food processing firm is a potential incidental additive. And so is the pesticide spray that is applied to farm crops in the fields where produce grows, or in the cattle and dairy barns—if in fact a residue remains on the produce or the meat or in the milk. The processor who eventually prepares foods for human use, though he has limited supervision over raw goods before delivery, is the one who bears the responsibility for the presence of unwanted "additives" in the finished food.

Whenever one hears or reads of an additive in foods, the word *additive* should spark an immediate consideration of the type represented. Is it an intentional additive, one added to a food for some specific purpose, or is it an incidental (unintentional) additive? You may say, "Who cares, if it's in the food I eat?" You should care, because policy and planning and methods of control are everyone's business.

One last analogy. Remember the strawberry jam you were "canning"? What were you planning to use for a container? A glass jar? More than likely. And if you were freezing vegetables, you probably would store them in some form of plastic (film) "freezer-wrap." Food manufacturers, too, must put up their products in some convenient, *protective*, container such as a bottle, jar, plastic (rigid or film) or paperboard overwrap. The only difference is that every package is a potential food additive or, rather, has the potential to leach some of its components into the food it holds. And every compound from which containers are made and every reaction component in every step from the time a log starts through the papermill until *and after* it comes off the processing line as a finished food container, resplendent with glue (adhesive) to hold it together and ink to color and inform, is a potential food additive—an incidental food additive. The list of chemical agents alone necessary to construct paperboard and various plastics includes hundreds of different compounds. And that list is small in comparison with the number of flavor compounds arraying the pages of regulatory documents.

Understanding the "Additive" Language

Much of the difficulty in reaching a balanced viewpoint of the benefits and hazards of food additives stems from a failure to appreciate fully the meaning of terms. Indeed, the language barrier that exists between food scientists and consumers, between government and industry, or between two processors within the same industry is often the equal of that present between two persons of differing national and lingual origins. Microbiologists tend to speak a language and create a jargon unique to their discipline. Economists develop another mode of speech, food scientists still another. Not only is it hard for scientists to communicate with nonscientists, but it is almost as taxing for two scientists of differing specializations to communicate with each other. It is perhaps inherent in the nature of the human family that such matters often become the source of pride, not the incentive to find better ways to communicate.

In the science of food technology intentional additives might be better understood if they were referred to as food *ingredients*—which is what they actually are. Incidental additives, then, would better be termed *food contaminants*. Ingredients are the mix from which food is prepared; contaminants are the materials that get into food from some outside source during food production, processing, or handling anywhere along the way from farmer to consumer. As a group, incidental additives comprise some 10,000 known chemicals.

But there is a distinction to be made. Incidental additives are contaminants of known origin and anticipated use. These are chemicals applied in the unending war against animal and plant diseases. These are chemicals applied to maximize yields. These are the insect and weed killers, the antibiotics and drugs, the growth stimulants and nutrients and the miscellany of associated ingredients required to bind active agents together in usable form. All—including associated chemicals—are tested before use. All are researched to ascertain their levels of safety, methods of application, and specific controls. All must be approved by the government before they can be used. And although the system is not foolproof (controls break down, unanticipated problems arise, carelessness and accidents occur), the compounds are known to exist, systematic procedures of control can be set up, some semblance of orderly use and supervision can be maintained. Insofar as abundance and low food costs are desirable (or essential), the benefits of applying these chemicals are considered to outweigh the risks. Rough estimates by veterinarians and university specialists place the loss of poultry, beef, and dairy animals, in absence of drugs, at approximately 50 percent of present production.

A decade ago, amid the furor over pesticide use, a number of hard figures were trotted out by a besieged industry. Absent the uproar, let's scan them again: 50,000 known plant diseases, 6500 insects known to infect plants and animals, 2500 known ticks and mites, a 50 percent reduction in the cost of weeding through the use of herbicides, a 25 percent increase in the value of pastureland from the same, an animal and poultry disease cost saving of $2 billion annually, a $20 billion annual saving in food cost from use of pesticides. Conservatively, cost figures would be doubled today, crop and animal losses holding about equal. To a world struggling to maintain food reserves, the implications are horrendous.

Incidental additives—environmental chemical contaminants of anticipated origin—should be distinguished from *chance* contaminants, the dirt and dust, specks and fragments of insects, the bits, drops, and pieces of miscellany that happen into food by chance alone. A worker's button, a strand of human hair, the organic gardener's manure and compost.

Intentional additives (ingredients), incidental additives (anticipated contaminants), chance contaminants. Over and above those descriptive words that may distinguish different kinds of additives, the consumer must also become acquainted with chemical terms. Fear and hesitancy are usually associated with the unknown. You might assume sodium chloride to be a poison, while in fact it is table salt. Or, an even more dangerous sounding chemical, aceton-3-hydroxy-2-butanone, loses its power to alarm when you know it to be one of the characterizing flavors found naturally in butter.

Looking at it another way, how many people would knowingly drink wood alcohol, a well-known poison? Not many. Yet wood alcohol (methanol) is a natural ingredient of coffee. So is acetone—a paint remover. In eggs can be found chemicals as weird as zeaxanthine and lipovitellin. Bread flavor alone consists in part of some eighteen different acids, five alcohols, seven esters, seven ketones, and fifteen aldehydes, among them formaldehyde, the oft-used component of embalming fluids. In regulations for flavor labeling, the FDA acceded to the wishes of the food industry by allowing substitution of the phrase "artificial smoke flavor" for the chemical name "pyroligneous acid" on packages solely because of the unfavorable impact such a term can have on a public unschooled in chemistry.

Chemicals, therefore, are literally the stuff of existence. All things, living and unliving, can be broken down into a baker's variety of chemical compounds. And whether you see in these names poison or plenty depends upon your knowledge of chemistry. So coffee contains methanol and acetone, bread has in it a witch's potion of formaldehyde. How is it, then, that we consume these foods daily and manage to survive? It is a simple, yet unimaginably complicated, question of the amount consumed and the relative toxicity (poisonousness) of the individual chemicals.

Anything consumed in overdose is lethal. A book has been written on the lethal quality of "pure" food. Salt, the harmless chemical applied to hundreds of foods, is a compound of considerable toxicity, comparatively speaking. It is more toxic than piperonyl butoxide, a component of household insect sprays. But caffein in coffee is even more toxic than salt. It is three times as toxic as the dandelion spray 2,4D. And somewhere between salt and caffein falls the pain-reliever aspirin. For any chemical, be it food or drug, the question of illness or death depends upon the *amount* consumed and the *inherent toxicity*. Clearly, the best way to avoid food intoxication is to eat moderately of a wide variety of foods—except where allergies dictate exclusion of certain specific food items.

Data in Table 10 give evidence of the wide differences in toxicity of some fairly common products. Smaller numbers (dose rate) imply greater toxicity. Note that the list is headed by a poison produced by certain molds. These molds have the ability to grow on a variety of foods and feedstuff. Their toxins may cause illness as slight as mild nausea or as complex and lethal as cancer. For a food processor the alternatives can be very limited: chance the growth of toxic agents or add a preservative.

Table 10. Relative Toxicity (LD$_{50}$) of Various Products[a]
(in Milligrams per Kilo of Body Weight)

Product	Oral	Dermal	Use
Aflatoxin B$_1$.	0.56[b]		Food poison caused by mold growth
Phorate.	1	3	Insecticide
Endrin	3	18	Insecticide
Strychnine.	16		Rat poison
Kerosene.	50		Fuel
Lead arsenate	100		Insecticide
Caffein	200		Coffee, tea
Warfarin (coumarin)	325		Rat poison
2,4D	700		Weed killer
Aspirin	750		Pain reliever
Monosodium glutamate	1660		Flavor enhancer in food
Pyrethrum	1870	2060	Insecticide
Borax.	2600		Cleaning compound
Table salt.	3000		Food
Ethyl alcohol	4500		Beverage

Source: Most of the data included in this table were taken from *The Toxic Substances List*, 1974 (U.S. Department of Health, Education, and Welfare, Public Health Service, Center for Disease Control, National Institute for Occupational Safety and Health, Rockville, Maryland).

[a] LD$_{50}$ is defined as the dose at which 50 percent of exposed test animals will die. Rats are the most common test animal used. Differences may occur as a result of age or sex of the animal. Smaller numbers are indicative of *greater* toxicity.

[b] Day-old rats were used as test animals; data taken from A. J. Feuell, *Types of Mycotoxins in Foods and Feeds* (New York: Academic Press, 1969).

Some Additives That Have Been Banned

Over the years a number of substances, some naturally occurring, some synthetic, have been found to be or assumed to be sufficiently toxic to humans so that they are excluded as ingredients of food. Their numbers are very small in comparison with the total number of ingredients (additives) allowed in food, and they are evidence both of more widespread testing and of the apprehensive climate surrounding the use of additives generally. The partial list presented in Table 11, the first of its kind, reflects an FDA attempt to consolidate and make more readily available to the public information of this nature. It is included here for its informative character and as a means of putting in perspective the question of risk.

Compounds in Table 11 were all banned after animal feeding trials indicated their potential toxicity. As will be asserted or alluded to throughout this book, the extrapolation of data on animal toxicity to humans is a tenuous business at best. But some such attempt seems warranted if only as a way of expressing relative risk. Take the fifth compound in the table, cycla-

Table 11. Some Substances Prohibited in Food and on Food Contact Surfaces

1. *Calamus, oil of calamus, extract of calamus*—calamus is the dried rhizome of Acorus calamus L.; former use: flavoring compound.
2. *Dulcin* (4-ethoxyphenylurea)—sweet-flavored synthetic chemical not yet found to occur naturally.
3. *P-4000* (5-nitro-2-n-propoxyaniline)—sweet-flavored synthetic chemical not yet found to occur naturally.
4. *Coumarin* (1,2-benzopyrone)—chemical found in tonka beans, among other natural sources; former use: flavoring compound.
5. *Cyclamate* (calcium, sodium, magnesium, and potassium salts of cyclohexane sulfamic acid)—synthetic sweet-flavored chemical not yet found to occur naturally.
6. *Safrole* (4-allyl-1, 2-methylenedioxy-benzene)—natural constituent of the sassafras plant; former use: flavoring compound.
7. *Monochloroacetic acid*—synthetic chemical not found to occur naturally; former use: preservative in alcoholic and nonalcoholic beverages; still permitted in food package adhesives with specific limitations.
8. *Thiourea* (thiocarbamide)—synthetic chemical not yet found to occur naturally; had been proposed as an antimycotic for dipping citrus fruit.
9. *Cobaltous salts*; acetate, chloride, and sulfate—former use: foam stabilizer in malt beverages, also to prevent "gushing."
10. *NDGA* (nordihydroguaiaretic acid)—resinous extract of certain plants; former use: antioxidant for some foods.
11. *DEPC* (diethylpyrocarbonate)—synthetic chemical not yet found to occur naturally; former use: ferment inhibitor in alcoholic and nonalcoholic beverages.
12. *Flectol H* (1,2-dihydro-2,2,4-trimethyl-quinoline)—synthetic chemical not yet found to occur naturally; former use: component of food packaging adhesives.
13. *4,4-Methylenebis* (2-chloroanaline)—synthetic chemical not yet found to occur naturally; former use: polyurethane curing agent and as component of food packaging adhesives and polyurethane resins.

Source: Adapted from Federal Register. 1974. 29(185):34172.

mate. More than any other this was the additive which, when regulatory action was taken to prohibit its use in food, touched off the legislative inquiry that led eventually to a far-reaching review of all additives regarded as safe. Cylamate was prohibited when test data on rats and mice pointed to its potential to cause cancer. Using this data and extrapolating dosage rate to humans based upon the consumption of soft drinks (one of the major products to use cyclamates), we can draw the following relationship. At cyclamate levels of ¼ to 1 gram per twelve-ounce bottle, the intake per person needed to equal the toxic dosage consumed by the test animals amounts to 138 to 552 bottles per day. A similar relationship determined for oil of calamus, at levels formerly found in vermouth, would put equivalent consumption at 250 quarts of this alcoholic beverage per person per day. For safrole in root beer and hard candy, the daily intake would have to reach 613 twelve-ounce bottles and 220 pounds respectively to match the dose found cancerous in rats. (Still, such figures must be weighed against the possibility, as viewed by some cancer researchers, that there is no such thing as a "no-risk" level of carcinogen). The last two compounds in the table were once constituents of packaging material. Human intake of food containing these two chemicals at the minuscule levels that might have been expected to leach into the food would have had to approach 300,000 and 100,000 times normal consumption to achieve equivalent toxic doses.

Another interesting relationship can be drawn for the animal growth hormone diethylstilbesterol (DES), also in the news of late. In this analogy DES is assumed to be present in beef liver at a level of 2 parts per billion. (The assumption has some basis in fact.) At that level, further assuming 2 percent of the total diet as liver, a person would have to consume 5 million pounds *per year for 50 years* to equal the intake from one treatment of "day-after" oral contraceptive.

These figures are not presented to hint at a relaxation of our vigilance, but only to put the question of hazard analysis, as observed for these additives, into some kind of focus. At the time radioactive fallout increased strontium 90 to an "alert" level in milk supplies, the hazard posed by the amount of milk consumed in the average daily diet was the equivalent of radiation absorbed during two hours of television viewing (from the TV set, itself) at a distance of six or eight feet. Such hazards may be real, they may even constitute a risk within the average lifetime; only future research will tell. Nevertheless, an objective view of such hazards would appear worthwhile if one is to appreciate the question of risk as applied to food additives. It might also be useful to compare these risk factors with those posed by naturally occurring food toxicants (poisons) found in many common foods (see Chapter IX).

Classification of Additives for Regulatory Purposes

As a matter of regulatory convenience additives are usually grouped into two categories, regulated additives and additives generally recognized as safe (GRAS). The latter group, as well as the acronym GRAS, was adapted from language used in the Food Additives Amendment of 1958. This document refers to one category of food ingredients as being derived from any "substance . . . generally recognized, among experts qualified by scientific training to evaluate its safety, as having been adequately shown . . . to be safe under the conditions of its intended use. . . ." Thus a group of food ingredients (many of them foods in their own right) was arbitrarily set aside as being, to the best of scientific knowledge, safe to use almost without restriction, at least without those restrictions normally applied to regulated additives. When cyclamates, one such GRAS substance, were banned in 1969, the president directed the FDA to review the safety of all materials being added to food, particularly those generally regarded as safe. This review is currently underway and no doubt will continue indefinitely as research techniques improve and new findings are brought forth. In the meantime the distinction between regulated food additives and GRAS substances, for all practical purposes, will become more and more obscure.

Foods, Food Additives, GRAS Ingredients

Without the need for regulatory distinction there would be little or no difference between the meaning of the terms *foods, food additives,* and *GRAS ingredients.* Casein, the principal protein in milk, is a food while present in the white fluid called milk. Isolated in chemical form, sodium caseinate, the same protein, is an additive. Citrus flavors as they occur in oranges or lemons are foods, but when extracted from these foods and added back to other foods they become additives. Because they are commonly found in most diets these food/food additives are assumed to be "safe." Just such assumptions were the basis upon which a number of substances, following passage of the Food Additives Amendment, were categorized as GRAS. In practice, GRAS ingredients were controlled less rigidly than regulated additives, which required proof of safety and functionality and which, as a rule, were limited to a given use rate (tolerance) in specific foods. GRAS substances were generally considered safe at almost every "functional" level and were allowed to proliferate freely in those foods in which useful applications could be found.

Distinguishing between food and additive promises to be even more difficult for formulated foods where each ingredient—food though it may be as a component of some plant or animal source—becomes just another ingredient in a dairy or meat substitute (or in other such foods). Where and how do you draw the line? Obviously, this is a regulatory responsibility and action in this direction has already been taken. Here, it seems well to note that "food" and "additive" are regulatory distinctions, and many common foods, under regulatory guise, are called additives.

Some Legal Aspects

Non-GRAS additives achieve legal status through regulation, usually after prolonged, costly research has been performed to document their safety. On paper, the procedure seems simple enough. Anyone wishing to use or sell their product as an additive need only petition the FDA to establish a food additive regulation. But petitioning and gaining approval are quite different matters. In support of the petition, the FDA demands "proof" of safety, functionality, and appropriate use rate. Or the agency may require specific test procedures or data. The filing of a petition may be preceded by years of research that has cost millions of dollars—and all this without guarantee that a regulation will in fact be promulgated.

If the evidence warrants it, the FDA publishes (in the Federal Register) a proposal to establish the substance under consideration as a food additive. Time for comment is allowed and, if no untoward findings are revealed, a final order is published; an additive is born.

A food additive regulation often sets specific guidelines for use. The regulation may (1) restrict the amount permitted in food, (2) impose standards of quality (purity), (3) prescribe manufacturing methods, those methods known to assure purity and safety, and (4) require specific labeling information (usually indicating procedures for safe use). If an additive is not used as specified or is used without prior approval, the food is considered adulterated, the handler subject to prosecution or injunction. Under the law, adulterated food may be seized and even destroyed, if conditions demand such measures be taken.

For a substance considered extraordinarily safe, GRAS status may be granted. GRAS additives need not be legalized by regulation (though recent proposals would leave them little short of regulation in specifics of use), but the process for gaining legal recognition, like that for "regulated" additives, starts with a petition to the FDA. Again, anyone may file. Or one may choose to assume a product to be GRAS, use it as such, and await legal action. In this case disagreement between the FDA and the user ends up in the courts. At issue more often than not is the question of safety—relative harmlessness.

Harmlessness, an Impossible Condition

"To provide assurance that any substance is absolutely safe for human consumption is impossible." So maintains an introductory comment in regulations regarding eligibility for GRAS status. Agreement on this premise is essential to further consideration of food (or additive) safety. If such agreement is possible, safety can then be defined as a consensus among scientists formed after they have evaluated suitable information. A further delineation would then be necessary to establish precisely the meaning of "suitable information." This is generally the approach taken by the FDA. Suitable information is expressed as data secured by (1) scientific procedures and/or (2) experience based on common use. GRAS status, therefore, is achieved through scientific investigation or knowledge gained from long-term experience with any substance found or used in foods. Safety is "reasonable certainty in the minds of competent scientists that the substance is not harmful."

For the future the FDA proposes to define "scientific procedures," as also including "those human, animal, analytical and other scientific studies, whether published or unpublished, appropriate to establish the safety of a substance." Common use in foods will come to mean, definitively, "substantial history of consumption of a substance by a significant number of consumers in the U.S." Note the restriction to the United States as the only country from which such information will be considered valid. This important point indicates the impropriety of deriving consumption data for local use from foreign countries where food sources and eating patterns may be significantly different from our own. Similarly an investigation of food (ingredient, additive) safety requires consideration of sex and age differences as well as eating habits (within the United States). For example, if a person for some reason consumes only vegetables or, out of religious or cultist preference, restricts his diet to a narrow group of foods, he may take in relatively larger (or smaller) doses of food additives or natural food toxicants than does a person on a normal diet. Or in restricted diets an additive may offer some nutritive value that otherwise might be lacking and result in dietary deficiencies. It must always be remembered that many food additives are themselves foods or constituents of foods with some nutrient value. They are simply extracted from natural foods, concentrated perhaps, and then added back in quantities sufficient to obtain a consistent quality from package to package.

Evaluation of safety must take into account probable consumption, cumulative effects, and such bizarre possibilities as mutation (mutagenesis) and effects on the unborn fetus (teratogenesis). Further, the use of scientific procedures to affirm GRAS status will in the future require the "same quantity and quality of scientific evidence" as that for a regulated food additive. Likewise certain GRAS additives will be accepted for use only under those restrictions heretofore applied only to the regulated class. Such restictions exclude use in food categories other than the one(s) assigned as well as use for different functional purposes and at higher levels than those set by the regulation. As indicated earlier, GRAS and regulated additives will scarcely be differentiated species any longer.

The GRAS review, expected to generate a very large number of references (by one estimate, a half million), will apply both to substances added directly to foods and to food contact surfaces (packaging materials).

What Additives Do in Foods

During the review of GRAS additives the National Academy of Sciences surveyed the food industry to determine the use rate of GRAS substances. This survey and an attempt to classify compounds by functional effect have resulted in a list of thirty-two categories of direct food additive functions. Generally, additives currently used as food ingredients will fall into one or more of these categories. As now applied for regulatory purposes, this list, with official interpretation of individual categories, is presented in Appendix II. A broader listing including incidental additives as well may be found in Appendix III, which I compiled from the additives currently permitted (on whatever basis) in foods. A third tabulation, Appendix IV, shows the food categories proposed for regulatory control in the future (for tolerances or limitations of direct human food ingredients). A perusal of these appendixes should be helpful in appreciating the breadth of the commodities and the uses of additives that are covered by food laws and regulations.

A summary list (without definition) of the technical functions of direct food additives is presented in Table 12. Although some of these functions (or the names by which they are known) may appear highly technical, practically all of them are related in one way or another to the necessity for preparing foods of good flavor and texture and then keeping the food in that condition during the time between processing and consumption. This time span, this relatively imprecise period (and condition) when the food is in the supermarket and in the home, is often critical and makes the use of additives essential. A cake baked from scratch in the home might (depending on the cook's ingenuity) last a couple of days before becoming dry and crumbly. A homemaker appreciates that fact and bakes and handles a cake accordingly—either urging consumption while fresh or throwing the leftover away. A commercial company baking the same cake and marketing through a grocer or chain store must assure a moist texture and good flavor for that time required to sell it at retail and store it in the home before eating. Almost without exception, assuming average time lapses between processing and consumption (which are considerably longer than would be anticipated for home-prepared goods), the only way to achieve this end satisfactorily is through the use of additives (many of which the homemaker uses while calling them ingredients). Texturizers would not be added as extensively were it not for this fact, nor would preservatives (which help to keep the product fresh during retailing and home storage), humectants, stabilizers, lubricants, or firming agents. Additives are placed in food to maintain (and provide) nutritional quality, to prevent spoil-

Table 12. Classification of GRAS Substances by Technical Effect

1. Anticaking agents, free-flow agents
2. Antimicrobial agents
3. Antioxidants
4. Colors, coloring adjuncts (including color stabilizers, color fixatives, color-retentive agents, etc.)
5. Curing, pickling agents
6. Dough strengtheners
7. Drying agents
8. Emulsifiers, emulsifier salts
9. Enzymes
10. Firming agents
11. Flavor enhancers
12. Flavoring agents, adjuvants
13. Flour-treating agents (including bleaching and maturing agents)
14. Formulation aids (including carriers, binders, fillers, plasticizers, film-formers, tabletting aids, etc.)
15. Fumigants
16. Humectants, moisture-retention agents, antidusting agents
17. Leavening agents
18. Lubricants, release agents
19. Nonnutritive sweeteners
20. Nutrient supplements
21. Nutritive sweeteners
22. Oxidizing and reducing agents
23. pH control agents (including buffers, acids, alkalies, neutralizing agents)
24. Processing aids (including clarifying agents, clouding agents, catalysts, flocculents, filter aids, etc.)
25. Propellants, aerating agents, gases
26. Sequestrants
27. Solvents, vehicles
28. Stabilizers, thickeners (including suspending and bodying agents, setting agents, gelling agents, bulking agents, etc.)
29. Surface-active agents (other than emulsifiers, including solubilizing agents, dispersants, detergents, wetting agents, rehydration enhancers, whipping agents, foaming agents, defoaming agents, etc.)
30. Surface-finishing agents (including glazes, polishes, waxes, protective coatings)
31. Synergists
32. Texturizers

Source: Adapted from Federal Register. 1974. 39(185):34175.

age, and to make foods appealing to the senses. Families expect as much from the homemaker who cooks from scratch, and it should not be surprising to find the food processor striving for these same goals for the families who choose to buy and serve commercially prepared foods. Additives, and what they do in foods, make it possible.

Annual Intake of Food Additives

There is no way of knowing the number or amount of indirect (incidental) additives (food contaminants) consumed annually. With a list of potential contaminants reaching into the thousands and with the many complex routes by which such additives gain entrance into foods, an objective estimate must admit that obviously some are consumed. That the amount consumed is slight—though not necessarily risk-free (witness those persons with extreme allergic sensitivity)—and probably far less than the smallest quantities of direct additives sprinkled into foods seems reasonable to assume. As an assumption, then, which is about the best that can be made, the amount of incidental food additives consumed each year per person would have to be measured in amounts a fraction of the weight of a grain of salt. Weighed against a bountiful, low-cost food supply, many—though certainly not all—of these additives can be justified and can be considered acceptable risks.

It is perhaps simpler (though far from easy and subject to much individual variability) to estimate the amount of direct (ingredient) food additives consumed. Currently, a total of 1540 such compounds are under review by the FDA. Before long, the list of direct food additives may be finite and each chemical will have been reviewed and either approved or disapproved.

Hall (see Chapter III, ref. 2) has compiled and graphically illustrated the consumption per capita of the majority of direct food additives in readily understood form. Of 1830 compounds, the median level (that point at which half the values are above and half below) of additive intake was found to be one-half milligram. That is, 915 additives are consumed at levels above one-half milligram annually, 915 at levels of less than this amount, a half milligram being about the weight of one grain of salt.

Of equal if not greater importance to health is the additive consumed in greatest amounts—sugar. The annual intake of sugar has been increasing for a number of years and is now slightly over 100 pounds per person. Salt is next at 15 pounds, followed by corn syrup and dextrose at 8.4 and 4.2 pounds respectively. Another 32 additives are consumed in amounts of about 9 pounds per person per year (see Table 13). The remainder—some 1800 different compounds—are consumed in total at an annual rate of about 1 pound per person.

Reflecting upon these thirty-two additives consumed in greatest quantities, it is well to ask why, in fact, they should head the list. Their functions are clue enough. And if you are honest, you will have to admit that the list mirrors the society in which we live. If you ask in this case which came first, the chicken or the egg (cultural eating habits or additives), the answer resounds in favor of the former. As a nation we have nurtured the desire for convenience foods—easy to prepare, tasty, no after-dinner dishes—and even the sweet tooth. We like to eat, and we don't want to spend time in the kitchen,

Table 13. Some Widely Used Food Additives[a]

Additive	Function
Monosodium glutamate	Flavor enhancer
Mustard	Flavor
Black pepper.	Flavor
Hydrolyzed vegetable protein	Stabilizer (thickener)
Acacia	Stabilizer (thickener)
Modified starch	Stabilizer (thickener)
Yeasts	Leavening
Monocalcium phosphate	Leavening
Sodium aluminum phosphate	Leavening
Sodium acid phosphate.	Leavening
Sodium carbonate.	Leavening, acidity control
Calcium carbonate	Leavening, acidity control
Dicalcium phosphate	Leavening, acidity control
Disodium phosphate	Leavening, acidity control
Sodium bicarbonate	Acidity control
Hydrogen chloride	Acidity control
Citric acid	Acidity control
Sulfuric acid.	Acidity control
Sodium citrate.	Acidity control
Sodium hydroxide	Acidity control
Acetic acid.	Acidity control
Phosphoric acid	Acidity control
Calcium oxide.	Acidity control
Lecithin	Emulsifier
Mono and diglycerides	Emulsifiers
Sulfur dioxide.	Preservative
Calcium chloride	Firming agent
Calcium sulfate	Processing aid
Carbon dioxide	Effervescent
Sodium tripolyphosphate	Curing humectant
Caramel	Color

Source: Richard L. Hall, 1973. Food additives, Nutrition Today, 8
(4):20-28. Reproduced by permission of Nutrition Today, Inc.
[a] Consumed annually in amounts of 9 pounds per person.

so we have additives that acidify and leaven rolls at any moment between purchase and days or weeks later, when, for that quick dinner, a panful is removed from the refrigerator, popped in the oven, and heated. Thus two leavening agents allow commercial bakers, not unlike the rest of the labor force, to work eight-hour shifts. Flavors and flavor enhancers speak for themselves; we like to eat tasty foods. Starches and emulsifiers, which make up most of the remaining top thirty, are thickening agents no less necessary for commercial sauces, gravies, and dressings than for their homemade cousins. And, if you will, top off the list with some carbon dioxide, the fizz in a Coke or the effervescence in a Scotch and soda. By our own actions and desires and lifestyle preferences we have asked for the foods—and the additives—now lining super-market shelves.

The question of natural food toxicants is taken up elsewhere, but for purposes of comparison while discussing levels of additive intake, it is pertinent to note in a limited way the amounts of some natural poisons that are consumed regularly in the diet of most Americans. Bear in mind the median level of additive intake of one-half milligram. Potatoes, then, at 119 pounds annual consumption per person, provide a natural poison, solanine, at a level of about 9700 milligrams. Nitrates in lettuce account for 12,000 milligrams of toxicant, myristicin in nutmeg 44 milligrams. Hydrogen cyanide gas, a product of lima beans, is regularly consumed at levels of 42 milligrams per person each year from that one source alone. Mercury in tuna fish, not an additive but a chance contaminant of the environment in which tuna abounds, looks small by comparison at 0.25 to 2.0 milligrams per person per year.

Food Additives, in Balance

To provide some basis for a rational viewpoint on the issue of food additives, we should perhaps list both the good and the bad elements of additive functions. A listing of this kind has been compiled by a committee of the National Academy of Sciences. Here, paraphrased, are the significant items as seen by this committee. First, the inappropriate additive uses:

1. To disguise faulty or inferior processes.
2. To conceal damaged, spoiled, or inferior goods.
3. To deceive consumers.
4. To gain some functional property at the expense of nutritional quality.
5. To substitute for economical, well-recognized good manufacturing processes and practices.
6. To use in amounts in excess of the minimum required to achieve the intended effect(s).

Now for the appropriate uses:

1. To improve or maintain nutritional value.
2. To enhance quality.
3. To reduce waste.
4. To enhance consumer acceptance.
5. To improve keeping quality.
6. To make the food more readily available.
7. To facilitate food preparation.

It is of course understood that any additive applied for whatever reason should not constitute a significant health or environmental hazard. Consumer needs must be recognized and given just concern, as should the essential facets of food processing and legitimate needs of processors. Underlying all this are economic and energy considerations that bear directly on food costs and total output. A weed killer that requires a pass over a field of corn by a single tractor or plane, as weighed against two or more trips with the harrow to remove those same weeds, must be viewed in light of present-day urgencies. And in that list of appropriate additive uses can be found at least three items of critical importance to a world suffering from food shortages.

In balance, food additives appropriately used have a valid, meaningful place in food for man. If there is hazard associated with their use, there are likewise hazards involved should they not be used, the latter primarily microbiological and nutritional in nature. Food spoilage (waste) and food poisoning are consequences of the one, malnutrition or natural food poisoning products of the other. These are ever-present hazards to health, and are areas where research and regulatory activity can be of continuing benefit. Clearly, the major needs of the world today—and in the foreseeable future—are more food and better nutrition. Well intentioned though they may be, those who would divert science and government from these ends do so at the nutritional expense of the family of man.

IV

Food Flavor

Natural Flavor (or natural flavoring)—*the essential oil,
oleoresin, essence or extractive, protein hydrolysate, distillate or any product of roasting, heating or enzymolysis
which contains the flavoring constituents derived from a
spice, fruit or fruit juice, vegetable or vegetable juice, edible
yeast, herb, bark, bud, root, leaf, or similar plant material,
meat, seafood, poultry, eggs, dairy products, or fermentation products thereof, whose significant function in food
is flavoring rather than nutritional.*

Artificial Flavor (or artificial flavoring)—*any substance,
the function of which is to impart flavor which is not derived from the sources indicated above.*

FDA definition

"Gee, that tastes good! It has such a good flavor." With that simple statement of pleasure, heard so often, unfolds a story of humankind's searches
and researches almost from the time our early ancestors discovered fire and
the sizzling good taste of a cave-side meal. And the search goes on, today as
never before, while flavor, the chemistry of it, still remains much a mystery
to those scientists whose job it is to analyze and formulate something that
"tastes good."

Without becoming bogged down in technicalities, we should have some common ground of understanding in order to discuss flavor in an intelligent manner. First, flavor per se, as will be spoken of here, means specific ingredients (chemical compounds) present in or added to food. In reality, though, flavor is a combined sensation of the tongue (taste buds) and nose (olfactory nerves). A food tastes/smells good. To coin a new, perhaps more accurate word, we may say that a given food yields a pleasant *flavaroma*. In this book no distinction will be drawn between the two sensations, only between the senses and the products that activate the senses. Secondly, only four basic tastes exist: sweet, salt, sour, and bitter. Beyond these simple distinguishable tastes, food flavor—as we know it—is a master brew of a number of sensual perceptions. As might be applied to a particularly tempting dish, flavor sensation itself can probably best be described as "indescribable."

But I began by implying that the chemical makeup of flavor ingredients was complex. As a starter, then, just to stress the point, take a look at some of the chemical compounds of a seemingly simple flavor (aroma), apple, as as shown in the accompanying list. That's one flavor (aroma)—or about as close to it as science is able to take us. The combined efforts of more than forty research scientists working since 1920 were required to segregate the eighty-odd chemical components now recognized as belonging to apples. There may well be a number of components yet to be isolated. Cocoa is even more complex, consisting of over 125 basic chemical elements. French cognac has 96 chemicals, coffee nearly 200.

It is important to note that there are literally hundreds of different flavors, many of which are fully as complex as apple flavor, and that flavors—those ingredients in foods that combine to titillate the sense of taste and smell—are actually a complex mixture of chemicals, with each chemical found in just the right proportion. The cook who adds a pinch of this and a pinch of that to produce a *flavor supreme* fashions, in chemical consistency, a minor miracle!

Natural versus Artificial Flavor

So you're a *natural* flavor buff. To each his own; no one will argue your right to prefer natural over artificial flavors. But are you sure of the grounds upon which your choice is made? Is it taste alone? Could you tell the difference—blindfolded—between two flavors, the one natural, the other synthetic? Or is your preference based more on an aversion to anything synthetic or artificial?

Chemical Compounds of Apple Flavor

Acetaldehyde	2-Hexen-1-ol, *trans*
Acetic acid	2-Hexenyl acetate, *trans*
Acetone	Isobutanol
Acetophenone	Isopentane
Benzoic acid	Methanol
Benzyl acetate	2-Methylbutanal
Butanal	3-Methylbutanal
Butanol	2-Methylbutan-1-ol
2-Butanone	3-Methylbutan-1-ol
1-Butoxy-1-ethoxyethane	Methyl butyrate
Butyl acetate	3-Methylbutyric acid
Butyl butyrate	Methyl formate
Butyl hexanoate	Methyl hexanoate
Butyl propionate	Methyl 2-methylbutyrate
Butyric acid	Methyl 3-methylbutyrate
Diacetyl	1-Methyl-naphthalene
1,-1-Diethoxyethane	2-Methyl-naphthalene
Diethyl ether	2-Methyl propanal
Ethanol	Methyl propan-1-ol
1-Ethoxy-1-hexoxyethane	2-Methyl propionic acid
1-Ethoxy-1-methoxyethane	2-Methylpropyl acetate
1-Ethoxy-1-(2-methylbutoxy)-ethane	2-Methylpropyl propionate
1-Ethoxy-1-propoxyethane	Nonanal
Ethyl acetate	Pentanal
Ethyl butyrate	Pentanoic acid
Ethyl formate	2-Pentanone
Ethyl hexanoate	3-Pentanone
Ethyl-2-methylbutyrate	Pentyl acetate
Ethyl-2-methylpropionate	Pentyl butyrate
Ethyl octanoate	Pentyl-2-methylbutyrate
Ethyl pentanoate	2-Phenylethyl acetate
Ethyl-2-phenylacetate	Propanal
Ethyl propionate	Propanol
Formaldehyde	2-Propanol
Formic acid	Propionaldehyde
Furfural	Propionic acid
Geraniol	Propyl acetate
Hexanal	2-Propyl acetate
Hexanoic acid	Propyl butyrate
n-Hexanoic acid	Propyl pentanoate
2-Hexanone	Propyl propionate
2-Hexenal, *trans*	2,4,5-trimethyl-1,3-dioxolane

Source: Adapted in part from Giovanni Fenaroli, *Fenaroli's Handbook of Flavor Ingredients,* ed., trans., and rev. by Thomas E. Furia and Nicolo Bellanca (Cleveland: Chemical Rubber Co., 1971). Used by permission of the Chemical Rubber Co.

Some readers will remember when the seal "Made in Japan" denoted a cheaply built, poor quality item or a substitute for a superior product, the label of which read "Made in U.S.A." Today, this distinction cannot be drawn. The numbers of Americans who own Japanese cars or cameras or snowmobiles—and swear by them—make it apparent that the distinction, if any, favors the foreign product. Yet in those who grew up with the inexpensive toys and goods from overseas, there still lingers the old prejudice of inferiority. It is a difficult prejudice to overcome.

The same is true of various synthetics. When they first appeared on American markets, they were often considered inferior. But today, many synthetic products—plastics, clothing materials, adhesives, rubberlike compounds, carpeting, insulation, etc.—are known to be superior in quality to their natural counterparts. Nevertheless the old bias, justified though it may have been, haunts us still. Synthetic (or artificial) still rings hollow, at least where food is concerned.

Since we cannot touch or feel or see or wear the synthetics in food, there is no direct way of noting quality and no experience to draw from, which, if favorable, would tend to lessen bias against a given food. Lacking tangible evidence, we perhaps need a speaking acquaintance with food chemistry to help us to "know" the flavor products being marketed, products which, in no way except through taste (and possibly food cost), can be known to us otherwise.

Many flavors are as common as the sour taste in lemons. In fact they may be derived from such everyday food products. And when they are, the flavor isolates are referred to as naturally occurring—that is, found in nature. One such flavor compound common to citrus fruits is citric acid. A share of the sour, puckery taste associated with these foods stems from the presence of this organic acid. But we can't feel or touch or see the shape of citric acid without the aid of a chemist who would tell us it looks (in two dimensional shape) like this:

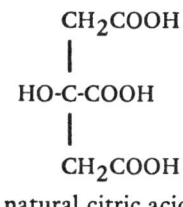

$$CH_2COOH$$
$$|$$
$$HO\text{-}C\text{-}COOH$$
$$|$$
$$CH_2COOH$$

natural citric acid

The flavor chemist who isolates this acid from citrus fruits has learned its shape and appearance with no less certainty than the blind man who knows shape and texture by "feel" alone.

Now, if our flavor chemist went into the laboratory and used his knowledge of chemistry to build from scratch a unit of citric acid, it would look like this:

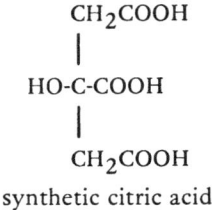

synthetic citric acid

Compare the two. Can you see a difference? Look carefully; examine every line and atom. Is there a difference? The answer of course is no. The two molecules of citric acid are identical, though one was produced in the sun and rain outdoors, the other in a chemist's artificially lighted laboratory. You can see no difference, you could taste no difference. Your body would utilize the one as it utilizes the other. The only conceivable difference between the two lies in the minute amounts of contaminants that may be associated with the sources from which the twin molecules were derived. And at present science is unable to prove that these differences are meaningful, although a number of research laboratories are pursuing this hypothesis as it relates to nutritive value. Certainly the two compounds taste alike; their flavor is identical. For purposes of distinguishing between the two, as would be necessary for twin children, we'll name them *natural* citric acid and *nature-identical* citric acid. Although the latter is synthetic, it is not termed *artificial*. Nor is the food processor required to label synthetic, nature-identical flavor as artificial. The question of dissimilarity comes down to this, Is water (H_2O) that falls from the sky any different from water formed in the following reaction (assuming both waters are distilled to remove contaminants absorbed from their respective environments)?

NaOH	+	HCl	$\longrightarrow$	NaCl	+	H_2O
sodium hydroxide		hydrochloric acid		sodium chloride		water
				(table salt)		

For that matter is the salt also formed in the chemical reaction any different from salt burning whitely in the salt flats of Utah? The one is natural, the other synthetic—nature identical.

The explanation above of course is an oversimplification. Lemon flavor consists of far more than the single acid identified in the example. But the chemistry by which citric acid is known, like the fact before your eyes when you spot a person you know, is valid. Chemists know flavor by chemical makeup. They are no less certain of their (known) facts than we of the facts we see. What if we went one step further, then, and extracted from lemons their flavor essence—the entire mix of chemicals that constitute lemon flavor? Would this not be, in effect, natural lemon flavor?

In practice, natural flavors are isolated from the varieties of foods in which they are found, then added back to foods to enhance flavor or to maintain flavor consistency. At present, though there may be some minor "flavor" differences, the law treats extracted flavors as "natural." On the food label they are declared as such. But there is natural and then there's natural. A quick review of the FDA definition of natural flavor reveals a number of terms or phrases all of which denote flavor compounds obtained from natural sources. The differences lie primarily in the methods used to extract or concentrate the flavor compounds. By FDA definition, the terms and meaning of natural flavoring substances are as follows:

Essential oil—aroma compounds obtained by hand or mechanical pressing (like squeezing the juice from a lemon) or distillation. Probably the truest flavors are pressed flavors. Distillation, while extracting the oils, may cause minor flavor changes. Some water soluble chemicals are lost and some heat damage occurs during distillation.

Oleoresin—the flavor compounds (usually derived from spices) extracted by percolation with various chemicals. Just as water percolating through coffee grains extracts flavors from the coffee, so do a variety of solvents leach out flavor compounds from food substances.

Essence or extractive—the concentrate of flavor or aroma compounds stemming from natural sources.

Protein hydrolysate—flavorful fragments of protein obtained by chemical or enzymatic splitting of the natural food protein.

Roasted (heated) flavors—flavors produced by heating. Sometimes the heating of foods produces off-flavors; other times desirable flavors are formed. The latter, of course, are the goal of flavor manufacturers.

Products of enzymolysis—flavorful products of enzyme reactions. Enzymes are natural catalysts with the ability to react specifically on food linkages, thus splitting larger food molecules into smaller, sometimes entirely different flavored compounds.

In the laboratory we'll now combine two chemicals in presence of sulfuric acid (a mineral acid) to form a product not found in nature:

$$\text{allyl alcohol + butyric acid (found in milkfat)} \xrightarrow{\text{H}_2\text{SO}_4}$$
$$\text{allyl butyrate + H}_2\text{O}$$

What have we got? Well, let's sniff it. Ummm, not bad. Smells like apple. Or maybe apricot. (It's hard to tell.) Anyway, *allyl butyrate* has an aroma similar to apple. Used in food, it would lead one to believe natural apple flavor was present—except that allyl butyrate has not yet been found in nature. It is not identical to any flavor/aroma compounds known to occur in apples. For that reason, and for that reason alone, it must be treated as *artificial*. On food labels it will be so designated, as indeed it should be.

No doubt many of us would turn up our noses or shrink from a food that contained allyl butyrate. It is human nature to do so. But we probably wouldn't take to dragon's blood either until we realized that it was a *natural* flavor compound derived from a palm tree of the Malay Archipelago. And even that might turn us off—natural or not. The same goes for dog grass, oak moss, schinus molle, and turpentine, all of which are natural flavor compounds.

Finally, to clarify a possible point of confusion, it should be stressed that the word *artificial*, as presently applied to food flavor, means synthetic or "unnatural" (not found in nature). It does not, nor should it, necessarily imply a nonnutritive character, as it does when applied to sweetening compounds. Artificial sweetener is low in calories or noncaloric. Artificial flavor is flavor not found in nature.

For a partial list of natural and artificial flavoring compounds, see Appendix V.

Flavor Usage and "Toxicological Insignificance"

Throughout this text reference is made to the concept of harmless amounts of food additives. Not always will such levels be known (or agreed upon) precisely. But the concept of "toxicological insignificance" is an important one, as it applies both to residues of crop chemicals and to additives placed intentionally in foods to achieve a given effect—for example, flavor.

The Flavor and Extract Manufacturers' Association (FEMA), which represents the flavor manufacturing industry, is a self-policing body active in evaluating the usage and safety of flavor ingredients. Members of the association are urged to submit new flavor chemicals to an independent scientific panel of experts for review and study; in turn, the panel carries out surveys of flavor usage. Most new flavors, by this means, are exposed to a broad screening program that helps to provide assurance of safety. Guidelines for evaluating safety are those recommended by the Food Protection Committee of the National Academy of Sciences/National Research Council. Among the guidelines are procedures for determining toxicological insignificance. The NAS/NRC views this concept as necessary for many food chemicals whose level of usage is very small. For these substances, the committee states, "It is justifiable to employ accumulated scientific experience and to recognize their structural analogy to other chemicals whose metabolism or toxicity is known." Toxicological insignificance can be established through "reasoning by analogy." The statement continues: "If a substance meets all the following criteria, it may be presumed to be toxicologically insignificant at a level of 1.0 ppm or less in the human diet. 1. The substance in question is of known structure and purity. 2. It is structurally simple. 3. The structure suggests that the substance will be readily handled through known metabolic pathways; and 4. It is a member of a closely related group of substances, that, without known exception, are or can be presumed to be low in toxicity." This safety category excludes compounds found to induce cancer or "any such serious condition" as well as compounds that "exert significant biological effects." These are the ground rules, then, by which relative harmlessness is established.

In their latest survey the FEMA independent expert panel reported 1249 flavor substances of GRAS status (FEMA GRAS status is to be distinguished from government GRAS status, though principles of establishing safety are similar). Of the 1249 GRAS flavor compounds, 831 were estimated at a *total* use level of 1000 pounds annually. The average maximum use levels fell below 10 ppm. The FEMA regards these data as indicative of toxicological insignificance.

What the Label Tells You

Information regarding flavor may be found in the name of the food and also in the list of ingredients on the package. The flavor designation tells whether a flavor is natural or artificial, whether more than one natural flavor is present, and whether or not a characterizing ingredient—peaches in peach pudding, for example—is present in amounts sufficient to characterize the flavor in and of itself, or, instead, if other flavors (natural or artificial) have been added to supplement the characterizing ingredient.

Natural flavors are often more costly than artificial flavors (though, depending upon individual taste preferences, not necessarily better tasting) and this processing expense will probably be reflected in a higher food cost.

What to Look for in Flavor Designation

Plum pudding! Now what is that? Certainly the words conjure up some meaning to us, a meaning tempered by age, past experience, and flavor preferences. But does that product contain plums? Is it flavored entirely, or only partly, with plums? Are other flavoring compounds present? These are some of the questions that can be answered by careful examination of flavor wording on a food label.

First, in a product like plum pudding, one would expect plums to be present, to be the "characterizing" ingredient. If the product has a standard of identity (there's no way of telling from the label), the amount of plums required independently to establish the food would be set by regulation. And if that amount of plums were present, the label would read simply "Plum Pudding."

Suppose, though, that there were not enough plums to flavor the product by themselves, or that a food processor wanted to enhance the flavor—and why not? Plums vary in sweetness and flavor, and if you, the customer, are familiar with a product and like it, you will want it to be the same each and every time you buy it; therefore additional flavor might be added. The flavor could be natural, stemming from plums themselves, or artificial. If natural flavor is added, the label would read, "Plum Flavored Pudding" or "Natural Plum Flavored Pudding." Notice in the latter that the words *natural* and *flavored* are both used. Those are the key words. There is flavor added, and it is natural flavor.

Now then, if natural flavor other than plum is present, the label would read, "Plum Flavored Pudding, with Other Natural Flavor," or "Natural Plum Flavored Pudding, with Other Natural Flavor." Note that specific reference is made to "other natural flavor." You can be sure it's a natural flavor, but one other than plum flavor.

Perhaps you're wondering why a processor would add other flavors. The reason, of course, is to create a more tasty product, or a unique taste, or to enhance the flavor in some way. For the same reason you might add lime juice to a food flavored primarily with lemon, to bring out the tart, lemony flavor.

Finally, if any artificial flavor—*any*—is added to the plum pudding product, the word *artificial* will be substituted for the word *natural* in the flavor designation. This will be true whether the amount of artificial flavor is a teaspoonful or a gallon.

In many food products no characterizing ingredient is expected; that is, you wouldn't expect vanilla beans in vanilla pudding, but simply a vanilla flavored pudding. Here again, information on the label can be useful in differentiating between foods and may provide insight into cost factors and certainly into flavor. When pure vanilla extract is used to flavor vanilla pudding, the food will be labeled "Vanilla Pudding." Vanilla, then, is the natural flavoring ingredient, a flavor extracted from the vanilla bean. The presence of other natural flavor is readily determined by the wording "with other natural flavor," tagged to the end of the food name. Likewise, the presence of artificial flavor poses no problem. The product will be labeled "Artificially Flavored Vanilla Pudding" or "Vanilla Pudding, Artificially Flavored."

There is one additional possibility where flavor is concerned—if you can imagine it—in which natural flavor is in fact labeled as artificial. This is not a common occurrence. But if, for example, natural lime flavoring were substituted for natural lemon flavoring, a consumer could be misled into thinking the product was, say, lemon pudding when indeed it was not. Therefore, flavor regulations require that if none of the natural flavor used in the food is derived from the product whose flavor is simulated, the food will be labeled either with the flavor of the product from which the flavor is derived or as "artificially flavored." In effect, this means you won't be deceived by the *origin* of the flavor or by the label, which must characterize the flavor origin of the food inside.

Since this may seem complicated, the following outline may help to simplify matters.

Flavoring	*Food Name*
Vanilla extract	Vanilla Pudding
Vanilla extract and other natural flavor(s)	Vanilla Pudding, with Other Natural Flavor
Natural flavoring other than vanilla, or any artificial vanilla flavor (presence of *any* artificial flavor that simulates the flavor indicated on the carton automatically requires designation as "artificial")	Artificially Flavored Vanilla Pudding or Vanilla Pudding, Artificially Flavored

When characterizing ingredients are expected:

Plums (enough to characterize the food)	Plum Pudding
Plums (not enough to characterize the food independently) and	
(1) Natural plum flavor	Plum Flavored Pudding or Natural Plum Flavored Pudding
(2) Natural flavors other than plum	Plum Flavored Pudding, with Other Natural Flavor or Natural Plum Flavored Pudding, with Other Natural Flavor

When Three or More Flavors Characterize the Food

Some foods consist of several different, separately identifiable flavors or are blends of flavors with no specific recognizable flavor. Mixed flavors of ice cream are examples of the former (chocolate, vanilla, and strawberry). Fruit punch is an example of the blended flavor, where no flavor is predominant over the others. In such products, flavor designation may be somewhat different, that is, not all flavors will necessarily be declared on the label. A specific generic term (*fruit punch*) may be used instead. However, the use of artificial flavors must be declared, and fruit punch so formulated would be labeled "artificially flavored fruit punch."

Exceptions

There are exceptions to most rules and flavor labeling is not unique in this respect. Perhaps the most notable exceptions in requirements for flavor labeling are to be found in those pertaining to frozen desserts—ice cream products. For many years the ice cream industry was regulated under a "three-category" system of flavor identification. Probably because confusion would arise if a new flavor declaration was implemented, ice cream (primarily) as well as butter and cheese are exempt from the 1973 flavor labeling requirements. Briefly, what this means is that they are exempt from the labeling requirements for artificial flavors *as such regulations are currently applied to other foods*. All three products, however, are standardized—are regulated by a standard of identity as previously described. The following summary describes the flavor designations as they apply to ice cream. The flavor descriptions also hold true for imitation ice cream products made with vegetable fat.

Flavoring	*Food Name*
Natural flavoring only (strawberries) or natural strawberry flavor	Strawberry Ice Cream
Natural and artificial flavoring, with natural flavoring predominant	Strawberry Flavored Ice Cream
Natural and artificial flavoring with artificial flavor *predominant* (or in entirety)	Artificially Flavored Strawberry Ice Cream

Flavor Designation in the Statement of Ingredients

Flavoring is the only additive that is associated with a food name and also declared in the statement of ingredients. Information regarding spices, color, and preservatives normally is found solely in the statement of ingredients. These additives, as well as flavor, are classed and declared collectively—all flavors as "natural flavor" or "artificial flavor," spice as "spice," etc. There are, however, some variations. For example, certain food components must be listed not as flavor, but by their common or usual names. The reason for this is that some products are more commonly thought of as foods than as flavors. Among these compounds are flavoring ingredients prepared by cutting, drying, or pulping fruit, vegetables, meat, fish, or poultry. If the flavoring is celery powder, it will be listed as such. The same can be said of granulated onions, garlic powder, etc. Salt, too, is a flavoring compound and is listed in the statement of ingredients. Chemically, salt is sodium chloride, but regulations require that it be listed by its common name, "salt." Monosodium glutamate, a common flavor enhancer, must, however, be listed as such—*monosodium glutamate.*

Smoked products have a characteristic smoky flavory that is hard to beat. Smoked cheese or smoked ham—just the words themselves—start the saliva flowing and create a unique flavor picture in the mind. Possibly you also envision a smokehouse out behind a farm with smoke rising from the chimney and a weathered wood interior reeking of dense smoke flavors of ham and bacon. This is one way flavor is imparted to foods. There is another way, though, which results in a tasty food, but doesn't require the expense and time associated with smokehouses. The method is simply to add "artificial smoke," a synthetic called pyroligneous acid. This acid is not listed by its chemical name on food labels, primarily because of public misgivings over chemical names. Instead, pyroligneous acid is termed "artificial flavor" or "artificial smoke flavor." What this says, then, is that the product has not been smoked by the natural process. True smoke flavor is not present, though doubtless few will be able to tell the difference.

On the subject of smoke flavor it should also be mentioned that true smoke flavor (as originates in the smokehouse) can be captured in liquid and added to a food. If this is the processing technique used, the ingredients statement will list the flavor as "natural flavor" or "natural smoke flavor," but not smoked. Sometimes the phrase used is "with natural smoke flavor" or "with added smoke flavor." But keep in mind: this doesn't mean a smokehouse treatment as such.

Spice. On the food label spice is treated much the same as flavoring is. Spices are grouped together and simply listed as "spice." Still, spices must be segregated from other types of flavoring, and for this purpose a legal definition is required. In processed foods spice is defined as "any aromatic vegetable substance in whole, broken, or ground form (except for products such as onions, garlic, and celery, which are traditionally regarded as food) whose main function is seasoning rather than nutritional."

Some common spices—though certainly not all—are listed below. They are spices within the legal definition of the word and would be grouped, for labeling purposes, under the single word *spice*.

Allspice	Cloves	Marjoram	Rosemary
Anise	Coriander	Mustard flour	Saffron
Basil	Cumin seed	Nutmeg	Sage
Bay leaves	Dill seed	Oregano	Savory
Caraway seed	Fennel seed	Paprika	Star aniseed
Cardamon	Fenugreek	Parsley	Tarragon
Celery seed	Ginger	Pepper, black	Thyme
Chervil	Horseradish	Pepper, red	Turmeric
Cinnamon	Mace	Pepper, white	

When an ingredient serves more than one function. When a flavor, spice, or coloring agent serves more than one function it is labeled either according to the various functions it serves or by its common or usual name. Spices such as paprika, turmeric, and saffron fall into this category. They are both seasoning agents and coloring compounds.

Flavor Enhancers (Potentiators)

No discussion of flavor can be considered complete without mention of a class of additives almost as important as flavor itself. Flavor enhancers, more aptly called potentiators because they make more potent the flavor of food without themselves contributing to flavor per se, after years of relative obscurity, have blossomed into full-fledged flavor counterparts with a future possibly as important as that of flavor itself.

In the broadest sense table salt may be considered a flavor potentiator, but only because, at very low concentrations, concentrations so low that the salty flavor is undetectable, other food flavors are heightened. By definition, then, salt fulfills the requirements of a flavor enhancer. But compared with the newer breed of potentiators salt is a relative weakling.

Monosodium glutamate. Speaking of the breed as new, though, is a mis-nomer of sorts. Monosodium glutamate (MSG), a powerful potentiator now sparking flavor buds at a worldwide rate of more than 150 million pounds consumed per year, came to light in the early twentieth century. But—and this is significant in view of suspicions that have been raised concerning its safety— it had been used for centuries by the Japanese who, unaware of the specific ingredient involved, commonly doctored their dishes with a dried mixture of seaweed. These naturalists of old, like naturalists today, prepared meals from nature's stores. And if the addition of a particular herb, spice, or seasoning compound improved the flavor, it became standard fare. Dried seaweed met the test. It was not until the turn of the twentieth century that a Japanese scientist asked, What is it in seaweed that enhances the flavor of food? He found the answer to be the compound MSG, a chemical agent by then well established in chemistry texts. There is nothing really new about MSG or, for that matter, *alginates, carrageenan,* or *furcellaran.* All are extracts of seaweed. All have technological functions in food processing. All have been a part of man's food for ages—except now we know them for what they are, and we call them by their first names.

Like so many food substances, MSG is a natural evolution of the centuries-old struggle to prepare a more tasty dish. The Japanese housewife added sea-weed; the food processor added a seaweed isolate, MSG. But seaweed is not the only source of MSG. Glutamic acid, of which MSG is the sodium salt, is a common amino acid of protein. Protein with significant amounts of glutamic acid is a potential source of MSG. At 36 percent each, wheat gluten and zein are rich sources of this amino acid, corn gluten is 25 percent glutamic acid, rice 24 percent, casein (milk protein) 22 percent, and wheat flour 20 percent. Green peas, spinach, canned mushrooms, and mother's milk all contain this amino acid, though in lesser amounts. From the glutamic acid in the foods we eat the body actually manufactures MSG during digestion. Protein is first hydrolyzed (split into smaller and smaller component parts) in the gut, then in the more alkaline environment of the lower digestive tract glutamic acid (if found in the protein) is converted to its salt, in this case MSG. This fact is im-portant. It is one criterion applied to evaluate safety. If an "additive" is known to be a food component or is a common constituent of the human body, the assumption that it is safe, at least at those levels normally found in the body, must rate high in probability. Certainly this kind of evidence must rank at least equal with the feeding trials of other species.

A few years ago, monosodium glutamate came under close scientific scrutiny. In his laboratory a researcher (in this case a psychiatrist) observed untoward reactions in immature mice which had been administered MSG. Brain lesions, obesity, and stunted growth were some of the results noted. All, certainly, are serious conditions. But almost as soon as the findings were reported, three other scientists, responding in the same technical journal, took issue not with the results but with the way in which the experimentation was carried out. Monosodium glutamate had not been given orally, rather it had been injected under the skin. Thus the observations were deemed "irrelevant." Animals do not react similarly to injections and mouth feedings. All previous research, that of several different investigators, which included feeding trials utilizing hundreds of test animals and thirty human subjects had cleared MSG of risk as an additive. In addition one investigator had found that MSG-fed rats had higher growth rates than did the controls and were more alert. Additional research has since borne out the basic findings. To date, MSG still has FDA sanction as an additive. In baby foods it has been voluntarily removed, not so much because of scientific evidence or belief that it is a health risk, but rather because it serves no critical purpose in these foods. It may be asked, then, why it was put there in the first place. The reason is simple, albeit of dubious merit, and derives from most mothers' habit of licking or tasting foods before feeding them to their infants. No doubt the habit is in part instinctive, a way of "testing" the quality, flavor, and temperature to ascertain purity and readiness. Food processors, aware of the tendency, added MSG in the not entirely altruistic hope of pleasing mother. And in all fairness, what other way would the processor have of attempting to please a child's palate? A satisfied mother means in all probability a satisfied child.

One reason for recounting the MSG story is that it exemplifies how circumstances converge at times to exaggerate an issue in the mind's eye. The MSG story broke at about the same time another additive, cyclamate, had come under fire as a potential carcinogen. These two chemicals should not be confused. Cyclamate is an artificial sweetener and, unlike MSG, is not indigenous to the body. Further, it is a synthetic. Found to induce a carcinogenic response, cyclamate was banned under provisions of the Delaney Clause, the exclusion clause for cancer-causing chemicals in foods which, although certain exceptions exist, demand deletion irrespective of the dose-rate required to initiate the cancerous condition.

At the time when the public was duly sensitized over cyclamates, an honest disagreement broke out among scientists regarding the safety of MSG. As we have already seen, the main issues were entirely different, for the two chemicals are entirely different. One is synthetic and unnatural, the other a natural component of human chemistry. One had been found to cause cancer under what were considered appropriate test conditions, the other had been found to yield a number of other bodily disturbances, but under test conditions considered inappropriate. The net result, shock on shock, could have been anticipated—public outrage; industry embarrassment; withdrawal symptoms in a research field critical to mankind's future. But how different the two settings. And if tempted to lay blame, remember: cyclamates spring forth from a natural human tendency toward overeating and a demand for foods that would allow us to have our cake and eat it too, i.e., a sweet taste and little or no risk of putting on weight. On the other hand, MSG is an additive designed to satisfy our general lust for flavor. As Pogo said, "We have met the enemy and he is us."

Although the decision to ban cyclamates was reasonably justified, it behooves us, considering the imperfect science of toxicity testing, to bear in mind that a carcinogenic response in test animals of other than human species and at levels of intake far in excess of those normally consumed by humans, proves only two facts—the effect of the compound on the animal tested at the specific level administered. Such evidence is far from a foolproof guarantee that man's response would be similar especially since he would consume cyclamates at vastly smaller rates. But with equal emphasis, out of deference to this deadliest of diseases, we should repeat: scientific opinion differs. To some scientists any evidence of a carcinogenic response is evidence that it can happen in man, and perhaps at any dose-rate, however small.

So far about the worst that can be proven about MSG is that it has potential as an allergen—a potential inherent in all foods. Evidence has been documented and the ailment named, one assumes facetiously though not without basis, the Chinese Restaurant Syndrome (CRS). Symptoms—burning sensation, facial pressure, chest pains, headaches—have been found to occur in subjects given MSG on an empty stomach. Yet subjects experienced similar symptoms when they were fed, at slightly higher dose-rates, both the L isomer and a racemic mixture of glutamic acid, the natural amino acid base from which MSG is derived. Because MSG is used amply in Chinese cookery and is present naturally in fermented soy products, a number of susceptible persons have reported distress following a meal of Chinese food. However, the ailment is in no way limited to Chinese foods. Allergies to beet sugar (or MSG derived from beet sugar) have long been known. As might be anticipated, persons allergic to beet sugar MSG are not necessarily allergic to wheat gluten MSG and vice versa. It is in the nature of allergies to be exceedingly specific. If you are allergic to MSG, check the food label. The compound, though not its source, will be listed. About 10 percent of the MSG used in food is derived from beet sugar, the remainder mostly from fermentation processes.

Other potentiators. If salt is a weak cousin of MSG, MSG is an even weaker relative of the 5' nucleotides, which are a hundredfold more potent, though just as common to well-known foods. Meat, yeast, pancreas, bonito tuna, and sugar beets head the list as raw products from which the potentiators may be derived. They can be identified on food labels by their chemical names 5' guanylate and 5' inosinate. A third member, xanthylate, being less potent than the other two, has not seen extensive use commercially.

Maltol is a potentiator with the ability to make sweet flavors sweeter. In effect, it reduces the need for sugar. Maltol is found naturally in the bark of young larch trees, pine needles, chicory, wood tars and oils, and roasted malt. Maltol finds its greatest utility in jams, fruit juices, and soft drinks.

The above is not a complete list and it is certain that new potentiators will be discovered, for it is in the coming struggle to ward off famines that the ability to enhance flavor may ultimately find its most noble end.

How potentiators work. Out of curiosity, the same curiosity that led to the discovery of MSG and its role in flavor enhancement, scientists would like to find out how potentiators work. Once the facts are known, new methods of potentiation may become apparent, resulting in cheaper, better ways of flavor potentiation. As yet the mechanism of flavor potentiation is unclear. Theories abound, including the following: (1) alteration of flavor sensors to yield peak response to stimuli; (2) promotion of salivation; (3) production of "feel" sensations that enhance flavor perception; and (4) intensified aroma perception.

Flavor in the Future

An estimated 10,000 new flavor formulations (or flavor combinations) seek to find their way into the food industry each year. Flavor houses, like perfume manufacturers, are masters of the art of concoction. And always the fruit of their ingenuity is a product that tastes good.

Most of this annual output of flavor is new only in the combination of ingredients blended, not in their uniqueness as individual flavors. Chemicals used are mostly of standard variety, natural or artificial, previously tested (if necessary) for safety. Only a relatively few (57 natural and 58 artificial flavors from 1965-1973) genuinely new flavors have been introduced. Still, 10,000 is a large number and can be taken as evidence primarily of the human tongue's insatiable desire for flavorful food, markets and profits being but by-products of this first cause. Of greater significance, however, is the basic question of palatability, and it is this food quality that looms large in the future of world food supplies. For, like it or not, the trend in the diets of people throughout the world, in both developed and undeveloped countries, is toward proportionately larger intake of cereal grains and vegetable protein. The day of cheap, plentiful animal products is already past.

Only recently has serious study been initiated to ferret out flavor components of grain products. Generally well accepted worldwide, grain flavors are of particular importance to palatability and, thereby, to the acceptability of food sources of less than good flavor. When the time comes that earth and sea are scavenged for food (and surely they will be), new products, perhaps entirely of unanticipated origin, can be made into overnight successes by the application of minute amounts of grain flavored chemicals. These flavors may be artificial. But because the search for flavor ingredients usually begins with natural products—after all, the flavors are certainly there to be found—the likelihood of the flavor chemicals being "natural" is better than average. And since these compounds will never before have been isolated and recognized for their flavor potential, they will be new. In reality, of course, they will be as old as nature. Yet it is important to realize that, new or old, natural or artificial, these compounds will be classified as additives for legal purposes. Their purpose, though, is directed at making palatable nutritive food sources, for both the affluent and the hungry. Already low-cost hamburger products, in which meat and vegetable protein extenders are blended in edible combination, are made appetizing through the use of meat flavors that mask unpleasant plant protein flavors. And the step from extenders (nutritive fillers) to nutritious *imitations* of traditional foods is not overly large. Some technological problems exist, but they are far from insurmountable. Giant strides in this direction have been taken in recent years. Imitation meat and dairy products, rare commodities in the past, may well be on the verge of wide-scale exploitation. Moreover, most of these products will not be called imitation, but will bear new names and will ultimately be widely accepted by consumers. That the movement is underway is evidenced by recent government action making it unnecessary to label as imitation any food substitute of equivalent nutritional value. A milk or egg substitute of equal nutrient content to its natural counterpart will henceforth be marketed as a product in its own right and with its own special food name. The trend has begun and is made possible by our profound knowledge of flavor chemistry.

On our present course a share of the world's population is doomed to starvation. If worldwide disorder does not somehow obliterate us first, Americans will probably survive. And if our luck holds, the worst we may be asked to accept is a vegetable product that tastes like steak!

V

Food Color

A color additive is "a dye, pigment, or other substance made by a process of synthesis or similar artifice, or extracted, isolated, or otherwise derived, with or without intermediate or final change of identity, from vegetable, animal, mineral or other sources" that "when added or applied to a food, drug, or cosmetic, or to the human body or any part thereof, is capable (alone or through reaction with other substances) of imparting color thereto."

FDA definition

No Rembrandt or Picasso ever dealt with artistic tribulations any more complex than those commonly encountered by food coloring specialists. Problems associated with coloring food encompass difficulties most painters, whether Baroque, Renaissance, or modern surrealist, never dreamed of. For there is much more to coloring food than the subtle mixtures of hues and oils, more than design and setting, more than the application of paint to canvas. Also involved are questions of solubility, color fastness, and the influence of heat. No canvas was ever subjected to retort cooking at temperatures above boiling. Nor are canvases frozen and thawed, or spray dried. There is the question of pH, too, since food acidity varies widely, and some colors react like chameleons to pH changes.

Other colors thicken or gel, which is all right for a canvas perhaps, but not for certain foods. Or a color may cause haziness or cloudiness, which is fine unless clarity is essential, as it is for a food like salad oil. Metallic ions may create problems as well, since their presence may cause food color to settle out in a sludge. Another problem is light, which to some coloring compounds is an anathema; they fade miserably. There is also a different medium, a different chemistry in different foods; the canvas is never the same. If the food contains reducing sugars, some coloring compounds cannot be used. Moisture varies widely and so does its influence on color. If protein and a high heat process are combined with one standard FDC color, the result is green chocolate pudding. Oxygen has a bearing on the problem, as do antioxidants (or reducing compounds), enzymes, protein, and surface texture. The most troublesome aspect of food coloring is that no single color reacts the same in all foods. And of course no painter ever had to confront the multiplicity of dangers associated with eating the canvas. His work may have caused indigestion, but he wasn't liable. The coloring of food is a complex matter.

The Makeup of Food Coloring

Food coloring chemicals fall into one of three categories: (1) synthetic organic compounds (the FD & C colors); (2) mineral or synthetic inorganic colors (iron oxide, for example); and (3) natural coloring of either vegetable or animal origin (vegetable or fruit juices or color extracts).

Just as there are natural and artificial flavors, so are there natural and synthetic coloring agents for foods. And like most other types of additives, food coloring compounds are not excluded from scientific and public controversies, as we shall note further on. But, first, a discussion of meanings and terms is in order.

To avoid possible confusion, the word *artificial* should be considered. Any time food is colored, it is generally spoken of as being "artificially" colored. Here the word *artificial* refers strictly to color that has been added over and above that which may be present naturally. However, food coloring can be either natural or synthetic and any reference to artificial coloring should not necessarily be construed as relating to color makeup per se.

Because government regulations apply to food *and* drugs *and* cosmetics, coloring compounds by legal definition are separable into one or more of these categories. More precisely, three separate listings are identified for legal use: (1) FD & C colors—coloring compounds approved for use in foods; (2) D & C colors (drug and cosmetic colors)—for internal use in drugs or cosmetics (not foods) which may come in contact with mucous membrane or which may occasionally be ingested (eaten) or swallowed (as in pill form); (3) External D & C colors—drug and cosmetic colors approved solely for external use, but not on lips or in contact with mucous membrane. Comments here will be confined primarily to the FD & C colors.

Legal Aspects of Food Coloring

Because they may vary in composition and, thereby, amount of known poisons, certain synthetic food coloring compounds must meet government specifications regarding content of impurities (in the same way drinking water is controlled through standards). Specifications are spelled out for each individual coloring compound according to known contaminants associated with the chemical. Limitations are placed on the following, among others: volatile compounds (at temperatures above 135°C); chlorides; sulfates; water insoluble matter; ether extracts; certain organic acids and aldehydes; chromium; lead; arsenic; certain iodine compounds; total coloring content; and other impurities to the extent that they are avoidable through good manufacturing practices.

Routinely, each batch of certifiable coloring is sampled and submitted to the FDA for analysis. Assuming the dye meets the necessary requirements, it is assigned a lot number and color percentage. The batch is then considered a marketable item.

Many coloring compounds do not require certification, but cost and technical considerations generally limit their use. In other words, certified colors predominate as coloring compounds in food because of cost and functional advantages. They form the basic colors from which a rainbow of various tints and hues is created by a blending process no different from that used by the artist—blue and yellow to obtain greens, reds and blues for crimson colors, and so on.

Provisional listing of food colorings came about as a result of the Color Additive Amendments of 1960 (Public Law 86-618). The law itself has an interesting history and evolved out of a need to clarify earlier legislation which required that coloring compounds be "harmless," i.e., incapable of producing harm to laboratory animals in *any* quantity under *any* conditions. Strict adherence to a regulation of this sort, as has been previously noted, could exclude food in general. No food is "harmless" under all conditions. To put it positively, all foods are potentially harmful.

The 1960 Color Additive Amendments gave the FDA the authority to impose limits on the amount of color used in foods, drugs, and cosmetics. They also required that all colors, synthetic and natural, be cleared for safety before marketing. To determine relative harmlessness, which is the most that can be done, the FDA was allowed to demand the retesting, under modern laboratory techniques, of any color compound about which the slightest question of safety had been raised. The burden of proof for safety fell on industry.

Because testing safety is a long-term project, the Color Additive Amendments of 1960 provided for the continued marketing of color compounds during the test phase by granting the color additive in question a "provisional" rating. If it passes all the requirements, its status is raised to the more secure "permanent" listing. This is not to say that the color additive will always have permanent status, but evidence—and odds—are against its losing this rating. To this day testing continues, as does the "provisional" status.

Why Food Coloring

Of all additives, food coloring is perhaps the hardest to justify as necessary, even though coloring does have a meaningful purpose. A strawberry flavored drink the color of water might be a case in point. Color, for some products, is a useful means of identification, and product identification by color has become a nearly instinctive reaction. For example, strawberry color and strawberry flavor go hand in hand. Color, too, serves to standardize the appearance of a product. Many natural foods vary from light to dark shades depending upon season, geography, or varietal differences. Sometimes—but not always—a uniform color may be desirable. If it matters to consumers, it matters to food processors. Were not most of us very selective regarding appearance, color would not be so important.

Color can also deceive. It may imply richness, flavor, or maturity where little—or less than we think—exists. As an indicator of quality, if, indeed, it can be considered as such, color (eye-appeal) must be related honestly to the product it attempts to designate, enrich, or standardize. To this end of course, laws, regulations, and enforcement agencies have been established, as have color-grade standards for many foods, whether organically or nonorganically grown, whether natural, refined, or processed. Until our eyes stop speaking for our stomachs color will remain an essential food component.

Naturally Occurring Color Additives

Many colors that flourish in the plant and animal kingdoms can serve as coloring agents for food. Table 14 lists most of the commonly used, approved, and regulated naturally occurring color compounds. Many of them are

Table 14. Naturally Occurring Color Additives Approved for Use in Food

Compound	Source	Restriction on Use[a]
Algae meal, dried	Algae	Chicken feed (to promote yellow skin and egg color)
Alkanet	Roots of *Alkanna tinctoria*	Sausage casings, oleomargarines and shortening
Annatto	Seed pods of *Bixa orellana L.*	
Beta-apo-8'-cartenal	Fruits, vegetables	Not to exceed 15 mg per pound or pint of food
Aztec marigold (Tagetes meal)	Flower petals of Aztec marigold	Chicken feed (to promote yellow skin and egg color)
Beets, dry powder	Beets	
Canthaxanthin	Fruits, vegetables	Not to exceed 30 mg per pound or pint of food
Caramel	Heated sugars	
Carmine (cochineal)	Female insects, *Coccus cacti L.*	
Beta-carotene	Fruits, vegetables	
Carrot oil	Carrots	
Chlorophyll	Green leaves	Sausage casings, margarine and shortening
Corn endosperm oil	Corn	Chicken feed (to promote yellow skin and egg color)
Cottonseed flour, partially defatted, cooked, toasted	Cottonseed	
Ferrous gluconate	Iron salt of gluconic acid	Ripe olives
Fruit juice	Fruit	
Grape skin extract	Grapes	
Iron oxide (synthetic)[b]		Dog and cat foods not to exceed 0.25% by weight of finished product
Paprika	Plant *Capsicum annum L.*	
Paprika oleoresin	Plant *Capsicum annum L.*	Not for use on fresh meat or comminuted fresh meat products except chorizo sausage and Italian brand sausage and other meats as specified by USDA.
Riboflavin (vitamin B_2)	Plant and animal sources	
Saffron	Stigmas of *Crocus sativus L., Iridaceae*	
Titanium dioxide	Found in the minerals rutile, titanite, anatase, brookite, almenite	Not to exceed 1.0% by weight of food
Turmeric	Rhizome of *Curcuma longa L., Zingiberaceae*	
Turmeric oleoresin	Rhizome of *Curcuma longa L., Zingiberaceae*	
Ultramarine blue	Mineral, lapis lazuli	Animal feed, not to exceed 0.5% by weight of salt
Vegetable juice	Vegetables	

Source: *Code of Federal Regulations.* 1974. Title 21, pts. 1-9.
  [a] Unless noted otherwise, there are no limitations or restrictions on use.
  [b] While the synthetic oxide is recognized as the food additive, iron oxide occurs naturally as well.

nutrients in their own right. Riboflavin is vitamin B_2. Three are carotenoids with pro-vitamin A activity. Some, like paprika and turmeric, are spices as well as coloring agents. Almost all have permanent status, even carmine which is a red extract from the female insect *Coccus cacti L.* found in Mexico and Central America and cultivated in the West Indies, Canary Islands, Algiers, and southern Spain. Though it may sound surprising, a yield of one pound of carmine takes 70,000 of these insects—female only. Regulations require that the finished color be pasteurized, i.e., treated to rid it of any germs, particularly salmonellas.

Mineral compounds included in the list do occur in the natural state, but synthesis is far cheaper and the end-products are more readily purified by such a process. The same can be said of the carotenoids, although they are organic, not inorganic. Nature annually manufactures about 400 million pounds of carotenoids. More than 100 different molecular species have been isolated and identified, though only three, beta carotene, apocarotenal (along with annatto), and canthaxanthin, are commonly used as food coloring. All three have permanent listing, canthaxanthin, the latest addition, being approved in 1969. None require batch certification. Carotenoid colors vary from light yellow to a dark, almost crimson red. In color and pro-vitamin A activity, synthetic carotenoids match their natural counterparts. As synthetics, they offer advantages in uniformity and availability.

Annatto coloring, also a carotenoid (bixin), is obtained chiefly by extraction from fermented seed pulp. Over the years, its use has centered primarily on butter and cheese products. Butter, in fact, was the first food authorized by Congress for color addition. The year was 1886. Butter could be marketed in noncolored form, as could margarine, but the one product would range in color from light to golden yellow, the other, unmasked, a pale bland white. No two products more clearly focus the issue of color additives (or additives in general). Margarine was anything but successful as a lardlike spread, and one must wonder even today (and in spite of continued price advantages) what its fate might have been as a colorless product. In view of rising food costs, and particularly the cost of animal products, we should ask ourselves, as consumers, how we might respond to colorless foods. Our honest answers to this and other such seemingly trivial questions will no doubt affect the cost of tomorrow's food, the aesthetic appeal of vegetable products, and ultimately the fate of those who hunger.

Just as "colored" margarine is the vegetable counterpart of butter, so too are there vegetable counterparts for meat (animal protein), the extenders.

Textured vegetable protein (made from soybean) which is added to hamburger is a prime example. A meat replacer of 100 percent vegetable protein is well within the realm of today's technology. But will it be colored like beef, or bacon, or ham? Will the consumer accept it if it is uncolored? Would the nation have accepted uncolored margarine?

Paprika and turmeric are both colors and spices. If their primary mission is coloring, they will be listed on the package label as such.

A household word, *chlorophyll*, denoting our planetary color is actually a mixture of two related compounds, chlorophyll -a and -b. No color could be in greater supply, though its commercial production stems primarily from lucerne, nettles, and grass. It is extracted with alcohol or acetone and is no different from synthetics in its content of minute amounts of impurities. The impurities (in this case extractants or diluents) are also additives and appear along with coloring compounds themselves in regulatory listings. Diluents are chemicals in which coloring compounds are dissolved or dispersed. They include such odd functional mates as castor oil and ethyl alcohol. Two other diluents cleared specifically for use in coloring food are dioctyl sodium sulfosuccinate and polyvinylpyrrolidone. Residue levels in foods are precisely regulated. While we are on the topic, it should be noted that an additional function of color is that of marking foods. Again, both coloring material and diluents are classed and recognized as additives.

As consumers' emphasis on "natural" foods increases, beet powder will no doubt gain in use. For years it has colored the tomato products in pizza. But until food scientists learn how to eliminate its flavor, its lack of color fastness, and its serious caking problems during storage, its growth in use will be at a snail's pace.

Naturally occurring pigments have a place in coloring foods, but their chief failing—color breakdown (fading, darkening, or hue change) in air, light, and heat or as a result of interaction with other food components—greatly limits their usefulness and explains why synthetic dyes predominate as colorings for food.

Synthetic Food Colors

Of a total of 695 coal tar dyes known to science in 1912, only eleven survive as legal coloring agents for food. However, three of the eleven account for almost 85 percent of coloring added to food in the United States. These three dyes are amaranth (FD & C Red No. 2),* tartrazine (FD & C Yellow No. 5), and sunset yellow FCF (FD & C Yellow No. 6). Their importance lies in

*Banned by FDA, February 12, 1976.

Table 15. Synthetic Coloring Compounds Approved for Use in Foods

Coloring Compound[a]	Common Name	Restriction in Use[b]
FD & C Blue No. 1	Brilliant blue FCF	
FD & C Blue No. 2	Indigotine	
FD & C Green No. 3	Fast green FCF	
FD & C Red No. 2	Amaranth	Proposed tolerance of 30 ppm in food as of July 4, 1972*
FD & C Red No. 3	Erythrosine	
FD & C Red No. 4	Ponceau	Maraschino cherries at levels not exceeding 150 ppm
FD & C Red No. 40		
FD & C Yellow No. 5	Tartrazine	
FD & C Yellow No. 6	Sunset yellow FCF	
Citrus Red No. 2		Skins of oranges at 2 ppm
Orange B		Sausage casings at 150 ppm

Source: *Code of Federal Regulations.* 1975. Title 21, pts. 1-9.

[a] All coloring compounds listed in this table are subject to FDA certification. Each batch is tested for purity by FDA. Lakes (salts of alumina or calcium) of these compounds are also permitted and are used primarily for oil suspension. Of these compounds, only the following have permanent listing as of 1974: FD & C Blue No. 1, FD & C Red No. 3, FD & C Red No. 40, FD & C Yellow No. 5, Citrus Red No. 2, and FD & C Red No. 40 lake.

[b] Unless noted otherwise, there are no specific limitations or restrictions on use.

the fact that they are primary colors, ones from which blends can be mixed to produce secondary colors. Not only do they provide color in their natural hue, they are needed to yield other shades.

The eleven coloring compounds legally permitted in foods as of July 1974 are listed in Table 15. All are subject to FDA certification for purity (see legal specifications). FD & C Red No. 2 has been proposed for restrictive use, at 30 ppm in food.* Red No. 4 is limited to the coloring of maraschino cherries at a level no greater than 150 ppm. An additional red compound, Citrus Red No. 2, is allowed solely for coloring oranges and then just the outside peel; it may not be used for other food purposes. The only application in food of Orange B is in sausage, not the meat but the casing, at a maximum concentration of 150 ppm.

Because of their high tinctorial value, only exceedingly small amounts of these dyes are required to color foods. For this reason their use is self-limiting. But another characteristic, their solubility in nothing except aqueous solutions, gives rise to a group of chemical relatives, salts of alumina or calcium which are soluble in oils. These derivatives are technically referred to as lakes. The aluminum salt of Green No. 3 is FD & C Green No. 3—aluminum lake. The basic ingredient remains the straight dye. While dyes are soluble, lakes are

*Banned February 12, 1976.

Table 16. Color Concentration in and Annual Consumption of Major Food Products
in Which FD & C Coloring Compounds Are Used

Food Category	Average Annual Consumption (lbs.)[a]	Color Concentration (ppm) Range	Color Concentration (ppm) Average	Maximum Color Intake per Day (mg)[b]
Candy, confections	10.0	10-400	100	7.0
Beverages	38.0	5-200	75	12.8
Dessert powders	6.7	5-600	140	6.6
Cereals	1.5	200-500	350	1.5
Maraschino cherries	0.85	100-400	200	0.4
Bakery goods	17.8	10-500	50	15.0
Ice cream, sherbets	15.0	10-200	30	5.0
Sausage	6.6	40-250	125	3.0
Snack foods	0.87	25-500	200	1.0
Miscellaneous[c]	1.85	5-400		1.2
Total				53.5[b]

Source: *Food Colors*. 1971. National Academy of Sciences.
[a] Figures represent approximate consumption and usage rates for the year 1967.
[b] These values are based upon maximum usage of color, the highest values in the concentration range indicated in the second column of figures. If, instead, average values are substituted, the total intake would be about 15 mg per day rather than 53.5 mg.
[c] Miscellaneous foods include salad dressing, nuts, gravy, spices, jams, jellies, and food packaging.

not. Lakes are—and remain so in food—discrete particles with the dye adsorbed to the surface. As a direct result, lakes are generally more stable than dyes.

An extensive report of the National Academy of Sciences entitled *Food Colors* (1971) draws together pertinent information on the uses of food color. Rearranged somewhat, data in Tables 16, 17, and 18 on the use, consumption, and allowable daily intake of food color were taken from this report. Findings of this scientific body indicate that 2.54 million pounds of food color were sold in the United States in 1967. With an annual food consumption of 1420 pounds per capita and a population of 200 million, the concentration of food color in food figures to be about 9 ppm. While this is indeed a small amount, other considerations, especially total dietary intake, are significant, nutritionally.

First, it should be understood that not all foods are artificially colored. The greatest share, including such major staples as bread, meat, potatoes and other vegetables, fruit, and most fluid milk products, are not. Those foods in which coloring has, over the years, been a common ingredient represent only about 10 percent of the various foods normally consumed. If maximum color addition is assumed, the total daily intake amounts to 53.5 milligrams (see

Table 17. Rate of Use and Estimated Maximum Daily Intake of FD&C Coloring by Various Food Categories

	Coloring Use Rate (1000 lbs.)[a]	Estimated Daily Intake (mg)								
		Blue No.		Green No.	Red No.			Yellow No.		
Food Category		1	2	3	2[b]	3	4	5	6	
Candy, confections	202.6	0.23	0.09	0.04	2.3	0.40	0	2.1	1.8	
Beverages.	563.4	0.36	0.05	0.01	6.4	0.02	0	1.8	4.1	
Dessert powders.	187.5	0.11	0.06	<0.01	2.2	0.30	0	2.1	1.8	
Cereals	105.8	0.01	<0.01	0	0.2	0.02	0	0.7	0.5	
Maraschino cherries.	34.0	0.01	0	0.01	0.1	0.04	0.13	0.1	0.1	
Bakery goods	177.8	0.31	0.06	0	3.7	0.81	0	6.6	3.6	
Ice cream, sherbet, dairy products.	92.0	0.14	0.01	<0.01	1.6	0.03	0	1.9	1.3	
Sausage.	164.6	0.01	0	0	0.7	0.09	0	0.1	1.8	
Snack foods	34.5	0.01	0	0	0.1	0.02	0	0.5	0.3	
Miscellaneous[c].	148.5	0.04	0.02	0.01	0.4	0.15	0	0.4	0.2	
Total	1999.1	1.23	0.29	0.07	17.7	1.88	0.13	16.3	15.5	

Source: *Food Colors*. 1971. National Academy of Sciences.

[a] Use rate is based upon sales of the respective FD & C coloring compounds during the first 9 months of 1967. Coloring compounds are those approved for use in 1973.

[b] FD & C Red No. 2 was placed under a temporary use rate tolerance of 30 ppm in foods on July 4, 1972. Therefore, figures in this column would currently be lower.

[c] Miscellaneous foods include salad dressing, nuts, gravy, spices, jams, jellies, and food packaging.

Table 18. Individual Acceptable Daily Intake and Safe Levels of a Number of
Food Coloring Compounds

Coloring Agent (FD & C)	Acceptable Daily Intake[a] (mg/kg body weight)	Safe Levels[b] (mg/day)
Blue No. 1	0-12.5	363
Blue No. 2	0-2.5[c]	181
Green No. 3	0-12.5	181
Red No. 2[d]	0-1.5	
Red No. 3	0-1.25[c]	91
Red No. 4	Not available	18
Yellow No. 5	0-7.5	363
Yellow No. 6	0-5.0	363
Others		
Annatto	0-1.25[c]	
Beta-apo-carotenal	0-0.5[e]	
Canthaxanthin	0-25	
Beta-carotene (synthetic)	0-0.5[e]	
Riboflavin	0-0.5	
Titanium dioxide[f]		
Turmeric	0-2.5[c]	

Source: Adapted from *Food Colors.* 1971. National Academy of Sciences. Data for apocarotenal, canthaxanthin, beta-carotene (synthetic), and turmeric are from Joint Food and Agricultural Organization/World Health Organization Expert Committee on Food Additives, 18th Report.

[a] Except when otherwise noted, these are unconditional levels established by short- and long-term toxicological studies and other biochemical data.

[b] Safe levels are calculated on the basis of 1/100 of the maximum no-adverse-effect levels established through long-term feeding trials using the most sensitive animal species and assuming daily dietary intake of 1814 grams for humans.

[c] This is a temporary value pending more information.

[d] Red No. 2 has been banned (see footnote, p. 96).

[e] This is expressed as total carotenoids by weight.

[f] Titanium dioxide is limited only by good manufacturing practices.

Table 16). However, calculations based upon average color addition reduce the total daily intake to 15 milligrams. But, as any dietitian will be quick to point out, averages can be misleading. What is truly significant is the kind and amount of foods regularly consumed by the person. Attention should be paid to the major food categories in which coloring normally occurs. As well, note in Table 17 the *normal* use rate by individual food category.

The beverage industry leads the list by far, followed by, at less than half the tonnage, candies and confections, then dessert powders and bakery goods. The popularity of these food categories makes it apparent that many Americans, especially children, probably take in higher amounts of food color than average figures would imply. And if, as one medical doctor seems to think (although his assertion is as yet unconfirmed and unacted upon by the FDA because of inconclusive evidence), food colors, specifically FD & C Yellow No 5, are indictable as causes of the childhood malady hyperkinesis (overactivity), it would be wise to learn about the eating habits of these children as a way of assessing total dietary intake of food coloring. Granted that some children might be more sensitive than others, but if their diets are composed primarily of soft drinks, fruit drink substitutes, candies, bakery sweets, and puddings, well. . . .

For reemphasis the following list summarizes food products in order of their significance in food coloring use.

1. Beverages
2. Candy, confections
3. Dessert powders
4. Bakery goods
5. Sausage (casing color)
6. Cereals
7. Ice cream, sherbets (butter, cheese)
8. Snack foods
9. Maraschino cherries

Other foods which collectively would be placed somewhere between bakery goods and sausage in rate of use include salad dressing, nuts, gravies, spices, jams, jellies, and food packaging material.

Where sufficient data exist, specific food colors are given an unconditional "acceptable daily intake." The levels for some widely used food coloring compounds appear in Table 18. It is important to note that levels for some of these compounds are temporary pending more research. Safe levels for man, also found in the table, are calculated on a factor of 1/100, the no-adverse-effect level established through long-term animal feeding trials.

A comparison of information in Tables 17 and 18 would strongly imply that acceptable daily intakes and/or safe levels for man should rarely, if ever, be exceeded. But this assumes "normal" eating habits and, to say it again, no one can be considered normal in eating habits. A diet favoring sweets could skyrocket the amount of color consumed. Thoughts automatically turn to children and adolescents who, by one USDA survey, ingest one-quarter of their daily caloric intake in snack foods alone. Again, these are averages.

Another important note for would-be natural food fanciers and other concerned individuals is the acceptable daily intake of naturally occurring food coloring material. Acceptable levels for carotenoids, riboflavin, and turmeric are no higher—in fact, they would average out lower—than levels for the coal tar synthetics. As so often emphasized previously, *all foods have their hazardous levels.* If we are to assume there is some risk from ingesting too much color in foods, we should balance it against the related risk, in this case, of excessive carbohydrate (sugar) intake. In its toll on American health, the latter has to be, at this time, the far greater of the two evils.

Food Colors under Suspicion

Time and laboratory testing have taken a greater toll of food colors than of any other category of additives. Upon passage of the Food, Drug, and Cosmetic Act of 1938, fifteen coal tar dyes were recognized as legal. The list has since decreased to eleven, but includes Red No. 40, a newcomer added in April 1971.

Through July 1972, when its use was limited, FD & C Red No. 2 was the most popular food color, its daily intake in the United States approaching 18 milligrams. The yellows, nos. 5 and 6, have never been far behind, their estimated consumption 15 to 16 milligrams daily. But two research reports and one unverified critical comment by a medical doctor have raised suspicions about Red No. 2 and Yellow No. 5. The research reports, by Russian scientists, suggested that Red No. 2 might cause cancer and might be embryotoxic (poisonous to the unborn infant). Yellow No. 5 and, by implication, other "artificial" flavors have been cast in the role of allergens in the nervous disorder called hyperkinesis.

The saga of Red No. 2 is typical of the agonizing decisions, the charges and countercharges, that engulf scientists, consumers, and governmental agencies in time of doubt. All the old, nettlesome questions rise through the dust of battle. How safe is safe? What is conclusive evidence? Is any test entirely valid? Are we, as one industry representative asked, simply "testing the test"? Is it true, as Goldberg laments, that, "by appropriate choice of dose, concentration of solution and frequency of administration, any chemical agent can be shown to be a carcinogen in the rat and probably also in other species of laboratory rodent"?

Or, in the same vein, is science engaged in a kind of mad pursuit of hazard analysis that places a premium on ingenuity and persistence? Goldberg asks, Does a chemical irrespective of evidence to the contrary, "remain under suspicion until such time as someone can discover an organism, devise a route of administration or achieve a sufficiently heroic dose to produce some positive biological results, however bizarre"? To this last question, an FDA advisory committee (on protocol and safety evaluation) seems to reply yes, in principle at least. It states, "The number of agents shown to be teratogenic in experimental animals is so great that the question arises as to whether all agents might not be teratogenic if applied at the critical time in the appropriate dose in the right species." Teratogenicity, you will recall, refers to the influence of chemical (or physical) agents on malformations of unborn fetuses.

The anguish of manufacturers is not without basis. Safety testing can run into hundreds of thousands, sometimes millions of dollars. And industry shoulders the cost, at least initially. The loss of a dye as versatile as Red No. 2 forces an immediate search for a suitable replacement. Even if one is found after years of safety evaluation, its application and formulation in various foods must be researched anew. The death of a dye involves the death of a library of facts. In the case of Red No. 2, this problem can be partly offset by the availability of Red No. 40, a somewhat more expensive, less versatile coloring compound.

So much for the problems of industry. There is the reverse side of the coin —public interests as usually promoted by consumer advocates and, directly or indirectly, by "natural" food proponents. Where doubt exists, these stanch activists usually opt for elimination of the additive. While their appeal tends to be emotional, who of us is so cocksure of himself that he is not comforted that another viewpoint is being heard? And though a government agency may be cursed by industry and advocate alike, it is appropriate that there be one attuned to toxicity research around the world and prepared to give ear to the single voice of a medical doctor.

Following publication of the Russian work on the possible effects of Red No. 2, the FDA undertook to repeat the research and confirm or refute the findings. This sounds simple, except that simple, indisputable answers are the exception not the rule in toxicity testing. What the Russian M. M. Andrianova observed was the presence of tumors in 13 of 50 male rats fed at a level of 2 percent Red No. 2. Even though 2 percent is a very large dose, control animals were reportedly free of tumors. Another Russian experiment, on a test group of female rats, found signs of decreased fertility among the animals, some stillbirths, and some fetal impairment. Each report must be examined separately, on its own merits and also in relation to previous findings. In this case, earlier extensive work on Red No. 2 had seemed to clear it of suspicion as a carcinogen. Knowing this, one is tempted to dismiss the Russian work as defective in some way. One hundred rats is a relatively small number. Factors other than a test chemical can lead to cancer. Control and test groups of animals do not always receive precisely the same treatment. There can be no certainty that controls and test groups were not divided, although innocently enough, into a "pairing" where one group was somehow predisposed to tumor development. There is room at least for doubt, though any finding of cancer or embryo pathology is bound to leave a lingering uneasiness.

In the second instance, where embryotoxicity was reported, it turns out that there was no previous work to fall back on for comparison. And though a cursory evaluation of the Russian work provides ample reason for suspicion (the number of animals was terribly small and only one male rat, conceivably the bearer of defective genes, was mated with the females), what other research can you point to and say, "look, there's good evidence to the contrary; the Russians must have goofed"? You can't do it. Neither could the FDA and the agency decided to investigate.

Initially, one well-designed experiment seemed to confirm the Russian findings on embryo toxicity; fetal deaths occurred in pregnant rats fed varying amounts of Red No. 2, the number of deaths increasing proportionately with increasing levels of dye intake. At this stage warning signals went up. Industry was placed on notice; Red No. 2 was being reevaluated, and regulatory changes might be in the offing.

Meanwhile other research brought forth some new evidence which conflicted with the FDA's findings. One study of pregnant albino rats fed Red No. 2 by gavage (tube feeding through the stomach) showed no increase in the incidence of fetal disorders. In another trial, using a different feeding procedure and rabbits as test animals, some differences were observed. Fetal deaths occurred in larger numbers in the test group, but deaths—as often happens with animals given to multiple birth—also occurred in the control group. The researchers were of the opinion that the death rate for the test group was not unusually high.

Still a third study involving four animals species, mice, rats, rabbits, and hamsters, and levels of Red No. 2 intake exceeding any as yet tried by the FDA revealed no difference between test animals and controls in either death rate or birth abnormalities.

Brought in to arbitrate, the Committee on Food Protection of the National Academy of Sciences/National Research Council, after hearing and studying the evidence, determined that the case against Red No. 2 had not been proven beyond reasonable doubt. Legal action against the dye was held in temporary abeyance. And, as we have already seen, it was not until July 1972 that restrictions were proposed for Red No. 2 and even then at use levels less stringent than those initially contemplated.

Nevertheless, work continued on the dye and further observations have been made. They are important if only in stressing the comparatively young and insecure nature of the art and science of toxicity testing. In one study, fertilized eggs injected with varying amounts of Red No. 2 produced high death rates and bizarre deformities in chick embryos. But this is a new, relatively untried technique, and critics will hastily point out that any foreign material implanted inside a fertilized egg can do no good; bizarre results are perhaps to be expected.

By June 1974 three separate laboratories were prepared to report on additional studies, including long-term feeding trials. Their results, first on gavage at 200 mg Red No. 2/kg daily feeding plus ad libitum feeding at 0.2 percent dye in distilled water, were all considered negative. The few adverse effects noted either were statistically insignificant or could not be pinned down specifically to the dye. A three-generation reproduction investigation by the FDA lab, with feeding levels ranging from 30 to 30,000 ppm Red No. 2, and another three-generation study by a second lab using 0, 1.5, 15, 45, and 150 mg/kg feeding levels both proved negative—no untoward results. And lastly, an independent study by a Canadian health agency supported in total these findings. It should be noted that not one of these experiments discovered any incident of cancer, which, based upon the one Russian study, touched off the flurry of research activity in the first place.

So what can be made of these conflicting findings? Out of one trial the findings seem to support an adverse effect, another set of findings seems to indicate the contrary. Two sets of "facts." You are the judge. What do you do? First it is clear that *all* findings must be regarded with suspicion, that the technique applied to uncover the facts—certainly the case here—is anything but foolproof. Instead of hard facts there are only soft "facts." Ultimately, one set has to be weighed against the other, taking into account improved techniques, the differences or similarities among the laboratories where the experiments took place, the numbers and kinds of test animals, the experimental design, and the statistical evaluation. All these factors must be weighed and a decision reached.

How did the three protagonists fare in this case? At this writing the final decision has not been handed down, but it appears that Red No. 2 will not be banned;* rather it will probably continue to be used on a restricted basis while more evidence and better techniques for experimentation are devised. This will not please the color industry particularly; much money has been spent, a good market has been drastically curtailed. On the other side, Anita Johnson, a spokesman for consumers, has called the entire episode "shameful,"

* See footnote, p. 96.

no doubt echoing sentiments of some consumers and more than one consumer advocate. Neither "side" is happy; the government agency sits squarely in the middle. Unfortunately, this is the way these kinds of issues often evolve.

What seems more important, however, than profit, indignation, or bruised pride is that something was done. Studies were undertaken. An effort, truly a serious, long-term effort, was made to clarify the risk of using Red No. 2. A government agency, aware that a legitimate health issue had been raised, took action. And although one might wish that research results could always be absolute, either black or white, and that no one ever had to deal with shades of gray, more is known today than was known previously. Red No. 2 is still around, but levels of intake, it can be assumed, are significantly lower. It also seems clear that had results of research been more supportive of a definite health risk at expected levels of intake, the dye would have been banned. And it may yet be;* certainly there are plenty of precedents for such action. (Ironically, at about the time research seemed once and for all to vindicate Red No. 2, Dupont Company, the sole supplier in the United States of sodium naphthionate, a basic ingredient of Red No. 2, announced plans to discontinue this chemical.)

As for Yellow No. 5's alleged allergenicity, the question remains unanswered. Supporting evidence has not been forthcoming, but the FDA has expressed its willingness to consider any findings of the scientific community and, when warranted, use them as a basis for regulatory action.

Specific allergies are difficult to pin down. Today, as never before, general environmental pollution adds greatly to the overall task of discovering the cause of allergies. Allergens have markedly increased in numbers and types and are widely distributed in the environment through the manifold routes of pollution. Isolating and identifying a single compound (or compounds) from among the myriad chemicals and pollutants which the individual sufferer comes in contact with is a monumental undertaking. But if it should happen that artificial colors and flavors are found to be offenders, another "food" may just come to the rescue. Observations of Dr. Robert C. Shnackenberg of the William S. Hall Psychiatric Institute provide evidence that hyperkinetic children respond favorably to, of all things, coffee. Stimulant though it may be for adults, on eleven youngsters who had been taken off the drug methylphenidate, a breakfast and lunchtime cup of coffee had an observable calming influence. The stimulant-tranquilization phenomenon is not new, only the "prescribed" medicine is.

A complaint—a single voice—has been heard; that is always good news. At

*See footnote, p. 96.

this writing, more research on color is being undertaken. As late as March 1975, at the annual meeting of the Society of Toxicology, two studies, one by the National Center for Toxicological Research, another by a commercial research laboratory, using four strains of rats as well as a group of beagle dogs, found no teratogenic hazard in Red No. 2. Thus have three independent laboratories now apparently agreed (quite remarkable in itself) on the teratogenic safety of this color additive. The question of carcinogenicity remains.* The Russian work has not been conclusively refuted or explained away (except to question whether or not the dye purity, as used, would in fact qualify as an FD & C color additive). In the meantime, any mother, any consumer, whether from private misgivings or recognizable symptoms of hyperactivity, can do much to reduce the intake of artificial food colors. Except for ice cream, butter, and cheese, which are exempt by law, artificial colors must be declared on the label. The information you need is there. It should also be remembered that 50 percent of Red No. 2 is consumed in soft drinks. Most other colors, including Yellow No. 5, are to be found in a variety of sweetstuffs. If these foods are decreased in the diet, whatever the possible effect of color, the results for the vast majority can only be beneficial.

*On January 19, 1976, after further work and a review of previous research, the FDA banned the use of Red No. 2 because of its concern that the dye might be a weak cancer-causing agent, because the dye's use in food is strictly cosmetic, and because the drug's safety could not be proven. Court action followed, which prevented the FDA from enforcing the ban until February 12, 1976. At this writing further hearings are scheduled and the status of the dye may yet be changed again. At issue, as always, is the question of proof of safety.

VI

Food Preservatives

A chemical preservative is "any chemical that, when added to food, tends to prevent or retard deterioration, but does not include common salt, sugars, vinegars, spices, or oils extracted from spices, substances added to food by direct exposure thereof to wood smoke, or chemicals applied for their respective insecticidal or herbicidal properties."

FDA definition

Strangely enough preservatives, especially chemical preservatives, are often regarded with suspicion. They are looked upon either as "chemicals" that should be avoided or as additives injected into foods as a cover for unsanitary practices in the food processing plant. As expected, a grain of truth undergirds such distrust. There are some potential health risks; a preservative can be used in lieu of appropriate routine sanitation. But there are health risks, too, if preservatives are not used. And a certain amount of poor sanitation in processing plants will exist whether or not preservatives are allowed. So it is the purpose, here, to examine the evidence both for and against chemical preservatives.

Since it is impossible for most of us to live directly off the land year-round, some method of food preservation is a must. Without exception, the choices— not the technology—available today are those that have sustained mankind throughout history. Food can be dried, frozen, salted (or sugared), spiced,

97

smoked, or pickled (fermented). A form of food preservation results. More precisely, one type of food spoilage is prevented, the kind resulting from the growth of microbes—bacteria, yeast, mold, and fungus. This is a good start, for many food-borne illnesses, not to discount a mountain of food waste, can be neatly eliminated by any method effective against microbial growth. There are, however, other forms of food deterioration. Without preservatives meat turns ghostly pale or unappetizingly brown. Peach and apple slices grow brown and mushy. Fats turn rancid, nutrients lose potency, flavors grow stale.

So it is that food preservation "preserves" in more than one quality area: (1) microbial inhibition, (2) rancidity (oxidation) prevention, (3) color retention, (4) nutrient maintenance, and (5) flavor retention. Each is a quality most of us would not be prepared to do without. Even color—or especially color, depending upon your willingness to consume pink sauerkraut or green bread or rusty-brown peaches—can hold its own as a valued food property, mainly because it is often the gauge, rightfully or wrongfully, by which people measure a food fit to eat. But it should be noted that food spoilage processes do not all proceed at the same rate. Color changes may take place far in advance of microbial spoilage or vice versa. Microbial spoilage is governed by the laws of living things. Color may be—and fat oxidation is—subject to different laws, the laws of chemical reactivity. The business of food preservation, needless to say, is not a simple matter.

When Preservatives?

Not all foods, even those like milk which are most susceptible to spoilage, contain preservatives. Other foods, rendered bacteria-free (sterilized), may contain compounds that provide a preservative function of one kind or another, though not one that is antimicrobial. Often, the presence or lack of these compounds depends on provisions of a standard of identity. Identity standards that provide for the addition of preservatives (by functional property) more often than not assure their use. And if such use is legalized, the assumption can be made that a definite need exists—or existed at the time a standard was written. Still, one is entitled to ask what, if any, rationale lies behind a policy to allow (or require) additives of this kind, whether through provisions of a standard of identity or through allowances for the preservation of nonstandardized foods.

Confining ourselves to technical requirements solely, we can say that the use of preservatives centers primarily on market needs (including consumer demands). Each food poses its own unique preservation requirements, which may have to do with microbial growth, color, flavor, or some combination of factors that cause spoilage. Ordinarily if a product can withstand—in *good* condition, because most American consumers will not accept less—the rigors of storage for such time as is necessary to market it effectively, the use of preservatives will be questioned if not excluded. By marketing is meant the movement of food from the point of production on the farm through various processing steps to the retailer and ultimately into the home, institution, or restaurant where it will be eaten. Throughout this marketing process are many unknowns. This is especially true in the consumer's home. How will the food be treated here? Will it be handled in a sanitary manner? Does it need refrigeration? What temperature is maintained in the refrigerator? How long can a given food be expected to stay in the home before being consumed? What unforeseen problems might there be—such as electrical failures, trouble with refrigeration, the handling of the food by children, insect infestation of storage areas? The list goes on indefinitely, and naturally the questions are unanswerable. It is possible only to anticipate the worst and guard against it with whatever methods are available.

Does this mean, then, that foods contain no unnecessary preservatives? Or that certain foods might not be more useful if preservatives were permitted? The answer, of course, is no. That would all be too tidy, and our world is not that kind of place. Some preservatives, nitrates for example, may be of questionable value. And it is, and should be, one of the prime purposes of industry and regulatory agencies to review continuously food products, the ingredients used, and the wisdom and justification for additives permitted.

But it should be recognized that there is as much reason to question what is lacking in a food that contains no chemical preservatives as there is reason to question what risk you may be assuming when you eat food that contains preservatives, however large or small the amount. If it is pertinent to question the use of preservatives, it is equally pertinent to question the health risk, nutrient value, cost, and availability of foods under a policy that prohibits their use.

Naturally Occurring Preservatives

In the natural evolution of some biological systems, substances have developed which are known to inhibit microbial growth. That man has chosen to eat some of these natural products alone distinguishes them from the remaining fauna and flora and makes of them something more than objects of interest or beauty. They are foods. If animals also take sustenance from them, they are feeds. Or at least man, in his self-centered existence, chooses to classify them as such.

Some foods as they grow or are produced in nature contain microbial inhibiting substances, whether by evolutionary purpose or strictly by chance. Since many microbes sensitive to these substances are not natural enemies or natural inhabitants of these foods, odds perhaps favor fortuity. But certainly man far exaggerates his own importance if he thinks nature evolved antimicrobial agents for human benefit alone. It is, of course, not the case, and it is the reason why nature falls short of producing a substance that might be considered an all-purpose preservative. Some foods harbor certain chemicals that may inhibit or retard—temporarily at least—the growth of one or more microbes. At the same time some foods contain certain chemicals that will, for a time, prevent oxidation (rancidity). But no food possesses one or a group of naturally occurring chemicals that will inhibit all microbes (or more than a few microbes) for even a relatively short duration. Sterility (complete microbial inhibition) is unknown. And only a very small number of food-borne human pathogens (disease microbes) are at all slowed in their phenomenal growth by the presence of these compounds in nature. It should be stated again: nature did not evolve solely for the benefit of mankind. Man exists with nature in a terribly fragile balance.

It is a fact, though, that foods are not automatically disease-free or spoilage-free, and some—some that are commonly eaten—are exceedingly poisonous as they come to us from nature's store. Processing alone renders them harmless and, in the end, nutritious. But this story is reserved for a later chapter. For now, let's look at some foods that contain naturally occurring microbial inhibitors or, as they are more frequently referred to, antibiotics.

In Tables 19 and 20 are listed a number of foods from which antibiotics have been isolated or in which they are known to exist. The list is not particularly long, but among the antibiotics present are some with the ability to inhibit certain potent human pathogens. Conalbumin of egg white is active against anthrax. An inhibitor in cow's milk is effective against the tuberculosis organism. A good number of vegetables and fruit contain antibiotics that can check the spread of an organism (*S. aureus*) that causes food poisoning. A casual glance at these tables, though, and the paucity of natural antibiotics becomes all too apparent. Likewise those organisms they inhibit represent an insignificant percentage of the total complex of organisms known to cause human disease.

This brings up a point important to individuals who, influenced by the idea of organic and natural foods, would consume only raw products, especial-

Table 19. Some Foods Known to Contain Naturally Occurring Preservatives[a]

Food	Naturally Occurring Preservative
Banana	
Green.	Antifungal compound
Ripe	Antibiotic
Cabbage	Antibiotic thought to be a carbohydrate with molecular weight of 10,000 or less. Content varies in different varieties of cabbage.
Cereals	
Rye, wheat, corn, barley.	Fungicide
Wheat, barley seeds, wheat bran.	Antibiotic
Egg white	"Conalbumin," a compound capable of inhibiting anthrax bacillus, antifungal activity
Garlic.	"Allicin" (diallyl disulfide oxide), an antibiotic. Vapors and juices inhibit some bacteria.
Honey	Long used to heal wounds; active agent was termed "inhibine," but found to be hydrogen peroxide produced by enzyme (glucose oxidase) acting on glucose. Perhaps other antibiotics.
Horseradish	Antibiotic with LD_{50} of 160 mg/kg in white mice
Milk (cow's).	"Lactenin," an antibiotic consisting of two fractions, L_1 and L_2. Pasteurization destroys 50% of L_1, none of L_2. Inhibits tuberculosis organism.
Onion.	Onion fumes and onion per se contain known antibiotics. Both S-methylcysteine-sulfoxide and S-n-propylcysteine-sulfoxide yield (in presence of certain enzymes) thiosulfonates, strong antimicrobial compounds.
Radish	"Raphanin," an antibiotic toxic to mice at 7 mgms or more, to guinea pigs at 50 mgms or more
Tomato.	Antibiotic glycosidal alkaloid called tomatine
Spices and essential oils	Long used as preservatives; various spices contain both antibiotic and antifungal agents
Strawberry.	Some wound-healing properties
Sweet potato	Three discrete biologically active materials in both the plant and the tuber

Source: Adapted from E. H. Marth. 1966. Antibiotics in foods—naturally occurring, developed, and added, Residue Reviews, 12:66-161.

[a] "Preservative" here refers to the presence of antimicrobial (antibiotic) compounds. These compounds may inhibit one or several different microbial species, but they are not "general-purpose" preservatives.

ly milk. Milk can be the source of a number of human disease bacteria, among them organisms that cause tuberculosis, diphtheria, septic sore throat, and scarlet fever. Not until modern times did man learn the technique to assure—absolutely—that milk would be free of any disease germs. Named for Pasteur, the man who discovered the process, pasteurization—the application of heat for a time sufficient to destroy every disease-producing microbe known to be

Table 20. Specific Antibiotic Activity of Certain Foods

Food	A. aceti	B. subtilus	B. thermoacidurans	C. sporogenes	E. coli	L. lycopersici	L. mesenteroides	S. aureus	S. ellipsoideus
Vegetables									
Asparagus					+			+	
Beets								+	
Brussels sprouts	+								
Carrots		+			+			+	
Cauliflower	+	+	+		+		+		+
Celery								+	
Chard									+
Chicory		+							
Cucumbers		+							
Green beans		+			+			+	
Okra								+	
Onions		+			+			+	+
Peas		+							
Peppers		+					+		
Potatoes								+	
Rhubarb		+	+		+			+	
Snap beans		+						+	
Spinach		+							
Sweet corn		+							
Tomatoes	+	+					+	+	+
Turnips		+							
Fruit									
Apples		+	+					+	
Avocados					+			+	
Cherries		+	+		+			+	
Grapefruit juice			+					+	
Grape juice		+			+	+	+		
Orange juice		+							
Peaches								+	
Pears		+		+					
Pineapples		+	+	+				+	
Plums		+	+	+				+	

Source: Adapted from E. H. Marth. 1966. Antibiotics in foods—naturally occurring, developed, and added, Residue Reviews, 12:66-161.

carried by or transferred to man through milk—is foolproof. One need only ask what might happen if we had to rely solely on natural preservatives. You can take all the antibiotics shown in Table 20, which is a fairly representative sample of those found in man's food, mix them together, add them to milk, or to any other food for that matter, and you still fall short of "pasteurization," the destruction of all disease germs. Certainly the result doesn't come close to sterilization, the treatment by which all organisms, disease or otherwise, are destroyed. Of course more potent antibiotics do exist, but they are not indigenous to foods normally eaten by man and they may even be harmful to man. Regulations exclude or limit their presence in food as they would any other potential contaminant.

Natural can be healthful; it can also be deadly. Nature does not do for us what we are readily capable of doing for ourselves, if we but choose to do so. In man's struggle for survival nature did not put forth germ-free, toxin-free fruit. Man had to learn—the hard way—what foods agreed with him biologically as well as how to "process" away natural food ingredients or germs that would otherwise do him in. That many succumbed during the learning process goes without saying. That it can happen today as easily as it did in the past is equally evident.

Food Processing as a Method of Producing "Preservatives"

Drying and freezing, two age-old methods of food preservation, in no way result in the production of chemicals with antibiotic significance. Food may be preserved (prevented from microbial spoilage) by these methods, but microbes that are present in the raw product initially may survive until the product is reliquefied (rehydrated) or thawed. Microbes differ greatly in their ability to tolerate the dried or frozen state. Some are readily killed; others live almost indefinitely. A recent drilling of Arctic subsoil uncovered microbes found to be viable (capable of growth and multiplication) after at least 10,000 years of dormancy. Some estimates placed their age at 100,000 years or more.

Drying or freezing as methods of food preservation *do not result in the death of all microbes.* They do not render food spoilage-free or disease-free.

A processing method does exist, however, in which "antibiotics" are produced and food made disease-free at least for a period of time. The process, fermentation, is nearly as old as man. Fermentation (souring, pickling), which can occur naturally or, with more consistent results, under controlled conditions established by man, is a sound method of food preservation. It is used in the manufacture of cheeses and cultured dairy products (cultured buttermilk, yogurt, acidophilus milk, etc.), pickles, sauerkraut, green olives, some soy products, vinegar, and the like.

In food fermentation man improves greatly on nature by providing essentially pure fermenting agents (bacteria, yeasts, and/or molds) and an optimum environment for their growth. Natural (uncontrolled) fermentations are inefficient (costly). Left to the whim of chance contamination, they are also susceptible to the production of vile flavors. Contrast the clean acid flavor of cultured buttermilk with the putrid, fruity, and unclean flavor of "naturally" soured milk.

Fermentation left to chance is wasteful. Tons of natural cheeses end up as processed cheese because, under the best conditions, the ferment sometimes goes astray; off-flavors are produced or compositional requirements cannot be met. By blending small amounts of such cheeses with perfectly good cheese and by further processing, capricious fermentations can be salvaged. Without this capability much natural cheese would end up as animal feed. The fate of other "wild" fermentations can be worse; the product is simply dumped down the drain.

"Commercial" fermentation is, as it has always been, a highly skilled art. Man improves upon nature, but only through the use of his best talents. It is still appropriate to speak of "master" brewers. The same term could be applied to cheese makers and pickle processors—or any of the skilled workers who oversee fermentation processes.

During fermentation some microbes produce antibiotics. Acids are also produced, and/or alcohol. Antibiotics, acids, and alcohol all inhibit certain microbes. Perhaps the best known antibiotic generated by fermentation is nisin, an end product of the ferment by which most cheeses, cultured milks, pickles, and sauerkraut are made. For some microbes nicin is lethal, for others it is fodder, and therein lies its limitations. Even so, it has been widely investigated for its utility in such foods as natural and processed cheeses, canned goods, meats, fish, and soups.

Publicity, hearsay, and legend surround the nutritive attributes of certain fermented dairy products. Among claims supporting them are minor health wonders and longevity. Examination of the evidence would indicate that, like the tall tales of fishermen, fact has been confused by fantasy. But who knows? For the more common fermented milks a few "facts" follow.

Lactobacillus acidophilus, for which acidophilus milk is named, produces antibiotic(s) capable of inhibiting some bacteria known to reside in the intestinal tract. Among these is a diarrhea-causing coliform. The milk product itself has also been shown effective against pneumonia organisms. Considered no less a miracle food, yogurt, the product of the combined efforts of *Lactobacillus bulgaricus* and *Streptococcus thermophilus*, boasts an equally impressive antibiotic (or antibioticlike) arsenal. Lactic acid, the natural end-product of all lactic acid fermentations, or other inhibitory agents present in yogurt retard or destoy a number of salmonella (food-poisoning) bacteria as well as bacteria that can cause diphtheria and tuberculosis. At least one pathogenic protozoan is also known to succumb to yogurt's health-giving touch. Kumiss, kefir, and other fermented milks belong rightfully to this class of "drug-producing" food processes, for they too contain inhibitory substances.

So what is it that fermented milks have that most other foods do not have? If anything—and scientific proof bears out the control mechanism if not the claims—it is the capacity to dispel from the intestinal tract a host of undesirable bacteria while implanting their own beneficial ones (with routine replenishment) as masters of the digestive system.

Before abandoning the topic, we should include a note of special significance to home canners of foods—*beware*! The acid that assists in the preservation of fermented foods is not common to all foods. Low-acid products (meats, poultry, fish, and all vegetables except tomatoes) require that special attention be paid to sterilizing procedures. For the risk of botulinum food poisoning is an ever-present possibility.

Naturally Occurring Antioxidants

Like other "preservatives" some antioxidants (or some chemicals with antioxidant powers) occur naturally. Perhaps the most common of these is ascorbic acid (vitamin C). Tocopherols of the vitamin E complex are also natural antioxidants, as is gum guaiac, a resin from the wood of *Guajacum officinale L.* or *G. sanctum L.*, Zygophyllaceae. Onions contain quercetin, asparagus rutin, and red wine anthocyanin pigments, all of which are antioxidants. Most GRAS approved antioxidants are either naturally occurring or derivatives of naturally occurring compounds.

For a number of reasons, including effectiveness, cost, and availability, synthetic antioxidants are commonly used. Were it not for such compounds, the variety and shelf-life of cake mixes and cereal products (among other foods) would be greatly reduced.

Antioxidants are also produced during some methods of food processing. Since they arise from natural food ingredients, they should perhaps be termed natural. For example, a good many sulfur-containing proteins, when heated, uncoil and expose chemical arms, called sulfhydryls, which serve as antioxidants. In the raw state (undenatured) these chemical arms remain encumbered and useless. If the heat treatment is not too severe—and it need not necessarily be—the protein quality, from a nutritional standpoint, remains unimpaired.

While "processed" antioxidants can be helpful, they are not sufficiently effective and widespread to be of more than limited value.

Chemical (Synthetic) Antioxidants

The three most important synthetic antioxidants are propyl gallate (PG), butylated hydroxyanisole (BHA), and butylated hydroxytoluene (BHT). To provide maximum effect the latter two are often combined. They appear on food labels in the abbreviated forms BHA and BHT. They are antioxidants solely. Chemically the three compounds are structured as shown in the drawing. PG is water soluble, BHA and BHT are fat soluble. This is an impor-

Propyl Gallate
(PG)

Butylated
Hydroxyanisole
(BHA)

Butylated
Hydroxytoluene
(BHT)

tant difference since an antioxidant can only be effective if it can be incorporated into a food. Some foods are aqueous systems, others fat or oil. An additional advantage of BHA and BHT lies in their heat suitability. They can be heat processed, or, rather, heated along with the foods they protect, without losing effectiveness. This is an essential consideration for the food manufacturer whose products require heat processing. Upon drying, though, these two compounds volatilize and a slight off-odor may occur.

Table 21 lists some specific food products in which BHA and BHT are al-

Table 21. Use of BHA and BHT in Foods

Food	Total Limits of BHT, Alone or in Combination with BHA (ppm)	Food	Total Limits of BHA, Alone or in Combination with BHT (ppm)
Dehydrated potato shreds.	50	Dehydrated potato shreds.	50
Dry breakfast cereals	50	Dry breakfast cereals	50
Emulsion stabilizers for shortenings	200	Emulsion stabilizers for shortenings	200
Potato flakes.	50	Potato flakes.	50
Potato granules	10	Potato granules	10
Sweet potato flakes.	50	Sweet potato flakes.	50
		Dry diced glaced fruit (BHA only)	32
		Dry mixes for beverages and desserts (BHA only)	90
		Active dry yeast (BHA only)	1000
		Beverages and desserts prepared from dry mixes (BHA only)	2

Source: Adapted from *Code of Federal Regulations*. 1972. Title 21, parts 1-119.

lowed, along with restrictions on amounts. To maintain a kind of perspective, because amounts may appear excessively large otherwise, it should be kept in mind that one part per million is an exceedingly small amount.

Safety of Synthetic Antioxidants—A Story Oft Repeated

Our story has its origin in two different laboratories where scientists are engaging in work of a similar nature, though on two entirely different chemical compounds. The work does not go on concurrently. There is no race to achieve an important finding in order to gain special recognition or prestige. These scientists are basically honest people carrying out exacting experiments in a field not renowned for yielding cut-and-dried results. They work with test animals in experiments so painfully routine and devoid of exciting "breakthroughs" that thought or hope of such coveted awards as the Nobel Prize is rarely if ever discussed, except perhaps as they relate to other people.

Uncomplainingly (or with no fewer complaints than would normally occur in such circumstances) the men go about the work of feeding the test animals (rats in this case) carefully formulated diets. Not a gourmet menu, the one diet contains controlled amounts of butylated hydroxytoluene (BHT). In the other laboratory, four or five years later, they will be experimenting with a formulation of nordihydroguaiaretic acid (NDGA). It is the common bond linking the two labs. Both are testing the safety of antioxidants usually used in commercially processed foods.

By now ground rules for affirming additive safety are reasonably well established. The LD_{50}, that lethal dose of "anything" which kills 50 percent of the test animals, must not reach levels of more than 1000 mg/kg of body weight. As well, the additive must not significantly influence animal growth over long-term feeding trials—one to two years—at doses of *100 times the amount* expected in human consumption. Comparatively speaking, the rats are receiving massive doses of the test compounds. Not only is a sound buffer between ordinary and test usage thus achieved, but any changes that take place are of such magnitude that they are more readily measurable. A number of responses are being measured: metabolites (breakdown products) excreted in the urine and fecal matter, growth rate, and, eventually, after sacrificing the animals, weight of various organs, especially the liver. Reproductive ability is also being studied.

Though working with rats, scientists in this field realize that dietary responses vary in different species of animals. Depending upon the chemical being tested, a rat might react more nearly like a human (it is often the case) or a rabbit, a dog, or, going up the evolutionary ladder, a monkey. And within the same species differences might even occur between sexes. Age, too, is a variable, all the way from conception (or shortly thereafter) on to the octogenarian, at whatever the time span needed to reach this comparative age in the animal in question. Rats reach old age in a fairly short period of time. It is one reason (along with cost and their metabolic similarity to humans) why they are so often used as "guinea pigs." Time looms critically important when research observations on offspring are desired. In this work such observations are a necessity somewhere along the way.

The experiments continue: measurements are made, the animals sacrificed, data collected.

The year is 1959. A group of researchers in a distant country have published an article indicating that rats fed a normal testing dose of BHA and BHT react adversely to the latter. On a BHT diet, baldness has been observed in Norway hooded rats. More seriously, three of thirty litters in one experimental trial produce anophthalmic (eyeless) young. Together, these findings suggest a possible teratogenic effect—the malformation of embryos during pregnancy. Deformed babies! The finger points directly at BHT. As word spreads, shudders begin to run up and down the spines of scientists, regulatory officials, and processors. This is a shocking discovery entirely contrary to previous knowledge.

As early as 1955 toxicity data on BHT had indicated no reason for concern. In dogs, rabbits, and rats, the latter over prolonged feeding trials, high dose levels of the compound caused no unusual responses, no observable pathological conditions. BHT had received a clean bill of health.

Obviously the word gets around about the alleged damaging effects of BHT. Consumers rise up in arms. Regulatory officials rattle their sabers. Industry throws up its hands in dismay. In a flurry of activity other laboratories take up the call. In one, the work just reported is repeated using rats and mice and immense levels of BHT. Another lab tests the effects of BHT on dogs. Two papers come from still a third laboratory where more detailed biochemical analyses are run; a fourth reports additional findings, followed by still another. By 1965, no fewer than nine independent experiments have been carried on which, in one way or another, repeat or enlarge upon the investigation in question. Ultimately some fifteen full-fledged research papers will be published. All dispute the one observation of pathology made in 1959. At that time honest results were scrupulously reported, but they have been found to be irreproducible. Yet shock waves of this sort have a way of hanging around for a long time. Today BHT is still somewhat suspect, all evidence to the contrary.

A single explanation for the reported disorders in rats suggests that similar dysfunction could have been induced by vitamin A deficiency. The basis for such deficiency, however, cannot be explained. And there, for this part of the story, the matter rests. A most interesting epilogue will be detailed later.

Let us jump ahead to 1968. A second laboratory engaged in toxicological studies on NDGA, another antioxidant, also makes an unwelcome discovery. A single breakdown product, orthoquinone, is found in the urine and kidney tissue of rats reposing on a diet containing 2 percent of the test compound. Theory has it that NDGA, metabolized by bacteria in the intestinal tract, is converted to the quinone, a known toxicant. In this case the rats become ill; obvious physiological changes occur, are measured, and reported. Again the word goes out. But this time, despite an almost complete lack of corroborating evidence, NDGA is struck from the list of GRAS approved substances, even though the study was restricted to rats, not necessarily the most appropriate test animal.

World society, it can be assumed, had long since been sensitized to such findings. The thalidomide tragedies, a cancer scare in cranberries, a case of mercury food poisoning, *Silent Spring*, PCB's in turkey meat, antibiotic residues, radioactivity in milk, phthalates, DDT, lead poisoning, botulinum food poisoning, FDA recalls—all combine to produce a gnawing sense of unease in most people's minds. The word *additive*, it seems, is spelled P-O-I-S-O-N.

(As a matter of interest not one of the cases or toxicants cited above involves a direct food additive. Thalidomide, of course, was a drug.)

Prohibited in the United States, NDGA still has approval for limited use in some other countries. Allowances are small, 0.005-0.01 percent, but the product nonetheless enjoys legal sanction. Interestingly enough, Austria, which permits use of NDGA, bans BHT.

So it goes. Partly as a result of its tainted history, BHT has spawned a great deal of research. Perhaps because of it, some intriguing findings have recently been reported in the press—among these an observation, not an isolated observation, that a test group of mice reared on a pinch (0.5 percent) of BHT outlived their control cousins an average of six and a half months. As mice measure age, this is nearly a third of a lifetime.

Quackery? Freak findings? Not necessarily. Evidence is fairly strong that aging, at least in part, involves an oxidation process within the human body. Lysosomes and mitochondria (cell residents active in life functions), clogged by peroxidized cellular products, are literally enfeebled. It is postulated that over time their end-products tend to accumulate in vital organs of the body. "Aging" takes place.

Such degenerative ailments as rheumatism, senility, lowered muscular (heart, too) strength may all be shown to be related to the peroxidation process which synthetic antioxidants so effectively inhibit. At the same time, some antioxidants are thought to lower the need for vitamin E! Either the antioxidant protects vitamin E until it can be absorbed into the body or it exhibits vitamin E activity per se. To anyone at all sensitive to shrinking world supplies of nutrients, the economic and life-giving prospects of this possibility are awesome.

Though inconclusive, evidence is mounting in favor of a number of beneficial side effects from the use of synthetic antioxidants. So intriguing are the scientific reports that those who ponder the consequences of overpopulation are beginning to worry lest we add to the problem of lengthening life through the widespread usage of age-preventive additives. In any event, the safety of those synthetics most often employed (BHA, BHT, and propyl gallate) appears for the moment to be above suspicion. But with due respect to the almost impossible task of obtaining incontrovertible proof, it should be noted that the issue of chronic toxicity is rarely if ever completely discounted. Other side reactions may yet be uncovered.

To top off this discussion, one final sidelight is in order. As recently as 1974 Utah State University researchers, appearing at an American Chemical Society meeting, offered evidence that peroxidized fats may enhance the carcinogenic potential of various cancer-causing chemicals. If true, antioxidants are perhaps serving a life-saving function. At any rate, wherever you may stand on the issues, select your own "poison." If nothing else, the alternatives are becoming more clear.

Nitrates and Nitrites

To even the most avid repudiator of evolutionary theory, a sense of kinship must kindle upon learning that test doses of some chemical or physical agent to which both man and another species are exposed induces in the other species, however remote, any form of cancer. Whether or not he is willing to admit to evolution, man is related to other animals, most would agree, if not on the surface then internally where life processes, like pyrotechnic displays, whirl and flare in chemical and enzymatic cycles. So it is that whenever a substance is found to trigger a carcinogenic response in a rat, dog, monkey, or guinea pig, an alarm sounds and the offending agent is scourged from its hiding place and instantly removed from the scene. At least this is the case most of the time. Certainly this was the intent of the Delaney Clause (of the Food Additives Amendment of 1958) which prohibits use in any amount whatever of any substance "found to induce cancer in man or animal." As defined, this clause refers not simply to direct food additives, but to food coloring compounds specifically, pesticides applied in production, storage, and transportation of raw agricultural goods, and all compounds given sanction before the 1958 amendment, including those admitted into commerce under both the Poultry Products Inspection Act and the Meat Inspection Act. It is a clause thereby applicable to all food. But it is a clause, too, presently under scrutiny for a number of reasons. No other additives better reflect the tormenting issues than do nitrates and nitrites.

Apparently, the origin of the use of nitrates in meats is lost in antiquity. Early users, it is assumed, not only knew nothing of the mode of action of these compounds, but presumably added them to meat as chance contaminants of salt, whose benefits were well known if not entirely understood.

For nitrate to be most effective microbiologically, it must first be acted upon by microbes, for it is in the conversion of nitrate to nitrite that a potent bactericide is produced. Quite understandably, the most significant present-day role of nitrates is the curing of fermented meats, like salami, where long-term microbial fermentation is required. In the past, generally poor sanitation, which caused heavy microbial loads, probably assured nitrate conversion. But the point is, it is nitrite, not nitrate, that is active against bacteria. Further, both chemicals have been in the marketplace a long time. USDA sanction, in the form of specific levels allowed, dates back to 1926. Today these levels are being reconsidered, alternatives are being sought, disallowances in certain products are under discussion. The reasons for concern are clear, the issues complex and not given to simplistic resolution.

How nitrites function in meat products. Nitrites perform three or four useful preservative functions. They are effective microbial inhibitors (even against botulinum, the most deadly of bacterial germs), they retard the development of rancidity, to some extent they enhance the flavor of certain meat products (primarily the beef-pork wiener, cured hams, and loins) and, reacting with myoglobin in blood, they serve to maintain the bright red color of meats.

No one disputes man's ability to survive something less than the finest flavor in cured ham. Neither does anyone strongly argue against man's ability to survive and perhaps, eventually, to enjoy a rainbow variety of meat colors other than red. Serious, well-founded doubts enter the picture only when microbial concerns are contemplated. And here, for valid health reasons, the issues get sticky; nitrites fulfill what can be an essential function.

In meat, which is slightly acidic, sodium and/or potassium nitrite salts break down to nitrous acid from which nitrous oxide gas (nitrogen monoxide), the curing agent, evolves. This reaction and other chemical reactions take place fairly rapidly. Two days after processing only about 50 percent of the original nitrite remains. Although the same processes occur in bacon, the reactions are somewhat slower. Still, significant amounts of nitrite are lost in a relatively short time span.

Depending upon the product and the initial concentration of added nitrite, only a few ppm or as much as 40 or 50 ppm may still be present after eight to ten weeks cf refrigerated (40-50°F) storage. Some 9 to 12 percent nitrite may be chemically tied up in cured meat pigment. Unlike the levels of most additives, that of nitrite is not a static phenomenon.

Foods with naturally occurring nitrates and nitrites. Like so many other seemingly man-made, strictly "unnatural" chemical agents, nitrates and nitrites do occur naturally in some foods. Most green vegetables contain nitrates, sometimes in amounts greater than those permitted for nitrate additives. Lettuce is a veritable nitrate storehouse at 1000 ppm. Levels in spinach can reach a point at which the food is actually toxic to infants. The same is true of natural water supplies. Infants are particularly vulnerable because of the low acidity of their stomachs. In acidic environments nitrates are readily converted to nitrites which, reacting with blood (remember, nitrites fix meat color by combining with hemoglobin), cause a blood ailment known as methemoglobinemia. One concern, that nitrates in *opened* cans of baby food vegetables would convert to nitrites, has now been disproven. Tests by the FDA on ten baby food vegetable products showed "no significant nitrite development" in opened cans mistreated in several ways.

Nitrate levels differ from one vegetable species to another and also in varieties within a species. Some parts of the plant are naturally higher in nitrate than are other parts of the same plant. In all cases nitrate levels are influenced both by stage of plant maturity and by environmental factors. Unless vegetables are badly abused, they are not significant sources of nitrite. Fortunately, the abuse necessary to yield nitrites usually renders the plant inedible.

Nitrites do occur naturally in some foods as well as in polluted air. And, believe it or not, nitrites are found in human saliva. You consume from 6 to 12 mg of nitrite each day from this source alone. To obtain as much from meat would require an intake of ¼ pound of cured meat having a residual nitrite level of 50 ppm.

We live, therefore, in an environment where nitrates and nitrites are common.

Toxicity of nitrates and nitrites. As a single dose, 30-35 grams (a little over an ounce) of nitrates can be fatal to adults. More potent, nitrites will kill at intakes of 22-33 milligrams per kilogram of body weight (about 2 grams for a 154-pound adult). Even at the lowest end of that range, the same 154-pounder would have to consume over 18 pounds of cured meat product containing 200 ppm nitrite to achieve a fatal dosage. Since actual residual nitrite levels are much lower than that, a fatal intake would actually require three or four times that amount—50 to 70 pounds of cured meat product at one setting.

The question of cancer. You will have noted that, so far, no mention has been made of cancer. Nitrates and nitrites per se are not carcinogens. But they *are* precursors; they are the starting material from which N-nitrosamines, described by some as among the most formidable and versatile group of carcinogens yet discovered, are formed. When nitrites react with secondary or tertiary amines, components of some proteins, spices, and flavors, the end-products are nitrosamines. The reaction looks like this:

$$(CH_3)_2N + NO \longrightarrow \begin{array}{c} CH_2NCH_3 \\ | \\ NO \end{array}$$

dimethylamine + nitrous oxide   N-nitroso methyl amine

The nitrous oxide stems from nitrite additives, the amines from meat protein.

Since their discovery in 1863, over a hundred nitrosamines have been isolated and identified. A whole family exists, varying only slightly in chemical makeup. In chemistry, however, minor disparities can mean the difference between life and death. Not all nitrosamines cause cancer; and the difference between those that do and those that don't is as small as a finger, a slight though distinctly different chemical extremity sitting like an outstretched finger on the chemical hand.

But the lethality of the 75 percent of so of this outlander chemical family that are toxic should not be denied. Those nitrosamines that cause cancer have been found to do so in *every species of test animal studied thus far.* Not only are they cancer inducing, they are also mutagenic (causing genetic damage) and teratogenic. Something as insignificant as a slight twist in the molecule, which results in a change in symmetry, will induce cancer in the esophagus rather than the liver, or vice versa. Thus do chemical makeup and physical shape both play their own deadly role in nature's scheme of things.

Man never improves upon nature in either beauty or deadliness. Nature simply stores her attributes; man rearranges and molds from them his own forms of art or poison. Nitrosamines are found in nature, in an African berry and in some saltwater fish. Very recently they were discovered in tobacco smoke. At the moment, the health implications of this new finding are not known.

Nitrosamines may also be present in prepared foods. FDA scientists, with the help of a highly sensitive broad spectrum analytical technique, are able to identify fourteen such nitrosamines. The years 1969 through 1973 saw the confirmation of trace amounts of nitrosamines in a number of different meat and fish products as well as in cheese. The amounts, it is emphasized, are minute, measured not in parts per million but *parts per billion.* That parts per billion are unimaginably small quantities is not necessarily expected to impress the reader. Where cancer is concerned one is correct in asserting that any amount is too much.

Still, it must be asked whether there is an amount of carcinogen, of whatever potency, so small as to require longer than the normal life span to produce its deadly effect. If the carcinogen is measured in parts per billion, two times the normal life span might be required, perhaps more. Professor Harden Jones of the University of California has estimated that, at an intake of 10 ppb nitrosamine in meat products, cancer would occur in man when he reached the age of 900 years! Since, in an environment capable of causing cancer, the great majority of mankind escapes this earth by some other route, the problem is obviously not all-pervasive.

Yet no one truly knows at this point what a "safe" level of carcinogen actually is. Evidently people vary in susceptibility, just as they do in allergic sensitivity. In kind, "safety" varies. Through 1973 (and it's only been since 1968 that suitable detection techniques were developed) nitrosamines have been detected in meat products at levels between 1 and 80 ppb, fish products to 26 ppb, and cheese at 1 to 4 ppb. None has been found in unfried bacon, but bacon drippings, by most comparisons, run high in nitrosamine content. Pan frying of bacon likewise produces significant (measurable) amounts of these compounds at nitrite levels of 30 ppm and above. That wiener you like so well, charring over a spitting, flaming grill, also yields nitrosamines. Oven cooking is less conducive to nitrosamine formation, microwave ovens least. As a general rule of thumb, higher temperatures and longer cooking times yield higher concentrations.

Convenience foods are a byword of most Americans. The use of "convenience" additives is a concept perhaps not quite so familiar. But it is a part of industrial operations. Allied industries specializing in additive technology do their best to make profitable and convenient the application of their products to consumer foods. Think of the demand for an additive mix that would transform a bland mix of pork chunks into an old-country-style sizzling pork sausage. Stir it in and, as advertizing glowingly tells the food industry, "prepare a delectable breakfast morsel each and every time you make a batch of sausages; consumer response will make your mouth water, your pocketbook bulge." Madison Avenue could do much better than that—and does. It is big business, however, and sometimes it is the only way a relatively small, unskilled firm (or a large firm with unskilled employees) can safely prepare a consistently tasty product.

Through food industry needs just this simple (and innocent) did meat curing "premixes" come into being. Premixes are the convenience additives for cured meat processes. As the package states, a given amount of additive can simply be added to each 100-pound batch and the pork sausage will be spiced, temptingly flavored, and preserved (with nitrites), all in one neat, no-fuss no-mess step. Very few food processors could long survive without the skills of these highly specialized artisans. Master sausage makers, like master brewers, master violinists, and artists, are always in short supply. Other highly skilled people, scientists in this case, evolve sophisticated methods for detecting carcinogens. Combine the labors of additive specialists and those of supersleuths with a nose for trouble and the following information comes to light: (1) Some spices and flavorings are sources of secondary and tertiary amines; (2) spices and flavorings mixed with nitrites (as readied for the sausage maker) produce nitrosamines; (3) acid (as in the baby's stomach) encourages the process; and (4) even buffered suspensions—specially formulated nonacid (alkaline) premixes—yield trace amounts of the carcinogens at least insofar as detection methods are able to discern.

These disclosures have already prompted action by the FDA and the USDA. Both unbuffered *and* buffered premixes have been banned for use in meat and poultry products barring proof of safety, i.e., absence of nitrosamine formation. Separately packaged premixes, nitrites apart from seasoning, are still allowed, though they must be accompanied by an appropriate statement warning users against mixing until just before use. Since homemaker and processor alike engage in the "art" of meat curing, the packaging procedures and warning message apply equally to both.

Where nitrates or nitrites are found, then, a potential for nitrosamine formation must be recognized. As already discussed, nitrates are common to a number of green vegetables. Mistreatment of these products does result in conversion of nitrate to nitrite. To date, however, green vegetables have not been found to be major sources of nitrosamines. Spinach, though, frozen, thawed, and refrozen a number of times, produces detectable amounts of the carcinogens. But that's not all. In our drug-riddled culture, as noted recently, tertiary amines are commonplace compounds, being constituents of some analgesics. One researcher, feeding rats a combination of 250 ppm of the analgesic amino-pyrine and 250 ppm nitrite, observed malignancies in every rat and a 100 percent mortality rate!

The alternatives. By this time the casual observer would probably have little difficulty selecting an appropriate course of action. This would be equally true were he additionally informed that certain scientists sincerely doubt the validity of present test methods used to detect nitrosamines, even those newer methods described earlier. It is possible, claim some investigators, that naturally occurring food components are being mistakenly identified as nitrosamines. I am certainly not qualified to judge, but, that being the case, perhaps it is wise to assume the worst. And may we also assume, as most meat processors state, that no suitable replacement for nitrites currently exists, that we are stuck with them and with nitrosamines as well.

What, then, should be done? The most obvious solution is to ban nitrates and nitrites across the board. Or (providing there are no legal restrictions) the processor might elect to do without nitrate/nitrite and take his chances. What this really means, though, is that one health risk is being substituted for another, both equally deadly. In 1963 four people died from botulinum food poisoning after eating contaminated smoked fish. An entire industry was nearly wiped out. Nitrites could have prevented the disaster. Or nitrites might have introduced the almost certain risk of nitrosamine formation through the combined action of nitrite, heat (smoking), and fish protein. (This fact, however. was scarcely known at the time. It is well understood today—or so scientists believe—and nitrite limitations, or exclusions, in smoked fish are expected shortly.)

Spurred by concern that the use of nitrite might be banned, researchers have been scrambling to gain a better insight into the potential hazard of botulinum growth in cured meats. All the answers aren't in by any means, yet the evidence is disconcerting. *Clostridium botulinum* will grow in canned ham. A light inoculum can be suitably controlled by nitrites at 200 ppm. A heavy inoculum leaves very serious doubts, to the extent that canned hams with 200 ppm added nitrite, have been found to contain botulinum toxin after seventy days of storage.

Much depends on the initial load of botulinum contamination, higher loads helping to assure growth, but it seems clear that the organism will multiply in cured meat products. Factors limiting growth include an acid environment (low pH), salt, low temperatures, and nitrite preservative. Acidity and salt levels in these products are not as high now as they used to be, leaving inhibition of botulinum primarily to temperature and preservatives. Botulinum toxin, the death-dealing agent, can be destroyed by suitable heat treatment, but obviously no one is willing to leave to chance the possibility that cured meat products will be suitably heated before consumption. Storage of these meat products below 70°F is another good way of retarding botulinum growth—in these kinds of meat products, *but not in other foods.*

Some bright spots. It is almost certain that a number of new restrictions will be placed on the use of nitrite in food products. In July 1974, after two and a half years of deliberation, an expert panel convened by the USDA recommended that maximum allowable nitrite residue levels in cured meat products be limited to a range of 50 to 125 ppm (in lieu of the 1926 standard of 200 ppm). Recommendations were as follows: cooked sausage products, 100 ppm; canned cured products, 125 ppm; canned sterile products, 50 ppm. In the meantime research for other alternative products or procedures goes on. Recent reports indicate sodium ascorbate (the salt of ascorbic acid, vitamic C) inhibits the formation of nitrosamines from nitrites (but possibly at the expense of nitrite preservative activity). Moreover the application of various ligands (N, N-diethynicotinamide, and methyl nicotinate) has resulted in some success in the formation of pink color. Other chemicals of promise include isoquinolines, pyridyl aldehydes, pyridyl ketones, and myosmine.

But what are we saying in reality? That, at a probability of about 99 to 1, the ultimate solution will involve the use of an additive, one whose safety will have to pass the test of time, just as nitrites had to do, in a technical world in which science's ability to detect, with ever more refinement, the many ways of dying will probably far outpace its ability to prevent death.

What the Label Tells You about Preservatives

Aside from specific amounts used, food labels tell all about preservatives, listing them by their common or usual name (BHA, BHT, sodium nitrite, sodium benzoate, etc.). A separate descriptive phrase indicates their function, whether "preservative," "to retard spoilage," "mold inhibitor," "to promote color retention," or the like. Amounts allowed, limitations, and restrictions are established by federal regulation. This information is provided in Appendix VI.

VII

Emulsifiers and Stabilizers
(Binders and Thickeners)

Without emulsifiers and stabilizers—the mortar and cement, the chinking and plaster and glue that bind a food system together—the cooky crumbles, cakes fall apart, puddings become thin runny syrups, processed cheeses melt to shapeless blobs. Without them there would be few vegetable toppings, no milk substitutes, no lemon, custard, or banana cream pie. Without them shortening would come as an oily lard, mayonnaise a greasy mess looking like wheyed sour milk, ice cream with the texture of crushed ice cubes. If that isn't bad enough, cheese slices would stick to the wrapper, both paper *and* icing would be peeled from a cake, pie crust would remain embedded in the pie plate, meat au jus would be served absent most of the jus, your beer wouldn't foam.

Homemakers have used emulsifiers and stabilizers all their lives, and the processor, if he has replaced you as chef in the household, uses them for you. Like all other additives they *are* composed of a variety of natural and synthetic chemicals. Some of them sport long, high-sounding chemical names. Taken together, they form a sizable family of ingredients. And yes, in the lot there's a black sheep or two.

Emulsifiers and Stabilizers—What They Do, How They Function

Unblushingly, I must admit that I could scarcely do justice to one, let alone several of the many physicochemical theories describing the function of emulsifiers and stabilizers. These are highly technical points best left to specialized academics. Nonetheless, a sketchy overview might prove helpful.

Emulsifiers, like stabilizers, serve several purposes. Chiefly, though, they bind fat to water or water to fat. In so doing aqueous and oil fractions are kept from separating. What this requires and what in fact emulsifiers have is a kind of split personality, an affinity for both fat and water. While their molecular arms grasp and tug at oily food ingredients, their mermaid's tail is anchored firmly in water. And like bees around a hive, they tend to swarm at that very spot (the interface) where fat meets water or water meets fat. There, like combatants in a tug-of-war, they hold their charges together.

In chemical makeup emulsifiers tend to be fatlike; many of them have their origin in fat. The monoglycerides and diglycerides, fragments of triglycerides (true fat), could be called the originators of the clan. From fat sources alone, an army of emulsifiers could be fielded.

In molecular flesh and blood emulsifiers are related to soaps and synthetic detergents. When fats are split into monoglycerides and diglycerides, the leftovers, free fatty acids, made into calcium salts, become emulsifiers. Common among them are calcium caprate, caprylate, laurate, oleate, and palmitate. The sodium salts of these same fatty acids would be soaps.

Milk and soybean harbor one of the most widely used natural emulsifiers, lecithin. Another natural emulsifier, which nobody likes to talk about, is heart disease's old nemesis, cholesterol. In the body, fats are dispersed by bile salts. To assure good whipping properties ox bile extract is used to emulsify free fat in dried egg whites.

Synthetics? Yes. And plentiful enough that one group is numbered to distinguish its various members. Of these, polysorbate 60, possibly with helpmates polyoxyethylene (20) and sorbitan monopalmitate, finds use in as many as sixteen different food categories. They are a versatile lot. Even the widely eaten food the pickle needs to keep its juices together with products like polysorbate 80. And shortenings, without polysorbates, stearyl 2 lactylic acid, and stearyl monoglyceride, would scarcely be recognizable.

Foods do need emulsifiers, the smoothers, lubricants, foamers, coaters, and wetters that put it all together—and hold it there. Table 22 offers a bird's-eye perspective of some of the more common ones.

Stabilizers could be characterized as swelled-headed. These compounds puff up, bloom, gel, and thicken. In the process they bind ingredients, thicken the mix, and smooth the texture. Without stabilizers, French and other salad dressings would be liquid and runny. Stabilizers put "body" in puddings, sauces, gravies, jams and jellies, icings and toppings. They hold in suspension the chocolate in chocolate drink. They stabilize the whip in topping, the foam in beer, and they add bulk (not calories) to special dietary foods. An

Table 22. Some Common Emulsifiers and Their Use in Certain Foods

Emulsifier	Some Specific Food Uses[a]
Acetylated monoglycerides	Shortening
Calcium caprate	
Calcium caprylate	
Calcium laurate	
Calcium oleate	
Calcium palmitate	
Cholic acid.	Dried egg white
Desoxycholic acid.	Dried egg white
Diacetyl tartaric acid esters	Blends of vegetable and animal fat
Dioctyl sodium sulfosuccinate.	Some cocoa products (solubilizes trace minerals)
Dipotassium phosphate.	Some cheeses
Disodium phosphate	Some cheeses
Glycerol-lacto stearate (and	
oleate or palmitate).	Rendered fat
Hydroxylated lecithin	
Monosodium phosphate	Some cheeses, meat (to decrease juices lost in cooking), canned ham, pork, bacon
Ox bile extract	Dried eggs
Polysorbate 60 (with possibly	
polyoxyethylene (20) and	
sorbitan monopalmitate	Whipped vegetable topping, cakes and mixes, shortening and vegetable oil, nonstandardized baked goods, icings, fillings and toppings, non-standardized dressings, milk substitutes, dough conditioner, gelatin desserts, nonstandardized confectionery coatings, various chocolates, chocolate syrups, protective coating on raw fruit
Polysorbate 65	
Polysorbate 80	Ice cream, frozen custard, sherbets, pickle and pickle products, canned spiced green beans (dill oil dispersing agent)
Sodium metasilicate	Egg white
Sodium stearate (and other	
salts of fatty acids)	
Stearyl 2 lactylic acid.	Shortening
Stearyl monoglyceride	Shortening
Trisodium phosphate.	Some cheeses

Source: *Code of Federal Regulations.* 1974. Title 21, pts. 10-129, subpart D.

[a] If no specific use is designated, the compound may be considered a general-purpose additive with rather broad application or potential application in a number of foods.

ambiguity surrounds them. In wines and beer they clarify, in fruit juice they hold juice solids in cloudy suspension. Some cheeses need stabilizers (as do certain frozen foods) to prevent "wheying-off," the separation of whey or liquid from solid ingredients. Stabilizers also prevent the formation of ice crystals in ice cream. They are even found in some processed meats, when a jellylike consistency is desired.

Many stabilizers are 100 percent natural products. Several are tree gums.

Gum arabic, sometimes called acacia after the tree genus from which it comes, is one. Ghatti, gum karaya, and gum tragacanth are others. Larch trees yield still another, arabinogalactan. Larch are common to the United States. Most other gum trees are native to foreign soils. Pectin is entirely natural, a product of fruit; gelatin is its animal counterpart. From seaweed stem four stabilizers of considerable importance: agar, algin, carrageenan, and furcelleran. Locust bean and guar are derived from seeds of leguminous trees.

To speak of natural, refined, synthetic, or artificial without precisely defining terms leads to confusion. Moreover, some compounds simply defy pigeonholing. One such group may be found in the listing of stabilizers. They are natural products in that they are not synthesized in the laboratory. But their original form, the form in which they exist in nature, is modified. Thus are there a number of cellulose (the stuff of which wood and plant tissue is made) derivatives, carboxymethylcellulose being one. There are starch derivatives and products that originate from microbial fermentation (*dextran*), as well as derivatives of pectin, algin, locust bean, and guar gum. These products may be called synthetic or refined or whatever, but there are other stabilizers not found in nature which are built from scratch in the laboratory. These can be labeled totally synthetic; they are stabilizers, they function as stabilizers, they are utilized primarily in nonfood products. Stabilizers most commonly used in foods are shown in Table 23.

Natural Is Better—Or Is It?

As previously mentioned, emulsifiers and stabilizers have their black sheep. Most recently in the news, because it is suspected of causing ulcers, is the one called carrageenan. Let's see how it went astray and what the outcome was.

Before the food processing industry came into being as we now know it, housewives had to make do with whatever ingredients were available when they cooked. If they wanted to make a pudding of pudding consistency, a thickener had to be found. So, how did they go about it? Probably they did exactly what the cavewomen did in their day; they left their abodes and scoured the countryside. Until modern times you can bet that this was what the housewives of Ireland were doing. And since they lived on an island, the seashore surrounded their hunting grounds. No one knows who first discovered it, but with persistence and ingenuity some Irish women (or liberated men), aware of the abundance of seaweed awash on their tidal flats, must have tried some out as food, perhaps in time of great hardship. They probably tore it from the rocks to which it was attached, brought it to the house, put it in a

Table 23. Some Common Stabilizers and Their Use in Certain Foods

Stabilizer	Some Specific Food Use(s)[a]
Acacia (gum arabic)	
Agar agar.	Clarify and stabilize wines
Algin	Bread mix and sauces, some cheeses
Ammonium alginate	Flavor stabilizer
Arabinogalactan.	Nonstandardized salad dressings, pudding mixes
Calcium alginate	
Calcium lactobionate.	Dry pudding mixes
Carrageenan and carrageénin	
Dioctyl sodium sulfosuccinate.	Gelatin desserts, fruit juice drinks
Furcelleran	
Ghatti gum	
Guar	
Hydroxypropylmethyl cellulose	
Hydroxypropyl sodium carboxy-methyl cellulose.	Foam stabilizer in beer
Karaya gum	
Locust bean gum (carob bean gum)	
Methyl cellulose.	Meat patties, French dressing
Potassium alginate, sodium alginate	
Propylene glycol alginate.	French and other salad dressings, ice cream
Sodium carboxymethylcellulose	
Sodium dihydroacetate.	Peeled squash
Tragacanth gum (more often called gum tragacanth)	

Source: Adapted from *Code of Federal Regulations.* 1974. Title 21, pts. 10-129, subpart D.

[a] If no specific use is designated, the compound may be considered a general-purpose additive with rather broad application or potential application in a number of foods.

pot, and boiled it. Upon cooling, the stuff gelled. Possibly this is the way the folks of Carragheen, Ireland, first learned of a use for the seaweed growing along their shoreline. Eventually it ended up in Irish puddings where, for hundreds of years, it served its thickening purpose admirably.

Like the Japanese researcher who wanted to know what it was in the seaweed along his shore that enhanced flavor, other men, curious and perhaps profit-motivated, wanted to identify and isolate the thickener in Irish Moss. From their effort came the emulsifier/stabilizer called carrageenan.

Extracted with hot water, carrageenan is purified and dried. It may end up commercially in pie fillings, baby foods, evaporated and flavored milk drinks, low-calorie foods, imitation dairy beverages, dessert toppings, and so on. It is particularly useful in stabilizing protein to keep it from separating into a curdy layer. The Food and Agricultural Organization/World Health Organization (FAO/WHO) Expert Committee on Food Additives places its acceptable daily intake for people at up to 50 mg/kg (about 3.5 grams for an adult male). Moreover, a hydrolyzed form (a smaller unit split off the parent molecule) has been utilized in the treatment of peptic ulcer.

So why is there concern about its use? Well, two decades ago researchers, injecting carrageenan under the skin of guinea pigs, observed the formation of small granules of collagen (collagenous granuloma). As already pointed out, skin injections are not considered comparable to feeding trials in ascertaining safety. But others, since then, fed animals hydrolyzed carrageenan and found it to be an ulcer-causing agent in the test subjects. This would be cause for worry except that similar studies of the natural product (unhydrolyzed carrageenan) had proven negative in rats, monkeys, and guinea pigs. And, over the years, *only the natural product has been permitted and used as a food additive*. It is still around, still performing, but a sudden turn of events leaves its future in doubt.

Just as it seemed that this additive would be removed from the GRAS list and given a regulated status, not because of any health risk but as an administrative maneuver to get it assigned to a listing where assurance could be provided against the use of hydrolyzed derivatives, the FDA announced that carrageenan would be given an *interim* food additive status pending further research. An interim basis leaves the final outcome in question. But the point is that the announcement came as a surprise to scientists who were knowledgeable in the area and who thought the issue of safety a closed one. What happened, apparently, was that further research by the FDA pointed to some adverse findings in rats—the presence of liver lesions (physiological changes not considered to be pathological) in rats fed at 2.5 and 5.0 percent carrageenan. The dose levels are very high, but the results were obtained through feeding trials, the method deemed appropriate for research of this kind. The on-again, off-again status of this naturally occurring food additive appears to substantiate further the relative uncertainties in the art and science of toxicity testing.

Label Information about Stabilizers and Emulsifiers

Not all, but most food product labels indicate if stabilizers or emulsifiers are present. Ice cream is probably the most notable exception, although the situation may soon change. Their inclusion on labels has to do with the standard of identity. If a standard of identity calls for emulsifiers and stabilizers as mandatory ingredients, they need not, by present law, be listed on the label. If they are optional ingredients, statutory authority exists to require their declaration. By a quirk of fate it happens that all ingredients in ice cream, the cream, the milk solids, the sweetening agents, and the flavor (stabilizers, too) are optional ingredients. Labeling of these substances could be demanded. If current trends continue and the public demands it, full ingredient listing on all products may shortly come to pass. For those few persons with extreme food allergies, this could be a lifesaver. At least one death, that of a ten-year-old boy with sensitivity to peanuts, has been recorded. In this case the boy knew of his allergic hypersensitivity, but ate ice cream containing peanuts, apparently unaware of their presence.

When declared, emulsifiers and stabilizers will be listed in the statement of ingredients. They will appear by name, their common or usual name, most of which can be found in Tables 22 and 23.

VIII

Toxic Metals in Food

It is a paper-thin veil on which mankind survives, the veil, like that worn to protect Moslem ladies' innocence, delicate, permeable, readily violated by rampant man. A few inches of soil—it is all that exists between mankind and oblivion. In and under this veil, in the earth's crust, are found a number of metals. Some exist free and pure; some, because of their excitable nature, exist only in combination with other elements. Here and there lie deposits rich in metal-bearing ore, laid down over aeons of time and mined in sizable amounts only recently. An isolated pocket of iron, copper, nickel: dug from the crust, the metal is separated, refined, purified, combined with the fruits of other diggings, poured into molds, and made into a thousand million things to be sprinkled like a great rain over a state, a nation, a continent.

A new veil is being formed, of a mix different from that which animal and vegetable life learned to coexist with through trial and error. And the reason still smacks of motives held by those pretenders to science, the alchemists of the Middle Ages, who, in their quest for a philosopher's stone, sought to convert one or the other metal—or all of them—to gold.

Besides those few ore-rich deposits in the earth's crust, metallic elements, as though mixed in a whirlwind, are scattered over the planet's surface in flinty bits of grain and dust. To be sure, this mix of metals would yet be changing had man not intruded. Earthquakes and volcanic eruptions generate upheavals as does the soil's microenvironment where bacteria, natural chemistry, and plant roots, in processes fully as miraculous as those ascribed to the philosopher's stone, transmute elements not to gold but to other forms, raise them up and lay them down in an unending cycle of birth and rebirth.

Meanwhile vegetable and animal life, feeding off soil and one another, consume traces of metal. If the environment includes a broad spectrum of metallic elements, none present in excess, life thrives. This has been the case for so long that metals have become a part of us and necessary for survival. Iron is found in our blood, cobalt in vitamin B_{12}, and copper in a variety of enzyme (life chemistry) systems. We arrive in this world chrome-plated, chromium serving an essential role in the utilization of glucose. Furthermore, unlike most metals, chromium decreases in the body as aging progresses and, should present hypotheses prove true, may be the reason why older adults are predisposed to diabetes.

Another metal, manganese, triggers a number of enzymes and seems to be involved in cartilage production. Molybdenum is a constituent of xanthine oxidase and aldehyde oxidase (both enzymes). Nickel appears to be essential, and so does selenium (as an antioxidant and in red blood cell formation), silicon (in bone and cartilage production), tin, and vanadium. Recent findings show that absence of zinc in the diet leads to dwarfism and belated sexual development. Man, then, is an alloy of sorts, so distributed with metals that it's a wonder he doesn't rattle.

Except in rare instances where soils are depleted, trace mineral requirements are met naturally in plant and animal foods. Plants absorb minerals primarily through root systems, herbivorous animals acquire minerals through vegetation consumed, and man and other carnivores, living off both plants and animals, derive mineral needs from them.

Basic to all questions of survival is the mechanism and health of plant life. It is the origin, the root source of our food chain. To the plant kingdom all else must yield.

All human foods contain trace minerals, although their amounts differ widely. To achieve a proper balance is one reason for varying the diet. Plants absorb minerals from their immediate environment (a significant concept as it relates to similarities between organically grown and inorganically fertilized foods) and deposit them, often in unequal amounts, in various parts of the plant. Animal and poultry products (eggs, milk, meat) also contain traces of minerals.

In proper amounts, certain minerals are known to perform unique life-serving functions. But *all* minerals can be dangerous when present in excess and some, particularly the heavy metals—lead, cadmium, and mercury—are exceedingly poisonous to life. Because, through man's activities, the mineral balance is being upset, serious issues need constant consideration. Thus far nothing has been said of environmental factors as they relate to food and health or of individual responsibility for conditions known to affect the food supply. The situation is clear concerning heavy metals: all of us, to the extent that we exploit the use of industrial goods and services, add our fractional bit to the potential for destroying life. Some examples should help to focus the issues.

Lead

Lead poisoning has been and continues to be a health hazard that requires monitoring and surveillance. Man is incapable of tolerating even minute amounts of lead and is particularly unable, once exposed, to throw it off. It sticks with him, small amounts adding to small amounts and building, ultimately, to toxic levels. If a few milligrams of lead are consumed each day, in three or four months dangerously high amounts of it can be traced in the blood.

Lead was one of the first metals to be recognized as a potential health hazard because its use in paints was causing extensive poisoning. Walls covered with flaking, lead-containing paint naturally attract animals and small children. Lead spewed forth from antiknock gasoline is a pollutant of air, a fallout pollutant of soil, and a source of lead in plant life upon which man's food depends. Lead absorbed in water supplies carried in lead pipes further adds to the dangers for some homeowners.

Had we been more clearly attuned to the hazards posed by lead, we would have been better prepared and perhaps averted recent crises involving mercury and cadmium. All metals—repeat, all metals—can become serious risks to health in the same way that lead spread its insidious, seemingly harmless pattern of engulfment. Lead is a perfect example.

In the earth lead is an insignificant risk to health; it cannot even boast abundance when compared with other metals. Spread thinly, it usually is associated with other metallic compounds. Only a few concentrations of lead occur naturally where plant life might take on elevated levels of it and transfer it ultimately to man. Lead is a minor constituent of earth and important only to man.

Taken from the ground lead becomes "gold" in its industrial application, formerly in paint but now in storage cells, mechanical parts, newsprint, gasoline additives, pencils, ammunition, industrial equipment, and toys. Like other metals lead is hoisted above ground, concentrated, and either cast away as refuse or made into something "useful." Through man's ingenuity it may rise into the atmosphere, hang as smog over cities, wash to earth in rains, and, where urban throngs live and breathe, concentrate its influence. Wherever this occurs food plants and animals take in small but steady doses; these chronic doses lead to ill health; they can cause death.

Symptoms and tolerances. Lead poisoning results in injury to the central nervous system and the kidneys. It impairs the body's ability to produce blood cells. One of the milder symptoms is hyperactivity. Though newer methods for reducing lead in the body are reasonably good, symptoms of the poisoning often remain indefinitely. Lead can be pulled from the body, but the damage already done may be irreversible.

As is often the case, it is the young who appear to be most prone to lead poisoning. They both absorb and retain lead in their bodies to a greater extent than adults do. Between the ages of one and three maximum tolerable intake—from all sources—is 300 micrograms of elemental lead *per day*. Adult intake should be restricted to no more than 3 milligrams (or 0.05 mg/kg of body weight) *per week*. Levels of lead in the blood of over 40 micrograms per 100 milliliters would suggest undue exposure.

Lead in food. Probably because the magpie is omnivorous, seeking food from any and all sources, *pica*, the name of the genus for this bird, has taken on the special meaning "craving for nonfoods." All children experience pica; mentally ill adults may have similar cravings. Today severe lead poisoning in the United States is almost totally associated with pica for paint, plastic, newsprint, or dirt. Nonpica lead sources, as previously mentioned, are air, water, and food. To these sources should also be added lead-containing pottery (ceramics) and enamelware suitable for the dispensing of liquids. If unsure, don't use that fancy ceramic pitcher you picked up during a vacation trip; it may contain lead, which is readily leachable into milk, juice, water, etc.

Though a potential contaminant of many foods, lead is primarily a concern in dairy products, not because of gross contamination, but rather because of the major role these foods play in the feeding of children. A six-month-old infant derives 59 percent of his diet from dairy products, a two-year-old about 36 percent. Over half the lead intake for both age groups stems from dairy sources. Both the FDA and industry monitor lead in evaporated milk.

The reliability and sensitivity of test methods restrict our ability to determine the levels of any poisonous substance and the potential for a food carrier to be a hazard. However, a limited survey of whole milk, using the latest analytical techniques, showed a lead content ranging from 0 to 0.069 ppm with the average at 0.021 ppm. These were significantly lower levels than had been reported previously. Industry and FDA surveys of evaporated milk (concentrated 2 to 2.5 times the original milk solids concentration) would put lead levels between 0.02 and 0.37 ppm, the average about 0.12 ppm. The later figures reflect an industry-wide effort to reduce lead contamination from one major source—the can. They are substantially lower than an earlier reported average of 0.35 ppm for evaporated milk. Improvements apparently have been made and an FDA "tolerance"—an allowable level of contamination when such contamination is unavoidable—of 0.3 ppm would appear to be well within industry capability.Lead in canned strained baby juice has likewise been lowered in recent years. Levels for 1972 and 1974 are shown in Table 24. All but one product (apple-apricot juice) showed reduced lead content. On the average, lead levels decreased from 0.32 ppm to 0.16 ppm between these two years.

To date, lead levels in food are not considered excessive—or perhaps it should be stated differently. More accurately, lead in food is always a hazard because any amount will be added to lead consumed from other sources. Indeed lead has localized significance confined primarily to older sections of large cities, those blighted areas where lead paint of a bygone era flakes and chips from walls within reach of the thousands of children who toddle there. Anywhere from 25 to 45 percent of these children show evidence of undue exposure to lead. Out of sight for only minutes, a two-year-old can chew and swallow many paint chips throughout the day. Add to that some lead in food and more in water supplies, and lead poisoning is a quick process. It happens all the time to between 6000 and 8000 children in New York City alone.

Cadmium

If you are in the mining business and find that after extracting a metal from its ore something is left over which, unless a use is found for it, is waste, you'll probably search for meaningful ways to utilize this material. The food industry operates in this way; it's what the recycling business is all about. It would be foolish to act otherwise. Not long ago cadmium was just such a "waste" product of mining operations.

Cadmium is not found in the free state. It always occurs associated with other ores such as zinc, lead-zinc, or lead-copper-zinc. The more zinc produced, generally the more cadmium produced. About 34 percent of the world's supply comes from the United States. With our insatiable appetite we have consumed 53 percent of the world's total production of cadmium since 1930. This metal, after being taken from its resting place in earth, travels the roadways in cars and trucks, the airways in planes, and the agricultural lands

Table 24 Lead Levels in Canned Strained Baby Juice
Products for 1972 and 1974[a]

Product	Lead Content (ppm) 1972	1974
Orange juice	0.23	0.14
Apple juice.	0.44	0.14
Orange-pineapple	0.39	0.15
Orange-apple.	0.34	0.08
Apple-cherry.	0.23	0.18
Orange-apricot.	0.25	0.14
Prune-orange.	0.45	0.10
Mixed fruit.	0.54	0.16
Orange-apple-banana	0.50	0.17
Orange-banana.	0.11	0.08
Apple-grape	0.29	0.16
Apple-pineapple.	0.26	0.16
Apple-apricot	0.16	0.21

Source: Data from National Canners Association.

[a] Overall average lead content of these products decreased from 0.32 ppm to 0.16 ppm between 1972 and 1974.

in corn pickers, grain combines, seeders, and tractors. It's now a part of hardware tools, textile equipment, business machines, radio and television sets. It's the stuff of which low-melting alloys find application as fire controls in office buildings and department stores. It's also a coloring for plastic. You would think that the continuing industrial revolution had evolved with the sole purpose of utilizing cadmium resources. It's everywhere—in paints and enamels, rubber, glass, printing ink, and ceramics. It's even used in medicine. You'll find it, too, in batteries (its second most important use) and in control rods of nuclear reactors. Mixed and refined, cadmium is being redistributed throughout the world. Where trash is burned, where metal refineries stand, where trucks pound the highways, cadmium concentrations build. High levels could also be expected around golf courses since it is a component of fungicides. Could there be any doubt that it would ultimately end up in food?

The situation is "lead" all over again. Plant life, where soil deposits accumulate, absorbs cadmium; if the plant is food or fodder, man takes it into his system. Plants vary widely in their ability to absorb cadmium without injury. Likewise they vary greatly in the amount they accumulate. Grown in a cadmium solution for three weeks, beet, bean, tomato, and barley leaves took in 280, 9, 1122, and 175 ppm of cadmium respectively. These are tremendous differences. Some plants will absorb cadmium only when the metal reaches a sufficiently high level in the soil. Lettuce leaves, broccoli leaves and roots, cauliflower leaves, pea pods and vines, and oat roots have been found to absorb cadmium at soil levels of 200 ppm, but not at those of 40 ppm. Radishes grown along a busy highway will yield higher cadmium levels than radishes grown beyond the "fallout" area—the cadmium in this case thought to be originating from the erosion of rubber tires.

Because it is important, a repetition is in order: Plants vary in their tolerance to cadmium and the levels of it they will accumulate. Whether to avoid excessive intake of heavy metals or to ensure adequate intake of trace minerals, the diet must be varied.

Cadmium levels in seafoods, meats, fats, and oils range somewhat higher than in fruits and vegetables. Such findings reflect either natural differences in accumulatory ability or dissimilar patterns of food intake, as animal life obtains and builds cadmium levels from the plant life on which it feeds. Both factors probably play a role. Although the accumulation of cadmium in food is not attributable entirely to pollution, higher than average levels of cadmium are found where pollution is greatest.

Humans take in cadmium through the respiratory system or the stomach (where it collects in the liver, pancreas, and kidneys). Symptoms of poisoning include shortness of breath, bronchitis, renal malfunctions, and, if exposure is excessive, death. Chronic intake of small amounts at a time is ultimately as serious as large doses consumed all at once.

A brief description of cadmium poisoning points out the varied places where trouble lurks. Ice cube trays have been serious offenders in the past. In one case, cadmium-plated trays leaked toxic levels of cadmium into lemonade, in another instance, into raspberry gelatin. Usually it is food acids that erode the metal. A leaking refrigerant has also been responsible for cadmium poisoning, the cadmium again stemming from ice cube trays. In this case sulfur dioxide leaked into the refrigerator and dissolved the cadmium, which in turn poisoned the ice cubes themselves. Mixed with punch and consumed, they caused illness. Iced tea prepared in a metal pitcher containing cadmium has led to poisoning, and one time twenty-nine children were made ill from popsicles frozen in molds plated with cadmium.

The incidents above were accidental, though preventable, occurrences. Industrial exposure is another matter. For sheer numbers and extent of illness, no single outbreak of cadmium poisoning has been worse than that endured by Japanese inhabitants of the Jintsu River basin. Heavily polluted drainage waters from a mining operation upstream were allowed to run into the Jintsu River which served as irrigation waters for rice paddies along its valley floor.

Quite literally, rice was grown on a cadmium bed. The diet of many inhabitants of the area consisted mainly of rice, and it was later determined that the individual intake of cadmium ran as high as 1000 micrograms per day. Ordinarily, in pollution-free areas, individual daily intake averages 50 to 60 micrograms. Some 7000 persons were subject to exposure. After the problem was uncovered, one hundred deaths were attributed to excessive cadmium intake. About fifty-seven people were found to exhibit definite symptoms of poisoning, and another fifty or more were considered suspicious. In numbers of this magnitude, the problem can be considered an epidemic. Like other epidemics this one was named. The ailment came to be known as itai-itai (ouch-ouch) disease.

In summary, two facts convey their warning for the future: When released into the environment cadmium persists tenaciously. Since the 1930s its production worldwide has been increasing.

Mercury

If in the past the Japanese have been our enemies, we should now honor them as martyrs to industrialization.

In a pleasant bay overlooking the vast blue reaches of the Pacific, Japanese citizens go about the business of living and working in the industrious manner characteristic of them. Conditions are crowded and much is accomplished on every inch of available surface. The only truly wide-open space is the endless Pacific extending beyond Minamata Bay. In contrast to the mainland, it appears boundless, unchanging, and unchangeable.

Starting sometime in the early fifties a few people began showing up at doctors' offices complaining of visual disturbances and hearing difficulties, headaches and dizziness, irritability, depression, restlessness, and insomnia—not unusual problems in that or any other day and age. Presumably normal steps were taken to alleviate the conditions. As time went by, more persons voiced similar complaints and there were others with coordination problems, loss of balance, and/or a stumbling walk resembling drunkenness. The search for causes went on, while some patients exhibited signs of mental derangement and two prenatal deaths were recorded. The symptoms, common as they were to a number of diseases and ailments, proved difficult to track to their cause. Before this was accomplished, in the years 1953 to 1960, 111 people, 27 of which were unborn infants, were known to have suffered from poisoning. Forty-one persons died, all citizens of Minamata Bay. During the same time another 65 poisonings occurred in Niigata, Japan. Five resulted in death.

Mercury, also known as liquid silver or quick silver, is a metal of unusual qualities, used in medicine, explosives, scientific equipment, recorders of temperature, and fluorescent lighting systems. In the world of metals it is a nearly universal solvent. Only iron and platinum do not readily dissolve in mercury. Other metals do, and from them arise alloys—amalgams.

Mercury is everywhere in nature, found from a few feet to 2000 feet below ground. Both its utility and its venomous nature are historical facts. If it has killed the innocent surreptitiously, it has also—so word has it—taken the lives through suicide of more than one of history's notorious figures. Ivan the Terrible, possibly Charles the II of England, and none other than Napoleon are all thought to have ended their lives with some form of mercurial potion. Other serious illnesses and deaths, numbering in the hundreds, have been caused by accidental poisoning with mercury compounds, but the tragedies of Minamata Bay and Niigata far exceed in prominence those poisonings known to have evolved from environmental contamination. Ultimately it was food— too little variety of food in the diet and mercurous contamination of food to levels between 27 and 120 ppm—that nurtured the epidemics. To have blindly stumbled into a disaster of such magnitude seems stupid in retrospect, but this is usually the case.

Either we truly have or we have led ourselves to believe that we have real need for certain industrial goods. Industry manufactures these needed items and provides jobs in the process; people then buy these products, and, as the saying goes, "that's what makes the world go round." In recent years no industrial product has been in more demand than plastics. Any industrialized nation can cite the production potential of plastics and the limitless "essential" goods out of which plastics can be made and sold and profits realized. No one can seriously fault the objective. Profits need not be used for evil purposes and, as proponents of the puritan work ethic would have it, jobs never hurt anyone—healthful jobs, anyway. Although the Japanese might have another expression for it, certainly the work ethic is near and dear to their way of life. As an industrially advanced people, the Japanese were among the first to see the wealth of potential in plastics manufacture. With American assistance, postwar Japan leaped to the forefront in the manufacture of a variety of goods. In Minamata Bay a plant commenced to manufacture vinyl chloride.

Because of its expanse, any ocean appears inviolable. No one doubted that the Pacific could handle the waste from a chemical plant producing vinyl chloride. The wastes would not be excessive, yet one element would be mercuric chloride, a chemical more deadly than a rattler's bite. You can almost see the minds at work: For something so deadly, what better dumping ground than the sea, so vast, so long the unquenchable dispose-all for man's waste? The sea can absorb anything and its bowels cleanse and purify.

For less than a decade, a short time in the span of economic history, mercuric chloride was released into the bay of Minamata. Mercury levels, ordinarily ranging between 0.03 and 0.27 ppb in seawater (depending on depth, somewhat on location), jumped to 1.6-3.6 ppb. Yet it wasn't mercuric chloride as such that made people sick. Herein lies a story unique to mercury compounds.

Inorganic mercury will not poison you (although all forms of mercury are toxic). Run your tongue over that tooth you recently had filled at the dentist's. The metallic taste is an amalgam of mercury. You live with it, without ill effects, each day of your life. The only actual hazard posed by metallic mercury, as almost everyone is aware, is inhalation of mercury vapors. Lung damage, and worse, can occur.

The real danger from mercury in the environment rests in its conversion by microbes to organic (carbon-containing) mercury compounds, particularly methyl mercury. On land or in water, seawater as well, metallic mercury and mercury salts (such as mercuric chloride) are readily transformed to methyl mercury. Or, if the waste entering the environment is already a form of organic mercury, then both chemical and bacterial digestion will convert these compounds to metallic mercury which, in turn, is changed to methyl mercury. It is this organic form of mercury that enters the food chain and is detrimental to health.

Mercuric chloride dumped into Minamata Bay was converted to methyl mercury. Fish and shellfish feeding in the bay picked it up directly from the sea or from food sources on which they survived. Where the seawater contained 1 to 3 parts per *billion* of mercury, food fish and shellfish showed levels of from 27 to 150 parts per *million* (on a dry-weight basis). This is a phenomenal concentration of mercury and part and parcel to the problem of mercury poisoning. Fish accumulate methyl mercury from their environment. Large fish feed on smaller fish and those at the top of the pecking order (tuna and swordfish in saltwater and pike in fresh water), like a distillery, build greater concentrates of methyl mercury. For persons on a diet consisting mainly of fish—as was the case with some of the citizens of Minamata Bay—

poisoning becomes inevitable. As reported in *Time*, March 17, 1975, Chisso Corporation, the polluter of Minamata Bay, has paid out over the last quarter century $67.3 million to 793 victims of mercury poisoning. Another 2700 persons are awaiting medical confirmation of poisoning so that they too may become eligible for compensation. Company officials, conscience-ridden over the horrors they unknowingly unleashed, are to this day still striving gamely to pay off their impossible debt to humanity. But at last word this and another unrelated disaster were threatening to put Chisso out of business. If that should come to pass, 1500 persons will be out of work. In pollution, as in war, nobody wins. Of note is the question why people were subsisting primarily on fish. In this instance one suspects, as so often happens, that it was not entirely a matter of choice. Those who could afford a varied diet no doubt had variety.

Pollution, the primary problem. Where environmental mercury levels become dangerously high, pollution can be blamed almost exclusively. The plastics industry and chlor-alkali plants (chlorine manufacturing plants) have been the worst offenders. Mercury slimicides, mining operations, and the burning of fossil fuel also add their fair share. So far, serious mercury pollution has been a local problem, in Japan, Sweden, and Iraq, and other small, centralized areas. The potential for more widespread pollution is of course great, though the two or three major disasters have placed the planet on worldwide alert. Knowing that a dangerous situation exists wins at least half the battle. Even so, 1969 estimates placed mercury consumption in the United States alone at six million pounds, of which about 70 percent was lost to the environment. As the world's largest consumer of mercury compounds, this country is also the world's biggest environmental problem. Floor waxes, furniture polish, fabric softeners, air conditioning filters, laundry preparations, paint—products right out of television's priority list of commercial advertisers—are all mercury users and sources of environmental contamination. To these public enemies of the environment must be added the industrial waste from the manufacture of plastics and chlorine, products for which any number of "essential" uses can be readily conjured up.

Some oddities of fact. Having established pollution as the culprit in food poisoning from methyl mercury, look at four ironies: (1) Mercury content in the world's food sources has risen scarcely a jot over the last thirty years. (2) Fish living in the nonpolluted lakes of the world contain readily measurable levels of methyl mercury. (3) Food fish from the worst polluted lakes often show lower mercury content than do fish from less polluted waters. (4) The same food, shellfish for example, containing precisely identical poisonous amounts of methyl mercury, may, in one case, cause serious illness while posing little or no risk at all in another case.

Paradoxical as the statements above may sound, these are the facts exposed thus far through well-grounded scientific investigations. Reserving comment on the fourth for a later section, I will concentrate now on the first three statements.

1. As early as 1930 certain scientists analyzed the mercury content of such foods as meats, milk, fish, fresh vegetables, grain, fruit, eggs, and beer. Later on, in the forties, similar analyses were run on some of the same foods. In 1964 a third sample of data was gathered. Comparison of the three sets of values—findings of three separate investigators—shows essentially identical amounts of mercury. Although it should be admitted that analytical techniques have improved during that time, the latest figures are still all on the same order of magnitude as those established in 1934. The conclusion, then, is that man has not added significantly to mercury levels over this time span. That may well be true. Spot increases, those obvious local flare-ups like that in Minamata Bay notwithstanding, the overall mercury content of food remains unchanged. That's good news; we can only hope we're wise enough to keep it so.

2. The answer to this second riddle is simple, although for a time it caused some head-scratching among scientists flung into the middle of a worldwide controversy as explosive as this one. Mercury is naturally present in the environment, scattered over the earth and in the bedrock under any number of pure northern lakes. Whether put there by man or by nature, mercury will obviously undergo the same natural processes. Metallic mercury will be converted to methyl mercury; fish will consume and store it in their bodies; older, larger fish (primarily those which feast on other fish) will build concentrations significantly greater than those of other species or younger, smaller fish of the same species. Lakes separated by no more than a few hundred feet could (and do) produce fish of widely differing mercury content with levels sometimes in excess of government tolerances for food fish. This does not and should not imply abstinence from a sizzling walleyed fish fry during that once-in-a-lifetime trip to a wilderness lake. All it implies is that you would not want to camp there year-round and eat absolutely nothing but walleyed pike.

3. Eutrophic waters become green as a result of pollutants conducive to the growth of aquatic plants and vegetation. Phosphates have been incrim-

inated as the major contributors to lake and pond eutrophication. The eastern shore of Lake Erie is a good example of blooming water, though certainly not the only example. Upon analysis, it was found that fish from Lake Erie's most polluted areas contained dramatically lower mercury levels than did fish from other, less polluted parts of the lake or even from the clean, clear waters of wilderness lakes. Mercury was and had been an industrial pollutant of Lake Erie for a long time. It was in the water, yet the findings still stand. Such seemingly conflicting information has all the makings for a good science-fiction novel. The explanation is straightforward. The facts are indisputable. What happens centers on a single characteristic of eutrophic waters, and is the reason why eutrophication, if complete, drastically curtails or completely eradicates the fish population at lake bottom. It just happens that bacteria with the ability to convert mercury to methyl mercury are the same bacteria that lack the ability to grow *in the absence of oxygen*. And, as already mentioned, fish cannot and do not make this conversion on their own. Therefore mercury sits at the bottom of Lake Erie's greenest coastal waters awaiting the day when rejuvenation and cleansing may once again oxygenate the deep. Meanwhile those clean, unspoiled lakes to the north, which happen also to be naturally high in lake-bottom mercury, will continue to produce fish with high levels of the poisonous methyl mercury. It is the way nature would have it.

Other environmental risks at stake. Being a natural though less than abundant element in earth's crust, mercury is in a state of continuous cyclical movement. Bacterial conversion to methyl mercury provides a certain amount of upward mobility. Evaporation from land and water raises mercury to the atmosphere while rain and snow wash it back to earth again. The cycle would go on indefinitely had man not entered the picture to threaten the balance and tip the scale with erratic consistency.

Mercury is released into the air we breathe from the burning of coal. It is brought to the surface and eventually to earth's crust in crude petroleum, at a rate of a half million pounds a year by one estimate. It hangs heavy in the smog of Los Angeles and New York. And—take note organic gardeners—soil rich in humus may run higher in mercury (to 2 ppm) than does more granular soil. Not only that, but vegetation apparently tends to concentrate mercury when soil levels are low (less than 1 ppm) and to limit mercury intake at higher levels (10-100 ppm). Plant life responds in its own unique way no matter what the fertilizer. And always are fish, birds, and animals at the top of the food ladder the most endangered species. Predatory birds, predatory fish, and predatory man; flesh and blood smelters of mercury. Pike can concentrate mercury 3000 fold, dead owls have been found with as much as 270 ppm of mercury in their bodies, and a number of humans have already succumbed.

Agricultural uses of mercury have not been found to contribute significantly to mercury in the environment, marketbasket surveys note. Mercury-treated seeds, at one time commonplace in the seeders crisscrossing grainland America, added no mercury to the food we ate, but did promote sharp increases in the mercury levels of seed-eating birds. Mercury-treated seeds have been the cause of numerous accidental human poisonings and deaths, the worst occurrences taking place in Iraq in 1956 and again in 1960 when seed grain intended for planting was eaten instead. Another serious poisoning occured in a family in New Mexico after they ate a family pig that was fed grain treated with mercury. These accidental poisonings are real; they have taken place and were the reason for banning mercury as a fungicide seed dressing.

Until the time they were prohibited mercury-treated seeds were not contributing to the mercury in man's food. It is well to segregate and to recognize this fact. Mercury being released into the environment today does not stem from agricultural uses but from industrial goods and services rendered to each of us as consumers. Without question we will pay higher costs for goods or higher taxes to ensure low environmental mercury levels. If not, the price will be ill health.

Mercury in the food you eat. All food—it can't be escaped—contains minute amounts of mercury. Mercury levels in seafood run distinctly lower in cod, clam, crab, flounder, lobster, and herring than in tuna and swordfish. Pike from unpolluted water can run as high as 1.4 ppm methyl mercury. The content of methyl mercury in fruits, vegetables, grains, meat, and dairy products is very small. An FDA survey of ten basic foods found fish to pose the only potential hazard.

Safe level of mercury. The only standard now in effect limiting the amount of mercury in foods is for fish and shellfish. A proposal by the Environmental Protection Agency published in the Federal Register, October 25, 1974, would limit mercury emission to the air from mercury ore processing facilities and chlor-alkali plants to 2300 grams of mercury per 24-hour period. Emission from sludge incineration plants would be limited to 3200 grams per 24-hour period. The action level is 0.5 ppm of the product. At this level one could consume 60 grams of these foods each day and maintain a tenfold margin of safety under amounts considered toxic. Data on human toxicity are still being evaluated, but a sound historical basis exists for setting maximum intake at 0.4 mg/kg of body weight (about 0.03 milligram for a person weighing 154 pounds). This amount provides the usual tenfold safety margin.

For reasons of choice or lack of it, people's eating habits vary greatly. A National Marine Fisheries Service study found 1 percent of the individuals surveyed consuming 77 grams of fish daily; 0.1 percent averaged 165 grams daily. At the latter rate of consumption the safety margin is reduced to less than fourfold. Other observers have noted regular intakes of over 50 micrograms of mercury per person per day without apparent ill effects. As the FDA points out, one would have to consume 600 grams (1 1/3 pounds) of fish and/or shellfish contaminated to the full 0.5 ppm allowed by standards to reach a blood level mercury content at which symptoms of poisoning are known to have occurred. But the fact is, of course, that it can happen.

This is not the end, by any means, to the story of mercury in food. A most curious and intriguing epilogue, the apparent explanation of the fourth irony, is one which no doubt was handed down to us by our ancestors.

Selenium and mercury poisoning. How could it be that two groups of persons, both taking in the same amount of methyl mercury, might respond so differently that one group would react with typical neurological disturbances (and even death) while the second would show no signs of ill health? How could it be that when two diets containing equal amounts of methyl mercury —one of tuna fish as the source of mercury, the other a synthetic diet with methyl mercury added—were fed to quail, only 7 percent of those on the tuna diet died while the mortality rate of those on the synthetic diet was 52 percent?

These are questions that scientists at the University of Wisconsin recently asked themselves. Under the direction of Dr. H. E. Ganther, they had made the startling discovery of tuna's apparent protective effect against the ravages of methyl mercury poisoning. One question of course leads to another and the investigators were soon trying to determine whether tuna might not contain a factor that was somehow able to inhibit mercury poisoning.

Work by another investigator, G. Lunde, pointed up the relevant fact that sea fish to varying degrees concentrated the element selenium along with mercury. It was a good clue. Selenium was known to be an essential trace element, though its role was unclear in metabolism. It was thought to be linked to vitamin E and sulfur amino acids; it could play a part in red blood cell formation; it had even been suggested as a protective agent against human stomach cancer. Also selenium was known to occur commonly as a binary compound with heavy metals. Mercury and selenium would naturally be expected to reside in close proximity to one another. Possibly in the evolution of man's metabolic system selenium had entered the picture along with mercury and had found a place in stemming the latter's poisonous touch.

The researchers decided to find out. They analyzed different samples of tuna and found concurrent parallel increases in mercury and selenium content. The more mercury, the more selenium. This was good, but circumstantial evidence. Next five groups of rats were fed a purified ration devoid (or nearly so) of selenium and vitamin E. In their drinking water each group received a constant dose of mercury, but at increasing levels for the different groups. Amounts varied from 0 to 25 ppm mercury. Weight gains were noted for four weeks. As might be expected there was a steady decline in growth as mercury levels went up. The scientists reasoned that, if selenium could exert a damping influence on mercury toxicity, the addition of selenium to the diet ought to alter dramatically the growth in weight. Into the rat's diet went 0.5 ppm of sodium selenite (a salt of selenium). Now growth rates in the five groups of rats were found to be similar. Moreover, all rats fed a basal diet (lacking selenium) were dead at six weeks while the others—every one of them—were alive at that time. At a rate of 25 ppm, the highest dosage, even the rats given selenium eventually died. But they remained alive for a longer time than unprotected controls. What happens in the body as selenium exerts its counteractivity to mercury has yet to be learned, but it does reduce the toxicity; that much seems clear.

Going back then to the disaster of Minamata Bay, we need to give some explanation to resolve the nagging question of why selenium apparently served virtually no useful function in delaying the onslaught of poisoning of the Japanese. One possibility is that natural variations in selenium content do exist and levels could have been too low in the species of seafood ingested. Another hypothesis currently favored by the Institute of Food Technologists' Expert Panel is that selenium, at the ratio of mercury to selenium ordinarily found in seawater, is taken in in proportion to the amount of mercury, and at the unusual levels of mercury found in Minamata Bay during the waste discharge into the sea, mercury so outweighed selenium in the bay-sea environment that it was absorbed by fish and shellfish at a disproportionately higher level. It's only a theory, though patently logical.

If your mind runs to the pragmatic, you are probably wondering why all the hullabaloo. Why not simply find a way to remove methyl mercury from seafood? It's an American-type question, characteristic in its innocence and presumptuousness. Yes, why not?

Obviously, somebody in the seafood business is going to ask the same question and attempt to do something to answer it. Indeed, at the 1974 meeting of the American Chemical Society, Dr. Eugene E. Schrier presented a paper explaining a process for doing just that. His method involved addition of sodium borohydride, a chemical that converts methyl mercury to elemental mercury which can then be washed away along with any excess sodium borohydride. By this procedure, fish protein concentrate (a fish protein extract that, because of its concentration factor, increases fivefold the mercury in the original fish) can be produced well within the 0.5 ppm of mercury maximum imposed by the FDA. The only catch is—it increases the cost of processing by about 20 percent. It is this kind of cost I refer to when asserting that we will pay one way or the other for industrialization and our appetite for "things."

Interactions—A Complicating Factor

As the saying goes, nothing is ever as simple as it appears to be on the surface. Concerning the question of minerals in food, this is profoundly the case. If problems of toxicity were as simple as input and output of separate, distinct mineral entities, life would be much simpler. But that is not the case. Some minerals can substitute for others. Some are mutually antagonistic. Others are mutually beneficial. For example, increased molybdenum and zinc in the diet expand the need for copper. Molybdenum is a competitor of copper. Because it strikes out for the same residence in the body, it places additional demands on the diet for copper. On the other hand copper and iron work together, and the body needs the former to make proper use of the latter. When two essential trace elements interact adversely in the body, further complications arise. So it is with copper and zinc once they reach the intestines.

The need for "nonessential" minerals may become "essential" in a complementary function of the body requiring both. The nonessential elements boron, arsenic, and tellurium play enhancing roles in the metabolic activity of vanadium, zinc, nickel, and selenium. To the contrary, other nonessential elements can exert antagonistic influences on essential elements. This is one of the ways in which poisonous heavy metals act to exact their toll on human health. Cadmium and mercury both can substitute for zinc in some systems. Of course, even essential minerals can be present in excess and cause serious illness. This is the reason for the controversy over iron fortification of bread. Too much of a good thing can be harmful.

Other Metals

In our present industrialized state, all metals now being extracted from the earth are potential troublemakers, primarily because of the subtle, seemingly innocent way their ill effects tend to materialize. Although we might wish that it were an issue restricted to metals, it is not. Hundreds of industrial goods with dozens of potential poisons are introduced into our lives each year. So much a part of the background do they become that we look at them without really seeing them. But they're there and it's only a matter of time until, by one route or another, they gradually show up in food. It's small wonder that almost every time you pick up a newspaper you find some new, strange-sounding chemical—polychlorinated biphenyls, pentachlorophenol, dibenzofuran, vinyl chloride (to cite a few recent headliners)—indicted as a source of chemical pollutant in food.

These are names, then, to be on the look-out for in household goods, food, and industrial and domestic wastes around cities and mining centers. As for metals, aside from the ones discussed, those most closely watched by the FDA are antimony, chromium, tin, copper, beryllium, germanium, and thallium. Nickel, too, poses a threat, as do iron, zinc, arsenic, and vanadium. Arsenic, thought until recently to be a danger because of its possible concentration in the food chain, has now been shown not to increase in this manner. Dr. Edwin Woolsen, a researcher with the Agricultural Research Service, suggests that arsenic is stabilized as insoluble compounds during natural processes occurring in water (both fresh and salt) and in plant and animal life surviving in the aquatic environment.

And One for the Future

As you search the literature and read and analyze the question of food contamination, particularly as applied to heavy metals, you become aware that major (or even minor) disasters could and should have been averted. All the necessary information required to foresee at least the potential for serious food contamination has been apparent. Heavy metals are among the oldest of known poisons, yet we have persisted in poisoning ourselves with them. Now another metal looms in our future and has already become a sure bet for disaster. It is the deadliest of known toxic compounds—more deadly than lead, mercury, and botulinum toxin. Like other metals the United States and other nations have begun to stockpile this one. Its buildup in years to come is an absolute certainty—all a part of the industrialization process. As conferees of the Western Food Industry Congress were recently warned by an executive of the National Canners Association, "Be prepared." In a couple of decades, we may well be reading about a new outbreak of food poisoning due to a rare metal produced by nuclear reactors. Its name: *plutonium.*

IX

Poisons and Antinutritional Factors in "Natural" Foods

A food shall be deemed to be adulterated . . . if it bears or contains a poisonous or deleterious substance which may render it injurious to health, but in case the substance is not an added substance (i.e., a "natural" one) such food shall not be considered adulterated under this clause if the quantity of such substance in such food does not ordinarily render it injurious to health. . . .

FDA regulation

In an interview on a well-known television talk show an expert on natural foods was asked how he determined whether or not a new-found food was safe to eat. The ensuing conversation went something like this.

"I take the product to a nearby college," the expert replied, "and have it tested."

"How do they do that?"

"Oh, they feed it to animals—rats, rabbits, and guinea pigs."

"And if the animals live, you eat it?"

"No," the expert answered facetiously, "then I feed it to my three grand-children and if *they* survive, I eat it!"

There are some truths to be learned from this bit of fun. First, though it may seem obvious, not all growing things are safe to eat. Many plants are highly poisonous. Secondly, appearance alone is anything but a sound way of distinguishing good from bad. The berries of some shrubs may look red and juicy but be poisonous; a poisonous-looking mushroom, on the other hand, may be safe to eat; or vice versa. In any event, it should come as no surprise that "natural" toxins abound in nature and that most edible foods contain chemical substances known to cause illness in humans. What may be surprising is that several common foods, if held to the same rigid standards of safety required of synthetic additives, would not pass the test!

In a very real sense, synthetics aren't necessarily all that bad or natural foods all that good. What's more, in our attempt to *prove* the safety of synthetics and other additives, we establish highly unnatural and uncommon test conditions and apply safety factors unrealistic even for natural foods. No one is suggesting that we do otherwise, but only that we achieve a sense of balance and perspective. The extent to which we will be able to meet world food needs in the future will depend entirely on our ability to see facts for what they are.

Although the basic health problems are not the same, the transition from a discussion of heavy metals to a discussion of other naturally occurring food toxicants follows logically. Note the implication that heavy metals are also naturally occurring toxicants, as indeed they are. Pollution may provide their entry into food, but once absorbed into the plant system heavy metals fit precisely the definition of a natural food component. Only the amount varies, but this is true of all food components, including the nutrients.

A naturally occurring food toxicant can be defined as any chemical component normally found in basic foodstuff and known to cause illness in humans or animals. The definition excludes additives of any kind, food-borne microbial infections as such, and also toxins produced by microbes. The poisons now under scrutiny are natural ingredients of everyday foods.

A few prefatory comments will help to set the stage. First, the following information is not presented to alarm anyone unnecessarily. Though foods long common to our dinner plates do contain known toxins, obviously we have survived quite nicely for centuries. This does not imply that poisonings have not taken place. They have, and still do, but usually under extraordinary circumstances, the same as poisonings from "safe" additives.

Secondly, scientists have been engaged in the study of these toxins and growth inhibitors for many years. But the work lacks drama and excitement and is not the kind of research you are apt to find splashed across the headlines of newspapers. Still it goes on, in industry, government, and university laboratories. The findings, where possible, are incorporated into food processing lines, which, like home canning, are necessary to put food into tasty, edible form and to preserve it for later consumption.

Thirdly, the food industry, confronted with some proven hazards, consumer alarm, and a slow but sure trend toward "natural" or "organically grown" food, assumed a defensive posture in the early seventies and began to rethink product lines. Out of this general reappraisal have come a number of so-called natural foods. This is an outcome that might readily have been anticipated, since food firms, like any business, will always respond to strong consumer demands. As consumers, we can literally have what we want. We can only hope that we have the sense to know what is good for us. This chapter focuses to some extent on that very concern. A like concern on the part of a number of toxicologists and food industry representatives prompted the publication of several excellent reviews of toxins known to occur naturally in food just as it comes from the ground, pod, tree, animal, or bird. There is much to be learned from these articles, for the issues they crystallize, the balance they lend to perspective, and the insights they shed on research priorities.

What would you think of a regulatory agency charged with the responsibility for food safety that allowed on the market food known to contain cancer-causing agents or food components known to cause abortions and fetal resorption? What if that agency turned its head on studies proving abnormalities in rat offspring—called teratogenic defects—caused by food on supermarket shelves? What if it permitted the selling of food with a kicker—an hallucinogenic drug as potent as LSD or cocaine? What if that agency, respected for its policy of setting one-hundredfold safety factor guarantees on additives, suddenly disregarded its long-standing policy and dropped the buffer zone to tenfold or less?

Think about it sometime as you sit down to one of those good old American roundhouse meals favored so highly by the makers of Tums and Alka-seltzer. Think about it as you gobble down that cinnamon roll, or savor a steamy hot baked and buttered potato. Think about it as you eat a chunk of juice-dripping grilled steak. And think about it as you sip luxuriously an after-dinner liqueur served in a tiny goblet.

Yes, think about it, and beware of the danger inherent in those selfsame foods. The cinnamon harbors an hallucinogen, the potato and butter teratogens, the grilled steak a carcinogen, and the liqueur possibly a toxin which could make you as ill as if you drank antifreeze, for it contains a common antifreeze compound called glycerine.

Even knowing this, we are probably not going to give up eating these and numerous other foods which contain a variety of toxins. Moreover, few of us are likely to leap on a soapbox and start hollering for legislation. It would be useless anyway because these poisons can't be legislated out of existence. We could, though, and should be doing more about some of them than we are, not so much because of risks to health, but because the nutritional value of vegetable products which tomorrow will crowd the shelves of supermarkets should be maximized and hunger and famine the world over should be averted. It can be done; all it takes is commitment.

Life Processes, a War of Chemicals

Poisons are a part of our normal diet. Digestion and metabolism mobilize energy and nutrients and screen out impurities. At all times living cells swim in a sea of chemicals, managing at one and the same moment to utilize needed nutrients while rejecting or detoxifying most toxic compounds. It is the purpose of the excretory process to rid our bodies of toxins which in a matter of hours would otherwise do us in. Putting it differently, we can say that healthy bodies are in a constant, perfectly normal equilibrium with naturally occurring food toxins. Excess, accidents, and unusual eating habits can tip the balance. Some of us simply have built-in weaknesses which tend to tip the balance in favor of toxins or in favor of toxic effects from "nontoxic" foods. A host of dietary ailments running all the way from allergies through diabetes and heart disease to cancer flaunt their evil through inborn errors in human metabolic machinery. It is perhaps well to keep in mind that we start dying the day we're born. Life, even as it is maintained by food, is the losing battle waged against that end.

Following is a representative, though incomplete, list of the food toxins we have to live with. For no special reason, save order, they are taken alphabetically. For anyone interested, entire books are available dealing solely with the topic of food toxins.

Alcohol

Little need be said about alcohol. It can become addictive; it may, in heavy drinkers, predispose to cancer. The effect of heavy alcohol intake is mainly on the central nervous system. The lowest published toxic doses for man is 50 mg/kg, with toxicity manifesting itself in disturbances of the gastrointestinal tract. Death has been reported at an intake of 1400 mg/kg of body weight. For a 170-pound man, this amounts to about 4 ounces of pure alcohol.

Allergens

Allergens, those food components causing allergies, are as varied as individual sensitivity. Infants and children often react adversely to milk or eggs. Generally a protein or protein fraction is at fault. Adolescents may "break out" from eating chocolate products, seafoods, or pork. A child has died from consuming no more than a few peanuts contained in a dish of ice cream. Sensitivity—one's ability to withstand a given dose of allergen—varies greatly. What to one person yields a case of hives, to another may result in anaphylactic shock and death. Severe allergies may stem from milk, fruits, vegetables, cereals, fish, rice, nuts—just about anything. And the only way to determine a person's food allergies is by his eating and noting adverse responses.

Allyl Isothiocyanate

With a name long and perhaps hard to pronounce, this natural ingredient of horseradish and mustard is little more than a skin irritant. In high doses it will kill rats, the LD_{50} being 339 mg/kg. Interperitoneal injections in the mouse have caused death at a level as low as 4 mg/kg of weight.

Amines

Biological amines are as near to us as a healthy bite of aged Cheddar cheese and a bottle of Chianti wine. In cheese the culprit is tyramine, though serotonin, another pressor amine, is found in banana, pineapple, tomato, squash, cucumber, and pumpkin.

The least amines can do is elevate blood pressure. Ordinarily monoamine oxidase, an enzyme present in the body, detoxifies the amines before they can do any damage. Dangerous, even deadly symptoms arise only when we add, of all things, a tranquilizer. One in particular, parnate, coupled with tyramine in aged Cheddar and Camembert cheese or serotonin in wine, makes for a potentially lethal combination. Never consume these foods when on parnate.

Antivitamin Factors

Certain foods create a need for abnormal amounts of a given vitamin. In other words, symptoms of vitamin deficiency occur when these foods are eaten unless vitamin intake is supplemented beyond normal requirements. Mostly raw (uncooked) foods are involved, since heat destroys the antivitamin factor. A thiaminase of clams and raw fish produces thiamine deficiency while avidin of raw egg white causes a shortage of biotin. Unsaturated fatty acids found in fish and vegetable oils increase vitamin E requirements. In corn and other cereals an unknown element works against niacin.

There is no danger of mass vitamin deficiencies in the United States or most other affluent nations. The problem, however, does affect those peoples of the world who depend on vegetable protein for survival. If they depend on America as a source of protein, they're in for stiff competition, and antinutritional factors in soybeans and other legumes are the crux of their problem. Some rough figures may help explain their plight.

One acre yields forty-three pounds of edible protein in soybean meal fed to animals. That same acre supplies six hundred pounds of protein when taken directly, as vegetable protein. Eventually a point is reached where, to free vegetable protein for export, it becomes necessary to increase meat production per acre. Maybe forty-four, forty-five, or even fifty pounds of meat can be derived from that one acre if the efficiency with which feed grains are converted to meat protein is maximized. At the same time the nutritional value of that same vegetable protein can be increased, if we learn how to cope best with both animal and human antinutritional factors. Antivitamin factors are only one of several nutritional problems that must be dealt with. But let's carry the figures out to absurdity.

For every pound of meat increase per acre, fourteen pounds of soybean protein become available for other food use. When supplemented with necessary nutrients, that fourteen pounds of "nutritionally sound" protein will maintain one adult at normal protein levels for a little over four months. If the conversion efficiency of that meat protein could be increased by about two and a half pounds, enough vegetable protein could be released to keep one adult going a full year. About thirty-six pounds of protein would do it.

This is only one aspect of the overall problem, but it serves to focus attention on the role of antinutritional factors in the world's delicately balanced people/protein ratio. In reality ruminant animals, like people, consume more than one food in the diet. And unlike nonruminants, these animals can convert grass and forage to highly nutritive human food. During years of surplus crops (grains in particular), little research effort was concentrated on improving forage crops. Time slipped by, research lagged, and we now wish we had been forging ahead full steam. Certainly we must firmly direct future investigations toward ways of reducing, or eliminating when possible, dependence of animals on cereal grains for feed.

It is necessary that animals and animal agriculture best utilize land resources but we must be judicious in designating land for this purpose.

Quite possibly animal agriculture of the future may have to use less fertile land areas and poorer quality feeds than are currently employed. For if a soil can produce forage and grass for animals, that very land can produce those same crops for direct conversion to human food. Again, antinutritional factors must be overcome, but the possibilities are monumental. Alfalfa, that perennial forage of enormous growth potential, is an excellent source of high-quality protein. So rich in amino acids is it that it can be substituted for milk protein. The technology for extracting human food protein from alfalfa is developing. Among the nutritional failings as yet to be totally corrected are presence of a saponinlike substance, a vitamin E antagonist, and a number of estrogens. Extraction of the protein leaves a residue with yet some value as animal feed. The scientific study and exploitation of such residues awaits the motivation of man. Meanwhile, the protein of alfalfa, in amounts sufficient to supply the entire world populace, goes essentially untapped. And this total quantity could be grown on a land mass about half the size of Alaska.

Carbohydrates

Though we generally think of them as nutrients, carbohydrates (starches and sugars) are promoters of obesity. We get fat from them and experience numerous untoward side reactions. Children and adults get dental caries—tooth decay. These two factors alone are sufficient reason for placing carbohydrates in a category far higher than the majority of food additives on a diminishing scale of health detractants.

Capsaicin

Capsaicin is a skin irritant found in red pepper. It's a chemical we would not rub on our hands, but would think nothing of putting into our stomachs.

Carcinogens

Cancer-causing additives that afflict almost any species of test animal are quickly struck from approved listings no matter what the dosage rate. Wording of the Delaney Clause is binding in this matter (although not always invoked). Yet, as mentioned before, certain additives with cancer-promoting activity may be used at levels so low that they constitute no threat for several lifetimes, so some authorities say. Perhaps the best supporting evidence for this theory of an "extra life span risk factor" is the known or suspected carcinogens residing in a number of common foods which hundreds of thousands of people regularly consume during a normal lifetime. Presence of these compounds should at least make us pause to ask if additives and the human response to them could not be equatable in this respect. Either that or we must do without tea, nuts, and some fruits. We should also do without barbecue-grilled steaks and hamburgers; cinnamon, nutmeg, and sassafras; cabbage, spinach, and lettuce; lemon oil. Except for lemon oil, thought to be a cocarcinogen, all these foods contain compounds which have been found to cause cancer in mice. If we add to them the list of food harboring *suspected* carcinogens, we would literally have to throw in the towel and starve.

The carcinogen benzopyrene forms on the blackened surface of charcoaled meats. It can be found in measurable amounts in cabbage, spinach, lettuce, and tea. Although these amounts can be extremely small, so were the amounts of certain cancer-causing additives that have long since met their Waterloo.

Are you a tea drinker? Tannic acid, a component of tea, is reported to produce liver tumors in rats. Safrole of nutmeg, cinnamon, and sassafras is a carcinogen sufficiently potent to be banned as a flavoring of root beer (an FDA proposal would likewise ban sassafras tea). With root beer (and sassafras tea), as with other soft drinks, there is always the added danger that someone with a hankering for it will go too far and drink enough to sink a battleship. If an additive is present, the intake may likewise be abnormally high. Everyone should be prepared to admit that, when it comes to food, some of us are likely to go off the deep end. But who is accountable? Who can anticipate an insane desire for some food or groups of foods? We ask such judgment of industry and regulatory agencies.

You may reasonably question whether the evidence for carcinogenic properties in the food components discussed above is valid. The answer is yes, certainly as valid as the data on additives that have subsequently been banned. Documentation exists for all the toxins reviewed in this chapter. How good is it? How good is any documentation?

Cholesterol

After many years of controversy, research and media advertising so nauseating as to verge on lewdness, one is perhaps justified in asking if the relationship between cholesterol and heart disease will ever be resolved. I plead ignorance. Note, however, that the compound has been duly acknowledged here as a naturally occurring food toxin. More will be said about this compound in Chapter XI.

Enzyme Inhibitors

Trypsin, an enzyme of the stomach, aids in the breakdown and digestion of protein. If its activity or the activity of another related enzyme, chymotrypsin, is restricted, poor growth results unless the stomach is flooded with extra protein. Should a food be termed toxic if it reduces the activity of either or both of these enzymes? Since the inhibitory effect can be overcome, leading authorities on the subject shy away from a term so severe as toxic. But no matter what the label, the soybean, cornerstone of animal and human nutrition, like some other food products, has within its crusty shell a latent enzyme-inhibiting factor. It is so well known that researchers abbreviate it T.I.—trypsin inhibitor. When present in raw meal, it is another one of those antinutritional factors already mentioned. Out of deference to scientific accuracy, if you will, they should be referred to not as toxins but as factors that reduce an animal's ability to utilize fully essential nutrients. It becomes a situation similar to the presence of sludge in your car engine; you can get where you're going, but in doing so it takes a lot more gas. Our most important animal fuel is being stripped of its nutritional punch by some as yet unknown, unisolated sludger.

The effect of the trypsin inhibitor can be dramatically subdued for human consumption by a process known as "toasting." This is a misnomer of sorts, since the word implies the application of dry heat. Actually, moist heat is applied; about 10 minutes at 100°C (212°F) will do it. This is a processing function which enhances the utilization of protein (maximizes the protein efficiency ratio) while introducing new nutritional defects and/or altering functionality, i.e., the ability of the soy product to perform adequately in specific foods. But that's a whole new ball game. Essential here is the realization that putting the soybean into commercially recognizable, useful foods requires major processing steps to assure proper functioning and *nutritional adequacy*. Much is yet to be learned.

Certain foods also sustain antitryptic or antichymotryptic activity. Kidney beans, lima beans, Irish potatoes, and egg whites are a few. There are other enzyme inhibitors, too.

Cholinesterase, an enzyme functioning at the very nerve centers of existence, is a part of the sizzle and snap that allows nerve impulses to race from head to toe. Not only humans but many insects rely on cholinesterase activity for necessary life functions, and scientists have put this information to work in the form of a variety of insecticides whose lethal sting strikes directly at this enzyme. It's an important one.

Would you believe that a cholinesterase inhibitor capable of causing human toxicity lurks with less than a tenfold safety factor in a staple as common as the potato? This antienzyme activity can be found elsewhere—radishes, celery, broccoli, egg plant, sugar beets, asparagus, carrots, the Stayman apple, and Valencia orange—but only in potatoes does it apparently pose any real risk to health, and then possibly not much. But see for yourself.

The anticholinesterase factor in potatoes is called solanine. Other toxic compounds, possibly dozens (including arsenic), infest the potato, but only solanine has been identified as a slight (perhaps) though bona fide health villain. Like MSG (and other suspect additives of food origin), the potato, with its intrinsic toxin solanine, has been around for many years. Only blighted or stressed potatoes appear to be of major concern, although solanine is ever present to some degree.

Is it dangerous? In one experiment potatoes rejected by industry were fed to pregnant marmosets. Newborns evidenced behavioral defects. Cranial (skull) malformations had been noted in earlier trials. When pregnant rats were fed blighted potatoes or solanine or glycoalkaloids, skeletal and renal tract abnormalities showed up in offspring. On parenterally administered solanine, pregnant rabbits became prone to frequent abortions and resorptions. High doses caused the deaths of rhesus monkeys. Human deaths have been attributed to solanine in potatoes and the incidence of anencephaly and spina bifida (ASB) has, with regularity, peaked in years of severe potato blight. Some researchers state unequivocally that 95 percent of ASB cases are caused by potatoes. Not all scientists agree. There are a number of recent case histories in which pregnant women, on diets including little or no potato, delivered ASB children. Does this kind of conflicting evidence ring a bell? Remember the research on Red No. 2, in which embryo-toxic effects were noted and then dismissed, in what some might say was a rather cavalier manner, because scientists lacked experience with the method? Solanine injected into fertile eggs through the yolk-sac resulted in death of the embryos. Solanine is not an additive. It is present in potatoes as they come from the ground, more so under stressful growth conditions, but also whether produced organically or by the far more fruitful methods of present-day commercial agriculture. This holds true for a number of toxic metabolites originating from the potato. What should concern us is not how potatoes are grown, but how new varieties of potatoes—some yielding as much as 10 percent protein, a phenomenal leap in protein content—affect human health. New varieties will have to undergo rugged evaluation, and growth and storage conditions will have to be carefully monitored. Or we will have to find ways of squelching the production of solanine. One new method involves the application of a hot coating of paraffin, but if it came to that, would anyone care to estimate, for those whose diets depend largely on potatoes, the cost of detoxifying them this way?

Estrogens

Estrogens found in a variety of foods are not generally considered a human health problem. As a group of substances of known toxic potential, they are thought to be isoflavones. Thus far estrogenic toxicity appears narrowly limited to breeding difficulties in sheep.

Fatty Acids

The role of fatty acids in human and animal nutrition is well known (or well researched), but so poorly understood that we are actually ignorant of their function. Be that as it may, certain fatty acids will cause poor growth in some animals (and an increase in the death rate). The unsaturated oleic acid produces anemia in rats. On diets high in polyunsaturated fatty acids, infants develop symptoms of vitamin E deficiency.

Favism

It appears that a hereditary weakness is necessary in order to contract hemolytic anemia from the ingestion of the Faba bean. Persons lacking glucose-g-phosphate dehydrogenase (an enzyme) are sensitive to the disease. As though eating problems were not enough, sensitive persons also succumb from inhalating the pollen. It is important to note that the toxic substance has not been isolated, and it can't be destroyed by cooking.

Only occasionally do persons in the United States show symptoms of favism. It is, however, common in other parts of the world, particularly in the Mediterranean, Asia, and Formosa. The problem is that people have to eat and the Faba bean grows abundantly in many parts of the world.

Flavor Components

Toxins of natural flavors have been touched on in Chapter IV. In quantities found in nature, they pose no serious threat. Acids, esters, aldehydes, terpenes, and phenols are all poisonous, but only at extraordinary intake.

Glycosides

Glycosides form an important class of naturally occurring food toxins. They are related chemically. Symptomatically, however, their poisonous influence varies all the way from stomach irritation to cancer. Food variety and human eating patterns make the latter a rare occurrence in North America. Nevertheless many known cases of cyanide poisoning have been documented.

Cyanogens are glycosidic carriers of cyanide, the infamous poison of gas chambers. Tropical lima beans may yield between 210 and 312 milligrams of cyanide (HCN) per 100 grams of beans. That's a knockout dose when compared with normal levels of 14 to 16 mg. Were you ever warned as a child against the danger of eating choke cherry pits? I was. And choke cherries are still around, as are peach and apricot kernels which contain 3 percent amygdalin, a potent concentration of this cyanogen. Wild black cherries should be placed in the same category as choke cherries. The fruit, the leaves, and especially the seed are cyanogenic. None are more lethal than unroasted bitter almonds, a handful of which will kill an adult. As few as twelve have caused the death of a child. Cassava, linseed meal, black-eyed peas, garden peas, kidney beans, Bengal gram, and Red gram all contain cyanogens in decreasing amount by the order listed. No great hazards are implied, although in cooking cyanogenic foods such as lima beans it is well to remember that cyanide is highly volatile; it readily dissipates as a gas unless a lid covers the pan, in which case the compound condenses and drops back into the stew.

Paradoxes are legion in this world. So it is with food. The cyanogen from apricot pits, though unproven in worth, is nonetheless being studied for its value as a cure for cancer. Its use in the United States is banned at present, but in Mexico at least one clinic continues to test its curative properties. The product is called laetrile.

Another glycoside, of no great significance in this part of the world but important to natives of the South Seas, is a potent carcinogen. One wonders how its danger was first discovered and how the simple process of washing it was learned and handed down through the generations. The compound, inherent in the cycad nut and called cycasin, yields, upon ingestion, methyl azo methanol, the causative agent. Somehow natives have uncovered its lethal quality and readily detoxify this food by an easy washing process.

Proponents of natural food diets recommend honey as an "unprocessed," "natural" food of inherent purity. And so it is—except when bees gather nectar from mountain laurel, rhododendron, azalea, or oleander. From these plants originate a glycoside, a powerful heart stimulant that can be most dangerous. It will turn golden-amber honey into "fool's" gold.

Not many people eat unripe grapefruit and so are spared the unpleasant stomach pains associated with a naringin, a glycoside of the immature fruit.

You're probably thinking that glycosides pose no serious threat as far as most of us are concerned. This is true and it is also true for goitrogens, which are compounds that cause thyroid enlargement—goiter. Still, life would be simpler and less risky if they did not exist at all. Rapeseed, an excellent source of protein yielding anywhere from 40-45 percent protein, carries and must be purged of a progoitrin which, altered by an enzyme, activates the goiter-causing principle. If it is formed before it is treated by heat (which destroys the enzyme and prevents further goitrogen production), only hot water or alkali extractions of steam treatment can render the rapeseed free of toxin.

Other goitrogenic foods include cabbage, cauliflower, mustard seeds, rutabagas, and Brussel sprouts. To produce thyroid enlargement from the thiocyanate of cauliflower a person would have to consume twenty-two pounds of cauliflower per day. Peaches, pears, and strawberries all show some goiter-causing tendencies. In white turnips the causative agent is 1-5-vinyl-2-thiooxazolidone.

Where food is concerned, Americans can only consider themselves unimaginably blessed. Goitrogens are no great risk to us, nor are most other toxins inherent in what we eat. Only where food is scarce or diets greatly restricted do these poisons constitute health hazards. In certain areas of India, for instance, a high incidence of goiter occurs because the diet consists primarily of cabbage. Parts of New Zealand have experienced similarly induced toxic effects. Goitrogens can even occur in cow's milk, and have cropped up in Finland where cows occasionally browse certain fodder or forage. Rutabaga is especially high in the goitrogen 2-hydroxy, 3-butenyl isothiocyanate.

Again, there is no need for undue concern, but in the back of your mind, where you ponder the fate of your children, and grandchildren, and children of other races and cultures, where at times you try honestly and sincerely to wrench order from confusion, to understand issues and estimate the consequences of a policy or stand, particularly as it concerns food, remember that the soybean, on whose leafy, beany foliage so many lives hang in balance, is mildly goitrogenic. Only raw soy milk, however, would carry the factor to any great extent. Rendering it nongoitrogenic is simple enough. Like the iodine in iodized salt, all that is required is a sprinkling of the additive potassium iodide.

Gossypol

In the pigment gland of the cottonseed may be found a polycyclic polyphenol compound known as gossypol. It is toxic and has been known to poison calves, rabbits, mice, poultry, rats, dogs, and guinea pigs. Since cottonseed oil is used in a number of foods, there is some risk to man, although the dangers are considered negligible as a result of alkali refining and washing techniques employed commercially. But the problem doesn't end there. Cottonseed meal is an important poultry feed, and gossypol and sterculic acid, another component of cottonseed oil, either individually or when combined, have been discovered to interfere with reproduction in poultry. In trout hatcheries the two compounds increase the number and size of trout tumors, acting as cancer-promoting agents along with aflatoxins produced by mold growth in cottonseed meal.

Gossypol can be inactivated by heat treatment. In the process it becomes protein-bound and the content of an essential amino acid, lysine, is reduced, which results in a significant nutritional loss.

The toxin gossypol is of sufficient concern that scientists are attempting to produce a glandless variety of cottonseed. Their only worry, it seems, is that, should they prove successful, the new plant may well be highly susceptible to insect infestation. One advantage will be gained only at the expense of heavier demand for the use of insecticides.

Hemagglutinins

So important to food production has the soybean become that one in every six acres of agricultural land is planted with this legume—one in six. Yet the soybean contains antivitamin factors; a trypsin inhibitor capable of causing pancreatic hypertrophy and a reduction in nutrient digestability; some goitrogenic activity; antimineral factors which we have yet to speak of; an antithyrotoxic factor; a certain degree of allergenicity; some sterols, the function of which in human cholesterol metabolism is not known; estrogens, found at high doses rates to inhibit growth in rats, to increase the iron content of the liver, the zinc content in the bones and liver, and the deposition of calcium, phosphorous, and manganese; and a hemagglutinin, which is a substance causing agglutination—clumping—of red blood cells. Plant hemagglutinins are widespread, and since weight is lost if their levels in the diet are accumulated, they must be considered toxic. The hemagglutinin of the soybean accounts for a fourth of the growth inhibition in rats on soybean diets. It can be processed away, by heat treatment.

Other beans carry hemagglutinins and are capable of causing illness under conditions where, because of high altitudes, undercooking is apt to occur. How many survive on diets consisting mainly of bean nutrients? And only two laboratories in the world are conducting researach on hemagglutinin. Consider again, "one acre in six."

Hydroxyphenylisatin

This is the chemical name for the potent laxative found in prunes.

Lathyrogens

Can you imagine a toxin so potent that it could paralyze a person from the waist down? If you can, perhaps you can further accept the fact that it is present in peas of the genus Lathyrus. But is it possible for anyone to truly perceive as many as 35 million people afflicted at one time? It happens today. It happens in India during periods of severe drought and famine. It happens from eating a plant which is otherwise well balanced nutritionally, with anywhere from 24 to 26 percent protein.

Lathyrism is a difficult disease to study because animals other than man are resistant to its effects. A famine that can cripple a human population only favors the rat which is almost totally immune to the toxin of plants used for human food.

The crippling property of this lathyrogen was first discovered in the sweet pea, that delicate climbing flower adorning fences and hedgerows. The sweet pea is particularly toxic, even to the rat. But it is the varieties of peas named *Lathyrys sativus, L. cicera* and *L. clymenum* that pose the greatest dangers to humanity because of their food value. These peas can be detoxified by soaking them overnight, followed by boiling for thirty minutes (or roasting).

Since it is inherently difficult to do research on a compound that is a toxin which to a large extent affects only humans, it will perhaps come as little surprise that science is only now beginning to isolate the specific neurolathyrogens. One of these compounds goes by the chemical name beta-N-oxalyl-alpha, beta-diaminopropionic acid. It's a shame the name doesn't evoke images of a skull and crossbones or shock us with the obscene picture of 35 million people stumbling, crawling, and hind-sliding along over a dry and dusty drought-ridden land.

Lycopene

Lycopene is a chemical relative of carotene, the precursor of vitamin A. It is found in ripe fruit and tomatoes. It is toxic. Accumulations in the body lead to a disease of the liver. No one is likely to become ill from this toxin unless restricted to a narrow diet excluding almost everything but tomato juice. Illness has occurred when the drinking of tomato juice was regular and excessive. If one-half gallon were consumed daily, toxic effects would take several years to manifest themselves.

Metal Binding Factors

Like most naturally occurring toxins, metal binding factors are not considered a significant hazard to man because his overall consumption of them is low. Still it is important to recognize the double standard that exists, or the different connotations of "natural" and synthetic toxins. Even at small dose rates, synthetics tend to evoke strong feelings of apprehension. Yet a number of normal food components which seem much less risky to us can lead to improper bone development, early tooth decay, or possible gastric hemorrhage, convulsions, and even death in children. Who worries particularlarly about eating rhubarb, spinach, celery, almonds, cashews, currants, prunes, or beans? Yet all these foods contain metal binding factors which can tie up calcium and prevent absorption in the stomach. Some of them are listed in medical records as the cause of death of young children. Sheep by the droves are known to have died from excessive intake of similar compounds in sorrel, purslane, and marigolds. During World War I, when short food supplies necessitated the substitution of rhubarb for other green vegetables, innumerable cases of poisoning became a matter of military record.

There is a tendency—a natural tendency—to apply a double standard because we are familiar with common foods but fear the unfamiliar chemical names by which additives or synthetics are known. A second, easy pitfall is the common, generally innocent tendency to judge everything by our own standards. Suppose you were to learn that there would be little risk to you or your children from the presence of metal binding factors in foods if you had plenty of vitamin fortified milk to drink (a good supply of vitamin D and calcium). Is there not an inclination to ignore the risk and go merrily on your way? There is for me, I regret to say. I tend to react that way. But instead should we not ask some obvious questions? Are there people in the United States and throughout the world whose intakes of vitamin D and calcium are insufficient? How many are there? Are metal binding factors a risk to them? Would a school lunch program with milk in the diet serve a good purpose for the children of the United States who, through no fault of their own, suffer from calcium deficiency? Must we not consider the millions of half-starved people around the world whose intake of most nutrients is borderline and who can be made healthy by fortifying their diets with soybean protein, cheese whey blends, or other cheap high-nutrient products like these, as long as attention is paid to the proper processing techniques to eliminate a metal binding factor from isolated soy protein? The problem is that science knows practically nothing about this factor except how to cope with it by expensive, stopgap measures. The stumbling block here is *expense*.

Two compounds thought to be important metal binding factors are phytic acid and oxalic acid (or oxalates). But our ignorance in these matters is unbelievable.

Menthol

There is no need to worry about menthol and cardiac arrythmia unless you are a heavy drinker of liqueurs. Injected subcutaneously, the lowest published dose known to have caused death in rats is 2000 mg/kg.

Mushroom Poisoning

If we adhere to the arbitrary definition of naturally occurring toxins given at the beginning of this chapter, then mushroom poisons do not belong in this discussion unless they are toxins of "edible" mushroom varieties. I plead ignorance of these but include mushrooms as a striking reminder of the very deadly toxins that do occur naturally in the plant kingdom.

Myristicin

You will probably not get high on myristicin, an hallucinogenic chemical found in nutmeg, carrots, and celery unless you are a nutmeg, carrot, and celery nut.

Nitrates and Nitrites

Nitrates and nitrites, already discussed in relation to their presence as additives in meats, are capable of causing methemoglobinemia, a blood disease. These compounds are present in potentially toxic amounts in beets, broccoli, cabbage, cauliflower, lettuce, rutabagas, and turnips. At levels between 40 and 2100 milligrams per kilogram of spinach, they are known to have been responsible for illness in children. Nitrites are also cocarcinogens.

Protein

Ordinarily, protein is not considered poisonous. Quite the contrary, it is thought to be one if not *the* most important nutrient required for human nourishment. Nonetheless, the amino acids methionine and tryptophan, common protein amino acids, have been found toxic to weanling rats.

Proteins are frequently the offenders in food allergies of children. Moreover, 1 in 1000 persons is subject to a dietary ailment that prevents him from eating bread or bakery products, noodles, spaghetti, macaroni, pizza, flour-thickened gravies, ice cream cones (just the cone), pancakes, waffles, poultry stuffing (dressing), farina, cereals, and any and all other products made from wheat, rye, barley, or oats. The reason for this is a built-in intolerance to gluten. The numbers of susceptible persons are probably greater but for lack of diagnosis the disease simply goes undetected.

The disease, an intestinal disorder, is called celiac or coeliac disease. Individuals who fall prone to the condition must completely avoid the foods listed above. Were it not for the fact that food technologists have researched products and processes and made commercially available a number of low-gluten or gluten-free substitutes for these foods, celiac patients would face a long life of terribly restrictive eating.

Purines and Pyrimidines

These compounds are almost as common as plant and animal tissue. They are cellular constituents which, along with other factors, promote atherogenic conditions.

Salt

Needless to say, salt can be highly poisonous and cause death at relatively small intakes. More often, however, it is associated with hypertension. Taken orally, salt toxicity manifests itself primarily in increases in blood pressure. Salt is also toxic (and can be lethal) in rectal administration, through adverse effects in the central nervous system. The lowest published lethal dose by this route of administration on a human is 163 mg/kg body weight.

Shellfish Poisons

Some lethal poisons can develop in shellfish that consume certain algae. So deadly do these foods become that less than two ounces of mussel could kill as many as ten persons. The toxin is heat stable, i.e., it is not destroved by cooking.

Stimulants

When you want a lift, you drink a cup of coffee or tea or possibly a cup of hot chocolate. Or you may drink one of these beverages to keep awake. The coffee break has become a national tradition. We should not forget the stimulants that reside in these products and which often lead to hypertension—nervousness, insomnia, irritability, tension. Caffeine, theobromine, and theophylline are the chemicals responsible for this problem. Being components of such commonplace foods, they are easily dismissed as of no danger. Nevertheless they are stimulants and excess intake can lead to ill-effects.

Vitamins

The vitamin supplement business booms these days on a tidal wave of vitamania. Like overkill and megaton nuclear warheads, the terminology of vitamin supplementation has breached the vocabulary of measurement; the megadose is born. As a nation we eat too much and most of us pop too many pills, vitamins included. At only ten times the recommended daily allowance of vitamin A over as short a period as a few weeks, human infants may develop swollen fontanel or hydrocephalus—bulging heads. Given the same dosage, older children and adults can form swellings like tumorous growths, with illnesses ranging from nausea, vomiting, and lethargy to ringing ears and double vision. A continued large intake can lead to a shriveling of the optic nerve and blindness. At a daily intake of 25,000 to 50,000 I.U. of vitamin A, a not extraordinary dose, symptoms appear within thirty days.

Vitamin A is teratogenic. The same is true of vitamin D, which may cause mental retardation in infants whose mothers took excessive doses during pregnancy. By itself, this vitamin can cause vascular lesions. Combined with nicotine of tobacco, vitamin D has been found to produce arterial blood clots (arterial thrombosis) in susceptible rabbits. In humans the more common disorder is called hypercalcemia. This is a condition in which the vitamin causes excessive absorption of calcium from the intestine. The calcium may then be responsible for calcification of soft tissue. Of the resulting complications, injury to kidneys is perhaps the most hazardous. Infants are particularly vulnerable to hypercalcemia. The first sentence of a statement by the NAS/NRC Food and Nutrition Board reads, "An excess intake of vitamin D can result in serious toxicity."

These are serious indictments of so natural and essential a food ingredient as vitamin D or vitamin A. Yet their toxicity is well documented (over 200 research reports on vitamin A alone) and provides the basis for serious debate regarding the extent of fortification in food products.

Try an experiment sometime. Mound a handful of fertilizer on your lawn and see what happens to the grass around it. You'll find that it turns yellow, withers, and dies. Overfertilization with essential plant nutrients hinders and stunts plant growth. Overeating and oversupplementation of the human body similarly account for a wide number of harmful effects, even death. We *can* overdo, and overkill is not an imprecise term for the net result.

Recap

This seems a good place and time to pin down some points raised in the preceding chapters. Because they are germane to the issues and important to our understanding, a few new elements will be added, those dealing with concepts used by scientists in evaluating toxicity, whether of poisons native to food or those purposely or accidentally added en route to the consumer:

*All foods consist of chemicals.

*The human body maintains a natural equilibrium with toxins or toxic metabolites (breakdown chemicals) in all foods, even so-called natural foods.

*Toxicity may be defined as the ability to cause injury. It is an unalterable property associated with a specific chemical structure. Nothing can be done about toxicity of a chemical per se except to know it exists.

*Toxicity and hazard are not one and the same. Toxicity defines an injurious response. Hazard is risk, or the probability that an injurious response will occur. Toxicity is beyond human control. Hazard, with appropriate knowledge, is predictable and subject to control.

*There are safe and unsafe doses for every single chemical that exists, whether natural or synthetic.

*Many commonplace foods and naturally occurring nutrients are toxic to man.

*The normal, healthy person needs a well-balanced diet consisting of a variety of foods to minimize the hazards posed by toxic chemicals in food, from whatever source.

*As distinguished from synthetic additives, chemicals ordinarily present in the food man has consumed without harm over the ages are assumed to be safe at the levels normally found in those foods.

*Persons most prone to the harmful effects of naturally occurring toxins in food are those whose diets are inadequate both in variety and in nutrient intake. The poor and malnourished of the world are the most susceptible to toxic compounds in foods common to their homeland and to imported foods.

*The refinement of food is not necessarily equatable with the loss of nutrients. Processing functions—cleaning, grinding, extraction, cooking—are essential to remove or destroy naturally occurring toxic substances from many foods. The soybean, invaluably important to man's future, like most plant foods must undergo a refining process to reduce its toxicity and enhance nutritive value.

*An important concern in supplying low-cost food to the needy is that processing functions add to the cost of food.

*Breeding out (genetically altering) naturally occurring toxins does not appear to be a promising method of eliminating these substances in food, at least over the short term.

*Priorities: there is as much need to research improved methods of processing and enhancing the nutritive value of foods rated high in production efficiency as there is need to evaluate the toxicity of additives that are required to maximize rates of growth and to put food in finished, edible form.

*Processing technology must at least keep abreast with production technology. This is gravely critical because the world is moving inexorably to plant foods as sources of nutrients.

Table 25 provides a summary of the information considered in this chapter.

Table 25. Poisons and Antinutritional Properties, Diseases They Cause, and Sources of Naturally Occurring Poisons in Food

Poisons and Antinutritional Factors	Disease, Ailment, or Disease Potential	Foods
Alcohol	Alcoholism, procarcinogen in "heavy" drinkers	Alcoholic beverages
Allergens	Allergies	Cow's milk, wheat, soybeans, eggs, seafood, chocolate, numerous vegetables, fruits, nuts, etc.
Allyl isothiocyanate	Skin irritant	Horseradish, mustard
Amines, biological	Can elevate blood pressure with certain tranquilizers; can cause death	Natural cheese, Chianti (wine), yeast extracts, bananas, pineapples, tomatoes, pumpkins, squash, cucumbers
Antivitamin factor	Vitamin deficiency symptoms	Raw egg white, clams, raw fish, linseed meal, corn (and other cereals), oranges, raw soybeans, fish and vegetable oils, raw kidney beans
Carbohydrates	Obesity and related health problems, dental caries, lactose intolerance in Asians and blacks	Sweetstuffs, carbonated beverages, sugar, milk (lactose), and other sugar-containing foods
Capsaicin	Skin irritant	Red pepper
Carcinogens		
Safrole	Found to cause cancer in mice	Cinnamon, nutmeg, sassafras
Tannic acid	Known to cause liver tumors in rats	Tea, nuts, fruits

Table 25—continued

Poisons and Antinutritional Factors	Disease, Ailment, or Disease Potential	Foods
Benzopyrene	Known carcinogen	Cabbage, charcoal-grilled meat, spinach, lettuce, tea
Lemon and sesame oils	Cocarcinogenic properties	Lemon and sesame oils
Cycad nut (cycasin)[a]	Liver damage, malignant tumors	Cycad nut
Selenium compounds, chromium, cobalt, iron, capsicum, cholesterol, oxalic acid, goitrogens, thiourea derivatives, naturally occurring estrogens	Suspected carcinogens	Many foods
Cholesterol	Heart disease	Variety of fatty foods
Enzyme inhibitors		
Trypsin	Growth retardation in animals	Soybeans, kidney beans, lima beans
Chymotrypsin	Growth retardation	Irish potatoes, egg whites
Cholinesterase	Block transmission of nerve impulses	Radishes, carrots, celery, potatoes, broccoli, eggplant, sugar beet, asparagus, Stayman apple, Valencia orange
Solanine	Cholinesterase inhibition, teratogen, abortion, and fetal resorption agent	Potatoes (blighted or stressed)
Estrogens[b]	Reproductive difficulties in animals; suspected tumor-causing agents	Wheat, oats, barley, soybeans, rice, plums, cherries, apples, liver, egg yolk, vegetable oils (cottonseed, linseed, olive, peanut, corn)
Fatty acids	Certain fatty acids cause poor growth in animals, oleic acid causes anemia in rats, infants on high diet of polyunsaturated fatty acids develop vitamin E deficiency	Fatty foods, vegetable oils
Favism	Hemolytic anemia in persons predisposed to toxic factor[c]	Faba bean (*Vicia faba*)
Flavor components	A number of toxic acids, esters, terpenes, phenols, aldehydes, etc.	All foods
Glycosides		
Cyanogens	Cyanide poisoning	Lima beans, cassava, almonds, black-eyed peas, wild black cherries, kidney beans, peach pits, apricot

Table 25—continued

Poisons and Antinutritional Factors	Disease, Ailment, or Disease Potential	Foods
		pits, choke cherry pits, Bengal gram, Red gram
Cycasin	Carcinogen	Cycad nut
Goitrogens	Goiter	Soybeans, rapeseed, white turnips, cauliflower, peaches, pears, strawberries, Brussels sprouts, spinach, carrots, rutabaga, cabbage
Glycosides of some honey	Potent heart stimulant	Honey from bees obtaining sugars from mountain laurel, rhododendron, azalea, oleander
Naringin	Stomach irritant	Unripe grapefruit
Gossypol (and/or sterculic acid)	Poisoning of farm animals, reproduction problems in rats and chickens, cocarcinogen in trout	Pigment gland of cottonseed
Hemagglutinins	Growth inhibition	Soybeans and other legumes
Hydroxyphenylisatin	Laxative	Prunes
Lathyrogens	Lathyrism (disease of the nervous system, bones, and connective tissue)	Peas: *Lathyrus sativus, L. cicera, L. clymenum*
Lycopene (carotenoid)	Liver disease	Ripe fruits, tomatoes
Metal binding factors		
Phytates	Interfere with calcium absorption	Cereals
Oxalates	Interfere with calcium absorption	Spinach, cashews, almonds, cocoa, rhubarb, Swiss chard, beet tops, parsley, celery, currants, prunes, beans, soybeans
Menthol	Cardiac arrythmia	Liqueur
Mushroom poisons	Food poisoning	Inedible mushroom varieties
Myristicin	Hallucinogen	Nutmeg, carrots, celery
Nitrates (and nitrites)	Cocarcinogen methemoglobinemia	Spinach, smoked foods, green leafy vegetables
Protein	Methionine and tryptophan (amino acids) toxic to weanling rats. Some children cannot tolerate wheat gluten[d]—thought to be an adverse reaction to a protein component	Meat, milk, legumes
Purines and pyrimidines	Coatherogenic	Plant and animal tissue
Salt (sodium chloride)	Hypertension	Salt

Table 25 – continued

Poisons and Antinutritional Factors	Disease, Ailment, or Disease Potential	Foods
Shellfish poisoning	Lethal toxin of shellfish consuming certain algae	Shellfish
Stimulants (caffeine, theobromine, theophylline)	Hypertension	Coffee, tea, cocoa
Vitamin A	Found to cause disturbances in rat offspring	Cheese, butter, yellow vegetables, fruits
Vitamin D	Vascular lesions, mental retardation; vitamin D and nicotine – arterial thrombosis in susceptible rabbits	Butter, egg yolks, fish oil, liver, salmon, tuna

[a] Cycad nut is an important part of diet in the Tropics.
[b] These are probably a group of related isoflavones.
[c] This is a serious problem in Asia.
[d] This is thought to be an adverse reaction to a protein component.

X

Natural/Organic Foods

*God said, "Let the earth bring forth seed bearing plants;
let fruit grow on trees." And all these things were done.*

Genesis I:11

A whole new food business has grown out of consumers' misgivings about additives in foods. Classed generally as "health" foods, natural/organic products are booming, with gross sales predicted shortly to reach several billions of dollars annually. Looking back now on its origin and growth, we see that the natural/organic movement takes on all the characteristics of a revolution. It began as a revolt—a consumer revolt against the unknown in foods. The early proponents were zealots, willing to go it on their own at all costs, and as convinced of their cause as any true believer. The rebellion started, it caught on and spread.

When confronted with an ideological revolution, one must constantly be on guard lest the guiding principles be perverted. If it can be agreed that some justification for the movement existed (or exists), then, as the revolution settles in, it is time to take stock, to see where the movement is heading, and to reaffirm the original goals.

The initial impetus for natural/organic foods stemmed not so much from nutritional concern as from fears over the presence of pesticide residues and other chemicals in foods. Nowadays the movement tends to subsist more on nutritional claims, but that is getting ahead of our story. Having segregated the motives, let's examine the legitimacy of those original fears and the implications of pesticide use generally.

Candor demands that we first admit to scientific proof that some pesticides are carcinogenic. There are those that do cause cancer in test animals, but much work has been done to weed out the most dangerous of these, and conditions now are generally improved over what they may have been at one time. Studies continue to analyze the hazards of pesticide chemicals. More is known now than was known five years ago. A recent evaluation (by the Commission of Pesticides and Their Relationship to Environmental Health of the United States Department of Health, Education, and Welfare, with much of the original research conducted by the Biometrics Research Laboratories of the National Cancer Institute) of 100 present-day pesticides indicates that at least some of them are tumorogenic, i.e., will produce benign tumors in animals. (Benign tumors should be distinguished from malignant cancers in degree of pathogenicity. The former are generally less serious than the latter.) What such evidence means to man is questionable, especially since tumorogenic responses usually occurred in test animals only after administration of massive doses far in excess of the minuscule pesticide residues that are left in food.

A technical panel of the commission, taking a conservative stance as assuredly anyone must, concluded that an "important objective in cancer prevention is the elimination, or reduction to a minimum achievable level, of all substances in the diet of man proven to be carcinogenic in either man or animal." It went on to say that the situation is so complex and poorly understood that "no assuredly safe level for carcinogens in human food can be determined from experimental findings at the present time."

We can conclude, then, that hazards persist, and as late as August 1974 the pesticides aldrin and dieldrin were banned by the Environmental Protection Agency because of evidence incriminating them as human carcinogens. For a slightly different slant on the matter, let's turn to a report by the Committee on Occupational Toxicology of the American Medical Association. In 1970 this committee issued a number of disclaimers on speculation concerning the all-powerful insecticide DDT, among them a rejection that proof exists of this compound's carcinogenic effects on man or of its perversive influence on human reproduction. After years of raging public controversy, proof still does not exist, even against the chemical which, more than any other, bears responsibility for creating concern in the first place.

If for no other reason than objectivity, we should emphasize that not all pesticides bear evidence of carcinogenicity. Not all but only some of the 100 pesticides surveyed showed a tendency to cause tumors, benign at that. And so we must ask if *all* pesticides should be banned for the misanthropies of a few. Should we not promote those in the majority which appear to pose little or no risk to man or nature? Many present-day biodegradable compounds, which break down in soil to any of a number of harmless elements, in the ecological cycle, fall into that category. For those that do not, good sense would seem to dictate the use of tests for toxicity as a selection process to govern the extent of their use. But further, and far closer to the nub of the matter, an appreciation of science's potential to find safe chemicals or safe alternatives to pest control such as sex attractants and insect viruses, should impel us to a worldwide commitment to seek from every scientific corner those safest methods yet locked away in nature. A cancer of the fruit fly—does it exist? A botulinum of the khapra beetle? A degradable, harmless chemical as lethal to insect pests as DDT?

In the end, variety in pesticides is as desirable as variety in the food we eat. The more chemicals, the broader the array to select from and the greater the assurance that no single one will be pressed into multiple crop service. The world should look forward to having an arsenal of chemicals that would permit the selective use of pesticides on either *one* insect or *one* crop to promote minimal residues of any given pesticide on any single food or group of foods. If DDT, the herbicide 2,4D, or any other seemingly panacean pesticides lead us awry, it is because they create a tendency to relax the struggle, to yield up that extra ounce of emotional urgency to research new products and methods. Panaceas, pseudo though they often turn out to be, promote complacency. And the genetic superweapons of insect pests, hidden like nuclear missiles in underground silos, await with lunatic patience for some natural trigger to touch them off.

You may have wondered why you hear or read many fewer "exposés" on the health hazards of weed killers (herbicides) than on insecticides; an explanation is in order, so that the two categories of pesticides may be more clearly distinguished. Pesticides, in the popular mode of expression, tend to become a catchall for any industrial chemicals. We would do well always to use the generic for fear that we may aim at the moon when our target is Mars. In any event, herbicides tend generally to pose less of a threat because they are usually shortlived in the environment, they don't accumulate readily in tissues, and they have low human toxicity.

Clarity of understanding, like regulatory and constitutional authority, depends upon precise definitions. If the natural/organic food business fails to perform up to the standards of traditional foods, it will be because its proponents failed to define exactly the meanings of *natural* or *organic* or *health* food. If consumer interests are served by a definition of mayonnaise, cherry pie, or textured vegetable protein, whether under a standard of identity or common or usual name regulation, and if the label of breakfast beverages must precisely inform the buyer of the contents of the product, then there is something to be gained in similar control of natural/organic foods. The matter is complicated because of the ambiguity surrounding these terms. To date, we can only repeat some commentary and definitions born of the struggle to describe these foods for legal purposes.

Natural foods. Natural foods came originally to be known as food grown under conditions uninfluenced by manmade chemicals. Implicit in the term was the assurance that such food came to the consumer free of process-oriented additives (preservatives, nutrient supplements, emulsifiers, etc.) and with minimal or no further refinement.

As the movement spread, it was shaped and molded to meet local (and even individual) needs and concepts. The word *natural* is now applied to a host of foods, few of which fit exactly the definition stated above. Rather than purify the standards, the definition has been stretched. Under such circumstances a point is soon reached at which no generalized standard or definition applies; the standard becomes almost as specific as the individual's desires (or the supplier's imagination). Lost in the maze of specifics and sub-specifics is any hope of setting forth precise rules by which growers and processors can be regulated and consumer interests served. It is as though a law had to be established which allowed each of us to define what turkey dressing is. By past experience and personal taste, each of us would come up with a recipe to suit his own desires—this spice at that amount, another spice at another amount, no spices at all, oysters, no oysters, dry bake, moist bake, and on and on.

Let's look at eggs as an example and attempt to define and interpret what they might be according to the natural food ideology. First, "natural" eggs should be untouched by drugs of any kind (whether residuals occur or do not occur). At the expense of the health of the poultry there can be no antibiotics in the feed. Similarly we should eliminate pesticide-treated feeds: no home-grown feeds treated with either herbicides or insecticides can be used, and feed purchased from farms (and here, for lack of specific regulations, we must trust the supplier) known to employ treatment with pesticides must be excluded. This ban on pesticides should also apply to the birds themselves and to the pens in which they are kept. To this end no spray can be allowed inside the henhouse or on the hens themselves. Production notwithstanding, the environment and the birds must be left to suffer the consequences of lice, ticks, fly infestation, and the like. For small flocks with plenty of space, the results of such conditions need not be catastrophic.

These would appear to be acceptable requirements for the production end of the business. Our definition of a "natural" egg is so far reasonably workable and enforceable by regulation. Would that there were regulations, however, for without them (and at least token supervision) how can you be sure any of the steps above will be adhered to? Even with regulations, experience suggests that you can only be sure that some of them will be carried out some of the time.

Processing comes next and we run into our first snag. Should the eggs be washed? For the sake of fastidiousness and to eliminate the possibility of contracting a dose of salmonellosis, many will vote yes. All right, then, we'll further narrow our definition of "natural" eggs by requiring that they be washed. Now we must decide what detergents to allow, or whether we should permit them at all. They are synthetics and, therefore, unnatural. Instead, we'll opt for soap made from lye and animal fat.

And what about sanitizers? The whole purpose of washing is to render the eggs as germ-free as possible. Perhaps we should allow their use for reasons of health. But sanitizers really aren't essential, and all we're doing is placing in the egg environment another chemical that may contaminate the product. What chemical will it be, should we go along with it? Chlorine? Perhaps one of those later compounds, like quats—quaternary ammonium compounds. Any danger of their getting into the egg? Can they cause illness?

We'll leave eggs at this point. The issues should not get that absurd—except that they do. Sugar and honey are two other examples. Natural food proponents recommend the avoidance of refined sugar. Yet a molasses-coated (brownish-colored) *extra* refined sugar, prestigiously named Turbinado sugar, is accepted fare. It can be assumed that the brown color hints (or deceives) of rawness, the same as brown rice and brown-colored whole wheat bread. But more to the point, how would this sweetener be defined for legal purposes—as natural or traditional?

"Natural" honey is usually heated and filtered, although not all advocates of natural foods would agree to this degree of processing refinement. Suppose we should settle on a heat treatment less intense than that customarily applied to traditional honey as a way of preserving some of the heat labile nutrients. The question arises, What precise heat treatment will fulfill the requirements of a "natural" honey? You could come up with enough fractional degrees of heat and time periods to satisfy everyone's specific whim. Compare that possibility with the milk pasteurization regulation which requires a heat treatment of $161°F$ for 15 seconds (or its equivalent).

Then there is whole grain flour and bread which, by natural standards, must be ground between millstones rather than in commercial grinders (in order, advocates say, to retain molecular structure and nutrient content). Stone versus metal grinders becomes another subpart to our imaginary regulation.

Vitamin supplements pose further questions of definition. Natural food stores stock rose hip vitamin C tablets consisting of natural rose hips and synthetic ascorbic acid. Natural B vitamins may be found in which synthetic chemicals are added to natural bases. There is also vitamin E which has been extracted from vegetable sources that were treated with insecticides and fertilized with inorganic fertilizer; the nutrient is then encapsulated in gelatin with added antioxidant. Is this a "natural" vitamin supplement? If so, we must rework suitable legal descriptions to account specifically for products of this makeup. And it is apparent that we have come a long way from our original definition.

As the situation now stands, the mere presence of the word *natural* on a label provides no unique assurance of content, safety, nutritive value, or advertising claim. Lecithin, a standard natural emulsifier used commercially, when sold in a "health" food store becomes a cure for dry skin, psoriasis, and liver spots. Carob—the same carob mentioned earlier as a source for a stabilizer employed by the conventional food industry—flowers as saintly as the tattered robes of St. John who, in the wilderness, was purported to have lived off this "natural" food. The fodder of cows, alfalfa, with its deep-growing root system, claims, in its "natural" food guise, a higher content of lower strata trace elements and minerals though subsoils are known to be *lower* in mineral content than are topsoils. The depth from which the minerals are absorbed, we agree, may be indisputably lower than that of shallower rooted plants, but the implication that the mineral content is therefore higher is incorrect.

It might be useful to note another example. Strictly speaking, the freezing of foods is a processing function which, according to most subscribers of natral foods, prevents these products from being characterized as "natural." Unlimited instances could be cited. Focusing mainly on legal definitions and standards, we simply relate a fact: on just such minutiae do attempts at regulating this burgeoning business run aground. And the business is burgeoning.

Organic foods. "Organic" foods are a logical variation of a somewhat illogical theme and seem to have been introduced as the movement to natural foods spread.

In its strictest usage, "organic" refers to the production of crops in which the application of nutrients to the soil is restricted entirely to such organic fertilizers as manure, compost, and garbage. Inorganic fertilizers—those purified, balanced, inorganic nutrients—are excluded. Yet there is overwhelming scientific evidence (ten investigations spanning seven decades) that no difference in major nutrient value exists between organically grown and inorganically fertilized crops. Minor variations in the minerals of plants do occur as a direct result of the mineral content of the fertilizer, but organic fertilizers fall short in their content of phosphorus, a major plant nutrient. Inorganic fertilizers, when needed, are balanced with trace minerals. So there appears to be little advantage, and possibly some nutritional disadvantage, in organically grown food crops.

If "natural" implies growth without pesticides, it goes without saying that "organic" is similarly exclusive in intent. A quick scan of crop disasters of the early years of the twentieth century leads us to ask how in present times nontreated crops grow relatively unscathed by plant pests. Entire fruit crops were wiped out, potato crops were devastated by blight, nematodes, wireworms, aphids, and psyllid beetles; and rice paddies were turned into weed fields when clogged with barnyard grass. We must also wonder why 1 percent of American farms excel so thoroughly that they produce 60 percent of all vegetables, 45 percent of all fruits and nuts, and 35 percent of all poultry grown in the United States. The answer of course lies in the controlled application of pesticides and in the good possibility that the survival of organic crops depends to some degree on the low rate of weed and insect infestation made possible by the use of pesticides elsewhere on the land.

Untreated crops and livestock are an indulgence the world cannot afford, say such universally renowned experts as Norman Borlaug. The truth of his words is apparent in projected reductions in food raised without the use of chemicals. Losses would total about half the food now produced. Rice, the mainstay of half the world, would yield 20 to 50 bushels less per acre. Such reductions are simply beyond contemplation. The advocates of natural and organically grown foods certainly did not have in mind a worldwide conversion to their standards of production. However, when the matter is brought closer to home, you can find a basic conflict in satisfying the desires of those who are repulsed by news of food that has been contaminated with rodent hairs and insect fragments but who, at the same time, decry the use of pesticides.

Another very real issue is the cost of organically grown food. Because losses are more extensive, and since nitrogen (plant food) supplied by garbage or cattle manure runs 50 to 100 times more expensive than inorganic nitrogen, considerably higher food costs might be expected. Supermarket and health food store surveys have shown prices of organic foods to be about twice those of similar conventional food products.

A third concern involves the alternative uses for the organic grower's fertilizer. Manure can be recycled to produce methane gas, a fuel. Much research, money, and technological expertise are now being directed chiefly toward that end. A study by the Stanford Research Institute estimates the fuel production potential from manure at 75 billion cubic feet of gas by 1985. There is also a residual manure by-product of this recycling which could serve as a fertilizer. Another recycling alternative would extract the protein from manure and use it as animal feed. In total, this organic fertilizer could supply protein in amounts equal to that derived from all the soybeans grown in the United States—a truly impressive protein source. As yet, no one is recommending that the protein be purified, modified, and made into meat analogues or similar protein foods. But why not? The only limitation might be energy.

Always is it necessary to return to the one limiting factor—energy. If organically grown foods are costly energy users, then inorganically fertilized crops are only cheaper when petroleum costs are down. Inorganic fertilizers are based on petroleum and fluctuate according to its cost. Truly the time has come to ask who or what will win the competition for petroleum products— the auto, the bus, the truck, and jet plane, oil-using industries from New York to Bangkok that turn out items from steel to food, war machinery poised for instant annihilation, our "essential" goods and playthings? All these, or tractors and fertilizer?

Our original objective was to define "organic" foods in a way that might make an enforcement regulation possible. Limited to essentials, the best proposals so far put forth would first rearrange the grammar and then specify the conditions for growth and processing. Thus "organic" food becomes organically grown food processed with minimal refinement in the absence of process-oriented additives. Conceivably, "organic" food could be distinguished from "natural" food on the basis of growth conditions, the former being attended by a grower (gardener or farmer) using only organic fertilizers. But because the basic intent of both categories is to achieve a food source free of intentional or incidental additives and of anything more than minimal refinement, it might be more logical either to eliminate one category or to use the words interchangeably.

Health food. Clearly this is a misnomer, a misleading term implying that products so labeled have healthful benefits not inherent in nonhealth or conventionally processed foods. Because of this, reference to or implications of special health significance will no doubt be banned one day. Possibly another term, *ecology food*, as one professor of food science would have it, will catch on or be legalized into prominence.

To leave this topic without at least a passing remark about research that would prove special health benefits of naturally occurring, unrefined nutrients would be less than frank. Indeed, such research is being conducted and, as often happens, common objectives make for strange bedfellows. Thus one finds segments of the food industry, with vested interest in establishing unique, marketable qualities in fresh fruit and vegetables, working side by side or at least with similar single-minded resolve with scientist-converts to the natural food movement.

But honesty and candor demand further proof of the superiority of naturally occurring nutrients than scientists of either persuasion have to this time brought forth. Differences have been documented, using Kirlian photography and chromatographic techniques. Visible evidence has been recorded—photographs showing the flash and flare and flamelike qualities of the so-called life forces of natural nutrients as opposed to the dull hues and rounded, lifeless appearance of the synthetic or processed nutrients. Obvious distinguishing characteristics of color and shape can be seen. What this means in terms of absolute nutritional qualities remains to be determined.

In a world faced with imminent food shortages, one is somehow struck with the absurdity of much of this commentary. To spend time, to devote a chapter to what in fact may be petty irrelevancies, is almost unforgivable. Only the need for common understanding justifies the effort. Again, a rethinking of the issues more than ever bolsters the feeling that the natural food revolution may yet achieve its finest hour in the promotion of a less extravagant lifestyle. And so, at the risk of offending sensibilities, I repeat: Pesticides are not the only pollutants of food. Industrial sources ooze out their own unique species of poisons. The list is long and ever-growing. And we dare confess: we are industry.

XI

Nutrients, Nutrient Sources, and Label Information

Inclusion of any added vitamin, mineral, or protein in a product or of any nutrition claim or information, other than sodium content, on a label or in advertizing for a food subjects the [food] label to the requirements of . . . nutrition labeling.

FDA regulation

Only a country exceedingly blessed with food resources could conceive, much less promulgate, a regulation requiring nutrition information on packaged foods. In no place in the world where food is in short supply could one imagine a regulatory effort of this kind, especially one seemingly destined to serve only that 10 to 20 percent of the populace interested in reading or using label information in any way. All consumers will bear the expense so one can only hope that increased cost and a tightening of American food supplies may in some way excite enough interest to assure a reasonable nutritional return on the investment made. Anything short of that must be considered extreme self-indulgence.

In autumn 1974, even as doors to the historical World Food Conference in Rome, Italy, were thrust open to those who would begin together for the first time to analyze the immensity of the world's food plight, regimented packets of nutritional information were appearing on packages of a large number of food items on the shelves of American supermarkets. To the unknowing eye, the little box of nutritional data appears so boringly plain and uninspiring that it seems to have no relationship whatever to the food held within that $10 billion business wrapping. Certainly one is not inspired to regard that printed information as anything valuable enough to sustain life. But that is exactly what nutrition does and what the figures on all those food packages are striving to say. With this in mind, it seems more than justified to dedicate a chapter of this book to that box of data and to that first of all purposes of food—nutrition.

The United States RDA Concept

Minimum Daily Requirements (MDR), long the phrase used to express daily nutrient needs, has been replaced by the term Recommended Daily Allowance (RDA). If the latter is not to suffer misinterpretation as did the former, some explanation of its derivation is necessary.

The concept of the RDA applies to nutrients considered necessary for good, overall health. However, some nutrients known to be essential are not sufficiently understood to allow the setting of a specific RDA for them. Labeling regulations require (or allow) listing of only those for which RDA's are relatively well established by science. Others will be added as the evidence warrants. The point is that not all nutrients are listed on the label; in fact, most labels will provide a listing of only protein, five vitamins, and two minerals. At most, the label will declare RDA's for protein and nineteen other nutrients (vitamins and minerals).

It is of primary importance to recognize that RDA's are, according to Hegsted, designed as standards to "serve as a goal for good nutrition." The key word is *goal*. An RDA is a goal that might be set by anyone for daily living. By that interpretation, RDA is a guide to good nutrition, but a guide only. It cannot and does not account for major variations in individual needs. On the other hand, RDA's are set sufficiently high to provide reasonable nutrient intake for *most* individuals. A few, however, may not be well served by such standards. It is obviously impossible to set standards that are adequate for all individuals when needs vary depending upon a large number of factors. It is good to be aware of this fact, but such awareness should not detract from the major intent and usefulness of the RDA concept. It can and does serve the majority of people very well indeed.

As established by the National Academy of Science/National Research Council, RDA's had to be further simplified to be more meaningful as label information on packaged foods. In general, the four user groups recognized by the controlling regulation are infants under twelve months of age, children one to four years old, adults and children aged four or older, and pregnant and lactating women. The method used for choosing appropriate RDA's for each category of dietary need was to select in each case the highest requirement of the four groups. In this way it was assumed individual needs would be met. Some, of course, would be exceeded, but this should not endanger health.

Recommendations of the NAS/NRC are shown in Table 26. As applied to nutritional labeling of foods purchased generally for adult use, these recommendations reduce to those shown in Table 27. It is the nutrients and RDA values of Table 27 that are important to most consumers. RDA percentages on food labels are all determined using the RDA values shown in Table 27 as a basis. Listing of nutrients will always be in the order presented in this table, i.e., vitamin C following vitamin A and so on down the line. This list will always be headed by protein RDA, which is considered separately later in this chapter.

RDA's listed on food labels are not requirements, but guides to good nutrition. Individual needs as related to age, sex, general health, physical activity, body size, and other personal differences may vary. Some individuals could require larger nutrient intakes; many could survive nicely, with no nutritional deficiencies, on lower intakes than the RDA's expressed on food labels. It is necessary, therefore, to understand the RDA concept and to use it accordingly. (For more detailed information the reader is directed to *Recommended Daily Allowances*, National Academy of Sciences, 8th ed., 1974, Washington, D.C., and to the preamble of Nutritional Labeling Regulations described in the Federal Register, January 19, 1973.)

The Nutrition Label

Before long most food products will bear nutritional information even though present regulations require it only for foods to which nutrients are added or for foods advertised for nutritional properties. Competition, regulation, educational efforts designed to make consumers aware of new labeling standards, nutrient fortification of snack foods, and more industrial advertising of nutrition—all point to a greater awareness of food value and a wider variety of foods with label information. No matter whether the reason for labeling is to comply with regulations, to advertise, or to meet competition, the information contained there is consistent both in form and in content.

What the label looks like; what it tells you. The simplest labeling format lists those nutritional factors shown in Figure 3. In evaluating such information, a number of factors should be considered.

Table 26. Recommended Daily Dietary Allowances[a] of the Food and Nutrition Board, National Academy of Sciences / National Research Council, Revised 1973.

Group	Weight (kg)	Weight (lbs)	Height (cm)	Height (in)	Energy (kcal)[b]	Protein (g)	Vitamin A Activity (RE)[c]	Vitamin A Activity (I.U.)	Vitamin D (I.U.)	Vitamin E Activity[e] (I.U.)	Ascorbic Acid (mg)	Folacin[f] (µg)	Niacin[g] (mg)	Riboflavin (B₂) (mg)	Thiamin (B₁) (mg)	Vitamin B₆ (mg)	Vitamin B₁₂ (µg)	Calcium (mg)	Phosphorus (mg)	Iodine (µg)	Iron (mg)	Magnesium (mg)	Zinc (mg)
Infants																							
0.0–0.5 years	6	14	60	24	kg×117	kg×2.2	420[d]	1400	400	4	35	50	5	0.4	0.3	0.3	0.3	360	240	35	10	60	3
0.5–1.0 years	9	20	71	28	kg×108	kg×2.0	400	2000	400	5	35	50	8	0.6	0.5	0.4	0.3	540	400	45	15	70	5
Children																							
1–3 years	13	28	86	34	1300	23	400	2000	400	7	40	100	9	0.8	0.7	0.6	1.0	800	800	60	15	150	10
4–6 years	20	44	110	44	1800	30	500	2500	400	9	40	200	12	1.1	0.9	0.9	1.5	800	800	80	10	200	10
7–10 years	30	66	135	54	2400	36	700	3300	400	10	40	300	16	1.2	1.2	1.2	2.0	800	800	110	10	250	10
Males																							
11–14 years	44	97	158	63	2800	44	1000	5000	400	12	45	400	18	1.5	1.4	1.6	3.0	1200	1200	130	18	350	15
15–18 years	61	134	172	69	3000	54	1000	5000	400	15	45	400	20	1.8	1.5	1.8	3.0	1200	1200	150	18	400	15
19–22 years	67	147	172	69	3000	54	1000	5000	400	15	45	400	20	1.8	1.5	2.0	3.0	800	800	140	10	350	15
23–50 years	70	154	172	69	2700	56	1000	5000		15	45	400	18	1.6	1.4	2.0	3.0	800	800	130	10	350	15
51+ years	70	154	172	69	2400	56	1000	5000		15	45	400	16	1.5	1.2	2.0	3.0	800	800	110	10	350	15
Females																							
11–14 years	44	97	155	62	2400	44	800	4000	400	10	45	400	16	1.3	1.2	1.6	3.0	1200	1200	115	18	300	15
15–18 years	54	119	162	65	2100	48	800	4000	400	11	45	400	14	1.4	1.1	2.0	3.0	1200	1200	115	18	300	15
19–22 years	58	128	162	65	2100	46	800	4000	400	12	45	400	14	1.4	1.1	2.0	3.0	800	800	100	18	300	15
23–50 years	58	128	162	65	2000	46	800	4000		12	45	400	13	1.2	1.0	2.0	3.0	800	800	100	18	300	15
51+ years	58	128	162	65	1800	46	800	4000		12	45	400	12	1.1	1.0	2.0	3.0	800	800	80	10	300	15
Pregnant women.					+300	+30	1000	5000	400	15	60	800	+2	+0.3	+0.3	2.5	4.0	1200	1200	125	18[h]	450	20
Lactating women.					+500	+20	1200	6000	400	15	60	600	+4	+0.5	+0.3	2.5	4.0	1200	1200	150	18	450	25

Source: National Academy of Sciences. 1974. *Recommended Dietary Allowances.* 8th ed.

a The allowances are intended to provide for individual variations among most normal persons as they live in the United States under usual environmental stresses. Diets should be based on a variety of common foods in order to provide other nutrients for which human requirements have been less well defined.

b Kilojoules (KJ) = 4.2 × kcal.

c Retinol equivalents.

d Assumed to be all as retinol in milk during the first six months of life. All subsequent intakes are assumed to be one-half as retinol and one-half as β-carotene when calculated from international units. As retinol equivalents three-fourths are as retinol and one-fourth as β-carotene.

e Total vitamin E activity, estimated to be 80 percent as α-tocopherol and 20 percent other tocopherols.

f The folacin allowances refer to dietary sources as determined by *Lactobacillus casei* assay. Pure forms of folacin may be effective in doses less than one-fourth of the RDA.

g Although allowances are expressed as niacin, it is recognized that on the average 1 mg of niacin is derived from each 60 mg of dietary tryptophan.

h This increased requirement cannot be met by ordinary diets; therefore, the use of supplemental iron is recommended.

Table 27. U.S. Recommended Daily Allowance (RDA) Established for Nutrients
Considered in Nutritional Labeling

Nutrient	U.S. RDA	Nutrient	U.S. RDA
Vitamin A	5000 I.U.	Folic acid	0.4 mg
Vitamin C	60 mg	Vitamin B$_{12}$	6.0 mg
Thiamine	1.5 mg	Phosphorus	1.0 gram
Riboflavin	1.7 mg	Iodine	150 mg
Niacin	20 mg	Magnesium	400 mg
Calcium	1.0 gram	Zinc	15 mg
Iron	18 mg	Copper	2 mg
Vitamin D	400 I.U.	Biotin	0.3 mg
Vitamin E	30 I.U.	Pantothenic acid	10 mg
Vitamin B$_6$	2.0 mg		

Source: Adapted from Federal Register. 1973. 38(49):6960.

Serving size. The figures giving nutritional information are all based upon one individual serving, not the contents of the package. An attempt has been made to standardize the serving size within similar types of food. Nevertheless, when comparing two foods of like nature, check to be certain the serving size of the two is identical; otherwise, adjust the figures on the label accordingly. Serving size will always be expressed in common meaningful units, i.e., cup, slice, ounce, teaspoon, tablespoon, etc. The serving size of a meal replacement will be the amount required to replace a single breakfast, lunch, or dinner.

Serving size also relates generally to age, with all persons over four years of age considered as adults, and those under four years of age regarded as children or infants. For adults, the term *serving* may be taken to mean "that reasonable quantity of food suitable for consumption as part of a meal by an adult male engaged in light physical activity."

Servings per container. This information provides a means of determining the overall nutritional value of the food contained within. The total servings per container times the nutrient values per serving will yield the total nutrient content. This can be helpful information in setting up weekly, biweekly, or monthly menus.

Calories. Calorie is a way of expressing energy value. A more widely used unit is the joule. Future food energy values will probably be expressed by this latter term. One joule equals the enery that is expended when one kilogram is moved one meter by one newton. It is generally more advantageous to use kilocalories or kilojoules to express the energy value of foods. A kilocalorie is the amount of heat necessary to raise one kilogram of water from 15 to 16°C.

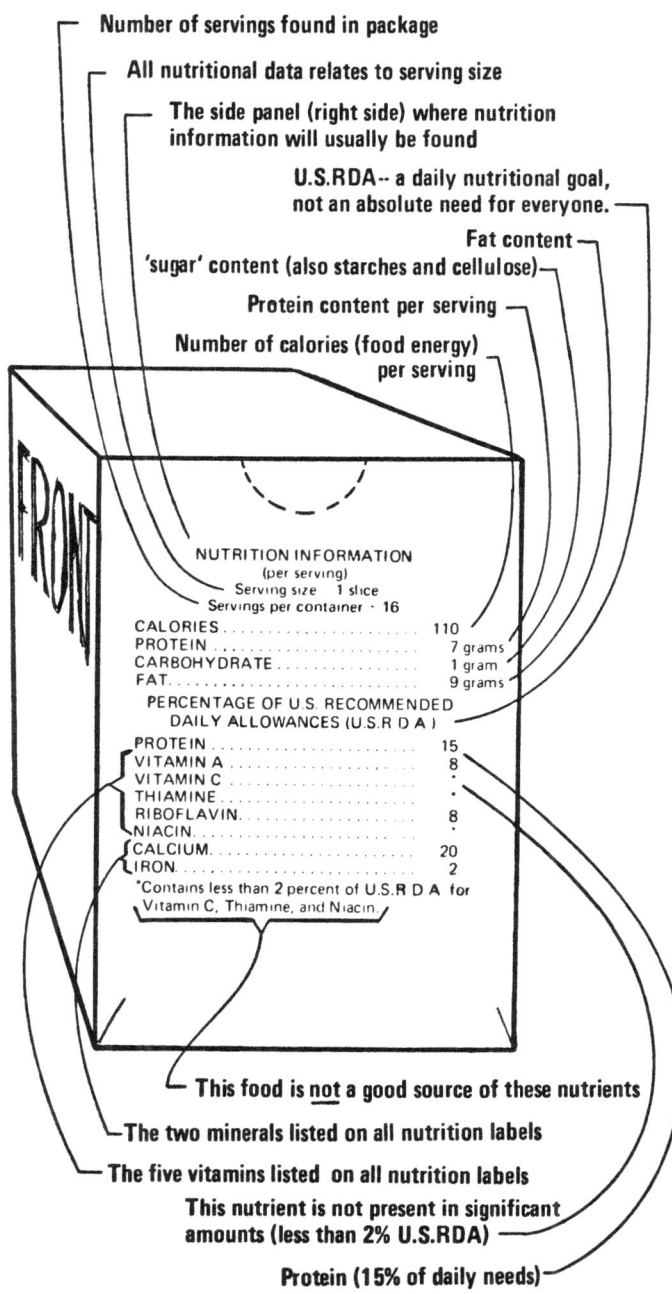

Number of servings found in package

All nutritional data relates to serving size

The side panel (right side) where nutrition
information will usually be found

U.S.RDA-- a daily nutritional goal,
not an absolute need for everyone.

Fat content

'sugar' content (also starches and cellulose)

Protein content per serving

Number of calories (food energy)
per serving

NUTRITION INFORMATION
(per serving)
Serving size   1 slice
Servings per container · 16
CALORIES   110
PROTEIN . . . .   7 grams
CARBOHYDRATE . . . .   1 gram
FAT . . . .   9 grams
PERCENTAGE OF U.S. RECOMMENDED
DAILY ALLOWANCES (U.S.R D A)
PROTEIN   15
VITAMIN A   8
VITAMIN C . . . .   *
THIAMINE . . . .   *
RIBOFLAVIN   8
NIACIN . . . .   *
CALCIUM   20
IRON   2
*Contains less than 2 percent of U.S.R D A for
Vitamin C, Thiamine, and Niacin.

This food is not a good source of these nutrients

The two minerals listed on all nutrition labels

The five vitamins listed on all nutrition labels

This nutrient is not present in significant
amounts (less than 2% U.S.RDA)

Protein (15% of daily needs)

Figure 3. A Nutrition Label

Kilocalories may be converted to kilojoules by multiplying by the factor 4.2 (or, more precisely, 4.184).

Food energy is derived from protein, carbohydrate, and fat. In Figure 3 you will note that these three nutrients are listed in grams per serving. According to label regulations, they should be listed to the *nearest* gram per serving. Knowing the energy value per gram of nutrient makes it possible to calculate the caloric content, which should roughly equal the number of calories listed. For this purpose, use 4 kilocalories per gram of protein and carbohydrate, and 9 kilocalories per gram of fat. In the product described in Figure 3, then:

$$
\begin{array}{lll}
\text{7 grams of protein x 4} & = & 28 \\
\text{1 gram of carbohydrate x 4} & = & 4 \\
\text{9 grams of fat x 9} & = & \underline{81} \\
\text{Total} & = & 113 \text{ kilocalories}
\end{array}
$$

The label shows 110 kilocalories per serving; our calculations yield 113. These two are close enough and well within the incremental and absolute tolerances allowed for calorie declaration under nutritional labeling regulations. The following summarizes the factors for energy conversion of food components:

Food Component	*Kilocalories per gram*	*Kilojoules per gram*
Protein	4	17
Carbohydrate	4	17
Fat	9	38

Why Energy?

We need energy for more than just physical activity. Work is required to digest and utilize the nutrients in food, to maintain body temperature, to grow, and, for females, to produce milk for the young. Energy is also necessary to run, walk, sledge rock, type letters, and paint a house. Energy needs are directly related to the amount of physical activity undertaken; thus a man weighing about 155 pounds would expend 540 kilocalories a day in sleep, and 2740 kilocalories a day in heavy work like felling trees, climbing mountains, and playing football. Therefore, physical activity, body size, sex, age, conditions of pregnancy and lactation, and even climate (colder temperatures requiring more energy to maintain body heat) determine how much energy an individual needs. It seems noteworthy to indicate that a food fortification policy allowing for nutrient balancing on a 2800 calorie daily intake is, because of the sedentary work habits of most of us, a more than ample calorie intake. In many instances, daily consumption of 2800 calories would lead to obesity.

Balancing energy sources. The combination of energy-producing carbohydrate, fat, and protein (also alcohol) a person consumes depends largely on his individual eating habits. The Food and Nutrition Board of the National Research Council now recommends as an average goal, considering all the obvious errors inherent in *any* average, that protein compose about 8 percent of the calories in a diet, with fat calories not in excess of 35 percent. These 1974 recommendations differ from those of 1968 which suggested that the intake of protein and fat calories be 13 percent and 42 percent respectively. The present recommended intake of carbohydrate calories (with protein at 8 percent and fat at 35 percent) would amount to 57 percent.

One complicating factor in the American diet is the consumption of alcohol. Many persons derive 5 to 10 percent of their total energy intake from alcohol. Others consume as much as 1800 kilocalories daily in alcohol, well over half their caloric needs. Since alcohol does not provide vitamins and minerals, substitution of this energy source for a large share of carbohydrate, fat, and protein energy leaves the body deficient in overall nutrient intake.

Protein

Following calories, the next piece of data on nutrition labels of foods is protein content (to the nearest gram). Proteins consist of various combinations of amino acids, which are nutrients used to manufacture protein for the body. The nitrogen of amino acids is also used in the building of other tissue constituents. All enzymes, the catalysts for most life processes, are protein. However, not all proteins are enzymes.

Protein differs from some nutrients in that its quality, i.e., its content of life-sustaining amino acids, varies. Since the body is unable to manufacture certain amino acids, they must be supplied from protein sources. A protein carrying essential amino acids is therefore considered to be of better quality than a protein lacking in one or more of these components. The nine essential amino acids are histidine, isoleucine, leucine, methionine, phenylalanine, threonine, troptophan, and valine.

Proteins that supply all the essential amino acids in those quantities needed to sustain life are termed *complete proteins.* Milk and whole egg fall in this category. An "incomplete" protein is one which lacks one or more of the essential amino acids. Gelatin is an example. Sandwiched between these two extremes in quality are those proteins which have all the essential amino acids but in such small amounts that they are inadequate to meet total body needs. These are often referred to as "partially complete." The protein of corn would fall in this class. Although a given source may be incomplete, protein requirements can be met by suitable intake from other appropriate sources.

In general, animal protein is of higher biological value than that of vegetables. An important need today is to find cheap, useful ways of upgrading plant protein, especially insofar as abundance and production efficiency make it a critical food source. High-quality protein may be found in meat, poultry, milk, fish, seafood, and eggs. A somewhat poorer quality protein, though still quite good, may be obtained from beans, peas, and nut products.

Protein needs. Like energy requirements, individual needs for protein vary with age, physical activity, pregnancy, and lactation. Stress also adds a greater demand for protein.

Recent evidence indicates that intake of protein in the United States may be more than necessary, if not excessive in some persons. Recommendations of the Food and Nutrition Board of the National Research Council were 10 grams per day *lower* for protein in 1974 than in 1968. This was for adults; children's levels remained the same. Levels for pregnancy were increased (over nonpregnancy intake) from 10 to 30 grams of protein daily. These recommendations are shown in Table 25 according to age, weight, height, and sex. It must always be kept in mind that these levels of intake are established as guides. They are not necessarily appropriate for specific individual needs, but are based on *average* requirements for overall public health and well-being.

Measuring protein quality. There are several methods for evaluating the quality of protein. Among these are the biological value, the protein efficiency ratio, the net protein value, the net protein utilization, and the slope-ratio methods. Defined, these methods are as follows:

1. Biological value (BV)—the ratio of nitrogen retained to that absorbed times 100.

2. Protein efficiency ratio (PER)—weight gained divided by protein consumed for each rat after 4 weeks of feeding trials with a test protein and a casein-based diet.

3. Net protein ratio (NPR)—weight gained by rats on a test protein diet plus weight lost by a second group of rats on a nonprotein diet divided by the protein consumed.

4. Net protein utilization (NPU)—the same as NPR except body nitrogen rather than body weight is used.

5. Slope-ratio method—utilization of three or four feeding levels of test and standard protein (lactalbumin), with only those levels which fall on the response curve's linear slope used to compute the assay value. The response curve slope for the test protein is then expressed as a percentage of the standard protein slope.

It is beyond the scope of this text to evaluate the various techniques. Instead, it must suffice to indicate that the system chosen for evaluating the quality of protein for nutrition labeling regulations was the protein efficiency method (PER), which is perhaps the most widely used of the five methods described.

No single procedure is without a failing, and scientists don't all agree on the merits of any one, but the most serious drawback of the PER method appears to be its inability to account for the protein needed for body maintenance. Growth *and* maintenance are both important functions of protein and PER measures growth almost exclusively. Some proteins may be poor promoters of growth and may have low to negligible PER ratios, yet they may function adequately in a maintenance role. Other factors that affect PER include age of the rat, sex of the rat, consistency of the diet, percentage casein in the reference diet, the duration of the experiment, influence of non-protein nitrogen and other biological factors that have a bearing on protein synthesis. These factors have been evaluated by one scientist (Hegarty, 1975) who has suggested a possible need to revise the PER of the reference standard (casein) upward in view of recent values reported in the literature. So this method is certainly not without its complications.

Protein quality and RDA. In nutrition labeling the RDA of protein is 45 grams when the PER is equal to or greater than casein (milk protein). When the PER is less than casein, the RDA becomes 65 grams. Any food containing protein of less than 20 percent of the PER of casein is not considered a significant source of protein, and the label must so indicate.

For children under four years of age and for infants younger than twelve months, labeling requirements set the RDA at 18 and 25 grams, respectively, depending upon the PER. Further, the source is not considered significant for this age group if its PER is less than 40 percent of casein.

Label information does not declare the quality of protein as such, nor is it particularly essential to know it, but it is possible to determine by simple arithmetic the quality present.

Calculating percentage of RDA of protein. Casein, the reference protein in the PER method, has a PER value of 2.5. If a protein source were found to have a PER of 3.0 (better than casein), the RDA would then be 45 grams. A food product containing 15 grams of this protein per serving would have 33 percent of the RDA per serving: 15 ÷ 45 = 0.33 x 100 = 33%.

Assume a PER value of 2.0 for a test protein and 13 grams per serving. The RDA, because the test protein is not equivalent to casein, now becomes 65 grams. Then 13 ÷ 65 = 0.20 x 100 = 20% RDA.

On a food label, protein quality is not declared as such, but can be determined from information provided. All that is necessary is to work backwards through the calculation above. The label declares the grams of protein per serving and the RDA of protein. Using the first example above (15 grams protein, 33 percent RDA), the arithmetic becomes 15 ÷ 0.33 = 45(+) grams of protein (or protein with quality equivalent to casein).

Carbohydrates

Following protein on the nutritional labeling list is carbohydrate. We might refer to this class of nutrients as sugars, although, chemically speaking, we would not be entirely accurate and it would be in error to think that sugar —granulated, powdered, white, or brown—is the only carbohydrate of food. Starches, the major constituents of bread, rice, potatoes, and cereal grains, are also carbohydrates. So is cellulose, the supporting tissue of many plants, particularly woody varieties. The breakdown product of gums must also be classed as a carbohydrate.

In the statement of ingredients, the presence of corn syrup, dextrose, molasses, maltose, honey, and even certain dairy products (dry milk, whey, skimmilk, and whole milk) can be considered evidence of the presence of carbohydrates. In milk the carbohydrate is lactose.

The simplest sugar consists of one sugar unit (molecule) and is called a monosaccharide. Cane sugar is a disaccharide (the prefix *di* implying two) linking glucose and fructose, two monosaccharides. There are also trisaccharides and tetrasaccharides, the latter important in legumes such as soybeans. Stachyose, the soybean "sugar," is of considerable scientific interest in its relationship to intestinal gas formation and flatulence (passing gas), one of several factors limiting the acceptance of soybeans.

Treatment of cane sugar (sucrose) with acid or enzymes splits the disaccharide into its separate components. This process, called inversion, results in a product referred to commercially as *invert sugar*. Invert sugar has special properties (better resistance to crystallization, higher moisture holding capacity) which are more useful than the original product in certain food formulations.

Polysaccharides, sugars linked together in long chains, are common food ingredients and have already been discussed in the chapter on binders and thickeners. Seaweed extracts and vegetable gums, which act as stabilizers for food, are complex (long-chain) carbohydrates. Along with cellulose, these ingredients differ significantly from other food carbohydrates because they are nondigestible and, consequently, nonnutritive carbohydrates. Even though they serve no nutritional need, per se, polysaccharide bulk seems to be helpful in aiding elimination.

Nutritive carbohydrates serve as human nutrients and are used mainly as energy sources. They help the body utilize fats efficiently and, by supplying energy, are able to release protein (also an energy source) to other, more important uses.

Reserve carbohydrate is stored in the body as glycogen, a polysaccharide of glucose. When needed, glycogen can be quickly converted to glucose by certain enzymes.

Carbohydrates and health. In the chapter on food additives it was pointed out that sucrose (cane sugar) is the "additive" that is consumed in the largest amounts. It has also been stated that the latest recommendations for balancing fat, protein, and carbohydrate intake would have the latter supply 57 percent of total calorie needs. Carbohydrates, therefore, are major dietary food components and are expected to contribute to a variety of nutritional needs of the average person. Evidence linking carbohydrates to heart ailments (namely atherosclerosis), while inconclusive, certainly warrants concern as an element in the more general problems of obesity. If it cannot be stated that carbohydrates are a direct cause of heart disease, then it can be emphatically stated that overweight persons are generally more prone to heart ailments. Insofar as carbohydrates lead to unneeded poundage, they contribute to the overall problem. It is possible, too, that the trend away from complex carbohydrates, such as the polysaccharides of potato, flour, rice, and cereal, to diets consisting of larger amounts of simple sugars may be a complicating factor.

Chief among specific dietary diseases induced by carbohydrates are diabetes mellitus, tooth decay, and, for many world citizens, an inability to digest lactose due to the lack of the enzyme lactase. Asians and blacks appear most troubled by intolerance to lactose, which becomes a significant nutritional issue where milk, the only source of lactose, is made a part of the diet. In the development of milk-based food products for aid programs, major research is needed in the study of this disease. Some work has been done, but scarcely enough to scratch the surface. The same physiological lack, that of enzymes capable of splitting disaccharides, occurs in kwashiorkor, the protein-deficiency disease of the starving, and in celiac disease, an ailment afflicting a significant percentage of nutrient-rich Americans. If there is an interrelationship between these nutritional logjams, no one knows.

Low-calorie, no-calorie sweeteners. Rightly or wrongly, the carbohydrates we know as sweets usually take the brunt of most person's attempts to lose weight. This fact alone can account for the great interest in "diet" foods, low-calorie foods that taste sweet. It was inevitable that low- or no-calorie food additives with the ability to sweeten would ultimately be incriminated as toxins.

If ever the statement "wanting our cake and eating it, too," more aptly applies than in the issue of nonnutritive sweeteners, it is difficult to imagine. What threatens to make any low-calorie sweetening agent a health risk is the sheer quantity of intake. Substitution of dietary sweets (colas, fruitlike drinks, dressings, etc.) for foods sweetened only with nutritive sugars presumes a significant total intake among the populace generally. There are comparatively massive intakes within certain segments of society (adolescents, chiefly). Coupled at any time with short supply and high cost, pressures to find and use both nutritive and nonnutritive sweetening agents other than sugar are bound to build. Indeed, the search is on in teeth-gritting determination.

Much impetus for the search for new nonnutritive sweeteners stems from the ban on cyclamates. (For a chronology of events, see Appendix VII.) It was in October 1969 that cyclamates were struck from the GRAS list; they were banned altogether the following year as a result of a study showing the presence of bladder tumors in mice exposed to cyclamate. Looking back, it is apparent that the ban grew out of both the research findings *and* a marked actual increase and a projected increase in the use of cyclamates. They had entered the market in force by the early 1960s. Between 1963 and 1967 total annual consumption jumped from 6 to 18 million pounds. Projections for 1970 placed the use of the artificial sweetener at 25 million pounds. A study suggesting a potential to cause cancer, a sensitized government regulatory agency, and a soaring intake of the additive all contributed to the cyclamate controversy. Perhaps this explains the FDA's reluctance to remove the ban some four years later when it was encouraged to do so by industry and when numerous scientific studies, including one by the prestigious National Cancer Institute, appeared to override earlier findings. To date, two studies of rats implicate cyclamate as a potential carcinogen; twenty-two other investigations involving several animal species contradict that potential. Additionally, evaluation of cyclamates has ranged over other health-related fields all the way from reproduction through heart and nervous system investigations, apparently without major findings of ill effects.

As yet unresolved, the question of the use of cyclamates now appears fated for more research and further evaluation by the National Academy of Science and other consultative groups. If it should happen that the ban is retained, if absolutely no future dietary benefit is henceforth derived from this former additive, then cyclamates may well have served a greater purpose in exposing to science and the world the tenuous nature of toxicity testing and the truly urgent need for developing more sure techniques for evaluating both toxicity and practical levels of additive use.

Saccharin, like cyclamates, is under scrutiny, with results equally uncertain and conflicting. A study by the FDA has found evidence of urinary bladder tumors when saccharin is consumed at high levels. One Canadian study disagrees. A further Canadian study indicates that saccharin, depending upon its method of production, may contain a number of impurities, some in relatively high amounts and some with the ability to induce cancer. A West German scientist, observing animals fed "extremely high doses" of saccharin or saccharin in combination with cyclamate, found no carcinogenic activity and no chronic toxicity over entire animal lifetimes. Hundreds of animals were used in the investigation and only one, fed a low level of cyclamate, developed cancer of the bladder. In 1972, saccharin was placed on an interim order basis in the United States, freezing its use at levels then being applied. The order was issued after the NAS/NRC panels had called for more studies, while acknowledging that "on the basis of suitable information, present and future use of saccharin in the U.S. does not pose a hazard." On January 9, 1975, after further study, an NAS/NRC report concluded that the question of its influence on cancer had not been conclusively resolved.

Not only are the two most commonly used nonnutritive sweeteners under some suspicion, but to some extent their future is interrelated. Ordinarily they are used in combination. Cyclamates particularly, without help from saccharin, would have to be used at much higher levels than has been the case previously. Because the level of use is related to risk, prospects for continued use (or for acceptance of new sweeteners) are linked inevitably to this technical consideration.

New sweeteners are being researched with a zeal that only necessity and business opportunity can foster. Nutritive and nonnutritive, natural and synthetic compounds are under study. The incentive to find new nutritive sweeteners lies in the hope of uncovering products more sweet than sugar, products that could be used at lower levels than sucrose while providing an equivalent sweet taste. One such compound, claimed to be twice as sweet as sugar, derives from fully ripened fruit. Proteins with this desirable quality are now known and are yet to be truly exploited for their sweet flavor.

A sweetener of seemingly great promise, which reached the FDA and was ultimately granted clearance for several different uses, has ended up entangled in much the same kind of controversy that has plagued other additives. This one has some food value and must be considered nutritive, but its sweetness, 180 times that of sucrose, allows real reductions in the level of use and, thereby, in caloric intake. As a combination of two amino acids (*L. aspartic* and *L. phenylalanine*) common to proteins, aspartame would have been expected to be quickly accepted, but no additive with the scope of use expected of sweeteners goes unmarred in today's cautious climate, and aspartame fell dutifully in line. At least one investigator found cause for opposing its use and filed an official protest during the interim provided for objections following the publication of an effective date. So it would seem that we have come full circle, to where we now opt for the certainty of metabolic disturbances that accompany excessive sugar intake instead of the uncertain and improbable potential for disorders that might accompany intake of sugar substitutes at the very low levels necessary to achieve sweetening.

Other sweeteners of more than passing interest include the dehydrochalcones, D-6-chlorotryptophan, glycyrrhizin, and extracts of serendipity berries. Depending on the source, the first of these ranges from 300 to 1500 times sweeter than sucrose, the second on the list about 1000 times sweeter. By comparison, cyclamates, saccharin, and aspartame would rank respectively 30, 300, and 180 times sweeter than sugar. Other products under study are acetosulfam (about half as sweet as saccharin) and N acetyl and N formyl kynurenine (with a sweetness 30 to 50 times that of sugar).

But much of the controversy and much effort would go begging were there a generally better relationship between carbohydrate intake and energy output. For many of us, though, such balance seems unlikely, and the search for sweeter "sugars" and calorie-free sweeteners goes on.

Fats

Information on fat content follows that concerning carbohydrates on the nutritional label. All such labels list fat content in grams per serving (to the nearest gram), but only food products containing 10 percent or more fat on a dry basis and at least 2 grams of fat per average serving will be allowed to go the one step further and declare the makeup of fatty acids. Anything less than these stated quantities is thought to be unsuitable in regulating the intake of fatty acids; thus, the restriction.

Fat and fatty acids—what they are, what they look like. In chemical terms, fat is glycerol plus three fatty acids. Since three fatty acids are needed to form a whole unit of fat, and since glycerol is the base from which fats are made, the chemical term, combining "three" and "glycerol," becomes triglyceride. In graphic terms a triglyceride looks like a block capital letter E. Each arm is a fatty acid linked to glycerol by a chemical glue, in this case a glue called an ester. The arms of the E representing the three fatty acids may be the same or different in length, depending upon the particular fat depicted. Fatty acids are chainlike and may vary in the number of chemical links. One very short fatty acid, butyric by name, is responsible for the pungent odor of rancid milk and the smelly, sharp flavor of blue cheese. Slightly longer fatty acids account for the unique flavor of goat's milk. Butyric acid has four links in its chain, and chains commonly increase in length two links at a time; thus chains of four, six, eight, ten, and on up to thirty or more links may be found. Not until chain lengths of eighteen links are reached is it possible to consider a major chemical difference between fatty acids, i.e., whether they are saturated or unsaturated.

Much like certain amino acids of protein, certain fatty acids are considered "essential." One, linoleic, cannot be produced by the human body and must be supplied through the diet. Arachidonic, the second of these essential fatty acids, can be formed from dietary linoleic acid, but otherwise it must be supplied from without. The reverse is not true; arachidonic acid cannot substitute for linoleic acid in the diet.

Linoleic acid has eighteen links in its carbon chain, arachidonic twenty. Both are unsaturated and to a greater degree than the simplest form of unsaturation in fatty acids. They are correctly referred to as polyunsaturated, i.e., containing more than one unsaturated linkage in the chain.

To better understand the concept of unsaturation in fatty acids, visualize a chain linked in the following manner: 0-0-0-0-0-0-0. Each pair of links is held together by a single thread or bond. Now imagine a chain in which two links along the line are connected by two threads (bonds): 0=0. This linkage, of two bonds holding together two units or atoms of carbon—C=C—is called an unsaturated bond. Fats (triglycerides) that contain fatty acids of this physical makeup are called unsaturated fats. Where fatty acids are linked together by single chemical threads, the term *saturation* or *saturated* is used to describe them:

C-C (saturated linkage)

C=C (unsaturated linkage)

Unsaturated fatty acids, thus unsaturated fats, are usually held in high regard as they relate to dietary needs and specifically to the management of cholesterol in the control of heart disease. Animals on diets deficient in unsaturated fats show poor growth, poor reproductive capacity, and less efficient calorie use. The condition of the animals' skin is generally unhealthy, the animals are less capable of handling stress, and the transporting of fat and fatlike compounds within their bodies is impaired.

Diets high in polyunsaturated fats have been shown to reduce cholesterol in the blood serum of both animals and man. What happens to cholesterol thus removed is not fully understood. After some fifteen or twenty years of highly publicized research, one can only be certain that while unsaturated fatty acids play a role in the overall problem, they are but one factor and one food component in an ever-increasing list of factors and nutrients now thought to be involved.

Unsaturated fats are more prevalent in vegetable oils than in animal fat (although arachidonic acid is found in animal fat, albeit at low amounts). Good sources of unsaturated fat and specifically the fatty acid linoleic include corn, cottonseed, sunflower, safflower seed, peanut, and soybean. Some plant fats, particularly that of the coconut, are highly saturated. Properties of coconut fat are such that it lends itself to specialized uses in foods. In the past, it found considerable application in imitation milk products. Olive oil, too, is a poor source of linoleic acid.

The latest recommendations of the National Academy of Science indicate a need for polyunsaturates in the diet to the extent of 1 to 2 percent of total fat calories. This is not a difficult goal to attain, as is apparent from data in Table 28, which lists the percentages of oleic acid (a fatty acid of one double bond) and linoleic acid in a number of foods. Linoleic, remember, is the essential fatty acid from which arachidonic, the other essential fatty acid, can be generated by the body.

If unsaturated fat has high biological value, it is also a weakness where food flavor is concerned. The same unsaturated linkage that apparently provides health benefits is the Achilles' heel of oxygen. In the presence of air (as oxygen) unsaturated bonds change in chemical makeup. The fat goes rancid or is oxidized (as distinguished from hydrolytic rancidity which is a splitting of fatty acids from glycerol to form the off-flavor characterized by butyric acid). Rancidity (oxidation) destroys flavor; it also destroys biological value.

Table 28. Fat Content and Major Fatty Acid Composition of Selected Foods[a]

Food	Total Fat	Saturated[c]	Oleic[d]	Linoleic[e]
Salad and cooking oils				
Safflower.	100%	10%	13%	74%
Sunflower	100	11	14	70
Corn	100	13	26	55
Cottonseed.	100	23	17	54
Soybean[f].	100	14	25	50
Sesame	100	14	38	42
Soybean, specially processed[g]	100	11	29	31
Peanut	100	18	47	29
Olive	100	11	76	7
Coconut	100	80	5	1
Vegetable fats—shortening	100	23	23	6-23
Table spreads				
Margarine, first ingredient on label[h]				
Safflower (liquid)—tub	80	11	18	48
Corn oil (liquid)—tub.	80	14	26	38
Corn oil (liquid)—stick	80	15	33	29
Partly hydrogenated or hydrogenated fat.	80	17	44	14
Butter	81	46	27	2
Animal fats				
Poultry.	100	30	40	20
Beef, lamb, pork	100	45	44	2-6
Fish, raw[i]				
Salmon	9	2	2	4
Tuna	5	2	1	2
Mackerel	13	5	3	4
Herring, Pacific	13	4	2	3
Nuts				
Walnuts, English.	64	4	10	40
Walnuts, black.	60	4	21	28
Brazil	67	13	32	17
Peanuts or peanut butter.	51	9	25	14
Pecan	65	4-6	33-48	9-24
Egg yolk	31	10	13	2
Avocado	16	3	7	2

Table spanning header: Fatty Acids[b] — Unsaturated (Oleic[d], Linoleic[e])

Source: *Agriculture Information Bulletin*, no. 361. 1974. U.S. Department of Agriculture. Agricultural Research Service. Washington, D.C.

[a] In decreasing order of linoleic acid content within each group of similar foods.

[b] Total is not expected to equal "total fat."

[c] Includes fatty acids with chains from 8 to 18 carbon atoms.

[d] Monounsaturated.

[e] Polyunsaturated.

[f] Suitable as salad oil; not recommended as cooking oil.

[g] Does not include the isomers of oleic or linoleic acid for which nutritional significance has not been established.

[h] Does not include small amounts of monounsaturated and diunsaturated fatty acids that are not oleic or linoleic.

[i] Linoleic acid includes higher polyunsaturated fatty acids.

These two reasons alone demand the use of a class of preservatives called anti-oxidants (see Chapter VI). Many unsaturated fats and oils on the market would be rank-tasting, nutritionally poor foods within hours without suitable additives to stem the attack of oxygen on unsaturated linkages.

Another characteristic of fatty acids may also be of biological importance. This has to do with the shape of the fatty acid chain. It has been pictured here as straight and kinkless, which is an oversimplification. Actually, unsaturated fatty acids have kinks in them, right at the spot where the unsaturated linkage occurs. Here, the chain may fold back on itself or it may merely step down, one step for each unsaturated linkage. Following are two fatty acids that differ in appearance:

$$(1) \quad \text{O-O-O- etc-O} \atop \qquad\qquad \overset{\|}{\text{etc O-O-O}}$$

$$(2) \quad \text{O-O-O-etc-O} \atop \qquad\qquad \overset{\|}{\text{O-O-O-etc -O-}}$$

Number (1) has a kink that folds the chain back on itself. This is called the *cis* form. Number (2) takes a step down, a physical shape referred to as the *trans* form. With more than one unsaturated linkage, *cis* and *trans* shapes may be combined. If the chain is depicted as a straight line, the various shapes possible are as shown on page 197.

The reason for showing these various unsaturated fatty acid forms is to help explain one scientific theory concerning fatty acids as they relate to the building of cholesterol in the body. Cholesterol is the compound thought to clog the arteries and create one kind of heart ailment. As advanced by Dr. F. A. Kummerow of the University of Illinois, the theory has it that *trans* forms of fatty acids lead to higher levels of cholesterol than do *cis* forms. In any one food, both the percentages of cis to trans forms *and* the amount of unsaturated fat may be important, according to Kummerow. If this is true, some altering of formulation and processing techniques may be needed to maximize the health stabilizing factors of products high in fat such as shortening and margarines.

Still, it must always be remembered that theories abound about any problem of such great interest as heart disease, a major killer in the United States. It is not the purpose here to evaluate theories, but to help explain them and assist the concerned individual in making appropriate use of label information when purchasing food.

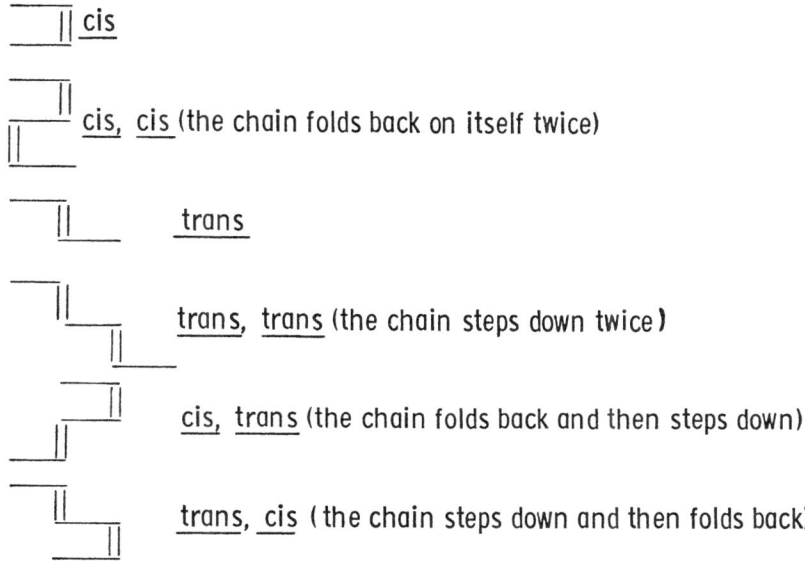

cis

cis, cis (the chain folds back on itself twice)

trans

trans, trans (the chain steps down twice)

cis, trans (the chain folds back and then steps down)

trans, cis (the chain steps down and then folds back)

Unsaturated Fatty Acid Forms

Cholesterol and Heart Disease

Cholesterol has been given such a bad name of late that it is almost impossible to think of it as an essential body chemical, which indeed it is. Problems apparently arise because the body is an efficient producer of this chemical and doesn't need as much from dietary sources as we tend to feed it. Cholesterol accumulates, therefore, and the well-known disease symptoms appear.

Cholesterol is found in fatty foods, in a number of animal tissues, and in blood. It is a *sterol*, as the name implies. Animal food products are major sources of cholesterol in the diet. Cholesterol (also phospholipids and fat soluble vitamins) is associated with fat because fats (triglycerides) are carriers of cholesterol.

Generally lost in any discussion of fats and cholesterol is the fact of their importance to human life. Healthy persons readily use fat, both animal and vegetable. Cholesterol, while not as necessary as fat from a dietary standpoint, serves major functions in the brain and nervous system. It is found in all animal cells.

Foods common to the American diet usually supply 600 to 900 milligrams of cholesterol daily. On low cholesterol diets, this level may be about 300 milligrams. Again, it is important to remember that the role of cholesterol in

Table 29. Cholesterol Content of Servings of Selected Foods (in Ascending Order)

Food	Amount	Cholesterol (in mg)
Milk, skim, fluid, or reconstituted dry . .	1 cup	5
Cottage cheese, uncreamed	½ cup	7
Lard	1 tablespoon	12
Cream, light table	1 fluid ounce	20
Cottage cheese, creamed	½ cup	24
Cream, half and half	¼ cup	26
Ice cream, regular, approximately 10% fat.	½ cup	27
Cheese, Cheddar.	1 ounce	28
Milk, whole	1 cup	34
Butter	1 tablespoon	35
Oysters, salmon	3 ounces, cooked	40
Clams, halibut, tuna	3 ounces, cooked	55
Chicken, turkey, light meat	3 ounces, cooked	67
Beef, pork, lobster, chicken, turkey, dark meat.	3 ounces, cooked	75
Lamb, veal, crab.	3 ounces, cooked	85
Shrimp	3 ounces, cooked	130
Heart (beef)	3 ounces, cooked	230
Egg	1 yolk or 1 egg	250
Liver (beef, calf, hog, lamb)	3 ounces, cooked	370
Kidney	3 ounces, cooked	680
Brains.	3 ounces, raw	More than 1700

Source: R. M. Feeley, P. E. Criner, and B. K. Watt. 1972. Cholesterol content of foods. J. Am. Diet. Assoc. 61:134.

dietary disease is not completely understood or absolutely established. For persons interested in the content of cholesterol in various foods, some comparisons are shown in Table 29.

What scientists say about dietary fat and cholesterol. Only one thing seems certain after all these years of study of heart disease: No longer is anyone willing to single out a specific food component or a specific cause of heart disease. Too many variables have come to light. Even diet alone, including all foods, cannot be blamed entirely, not when smoking habits, exercise, and high blood pressure are also taken into account. Some of the factors thought to be involved now include (in every and all combinations of individual dietary and living habits), but in no way imply a priority of order or a complete list, the following:

*General state of health.

*Food components (fat, fatty acids, cholesterol, sugar, sodium).

*Exercise, the lack of regular physical activity.

*Dietary fiber intake—a more recent addition to the list, thought to reduce (control, balance) blood lipid and cholesterol level.

*Soft water—a possible risk to heart victims, perhaps as a result of high-sodium content. Water softeners exchange sodium for hard water salts. Some waters are naturally high in sodium content. Evidence that hard water may be a risk has also been reported.

*Xanthine oxidase from homogenized cow's milk—thought possibly to be absorbed intact from gastrointestinal tract into the body, there to be transported by the lymph system to heart arteries where cell membranes are altered and atherosclerosis is initiated. As yet, this theory has not been widely accepted.

*Yogurt—found to reduce cholesterol level of African tribesmen on high cholesterol diets.

*Sucrose (sugar)—exposure (at weaning) of carbohydrate-sensitive rats to sucrose apparently leads to high blood cholesterol in adult rat population.

*Vanadium (a mineral)—significant lowering of cholesterol has been found in young men on a diet containing 125 mg of vanadium daily. Other elements may also be important in this regard.

*Hereditary influences.

*Emotional makeup.

A report by the Inter-Society Commission for Heart Disease Resources offers a number of recommendations helpful in preventing heart disease that can be started in early childhood: (1) consume calories in those amounts allowing for optimum weight; (2) reduce the intake of fat to 35 percent of total calories; (3) consume saturated and polyunsaturated fats in amounts of less than 10 percent each of total fat calories, the remaining fat to be mono-unsaturated (one double bond); (4) restrict daily intake of cholesterol to 300 milligrams or less; (5) control hypertension; (6) quit smoking.

Label information on fat, fatty acid, and cholesterol content. Not all nutritional labels will bear information on the content of fat, fatty acid, and cholesterol, as already mentioned. Those that do are usually found on foods that tend to contribute substantial amounts of fat to the diet, and they will provide the data according to standard form. Figure 4 is an example of a label showing fatty acid and cholesterol declaration following the standard format. Several facts should be noted:

*Fat is labeled both in grams per serving to the nearest gram *and as the percentage of calories.*

*Following the information on fat is the content of polyunsaturated fatty acid (labeled simply polyunsaturated) to the nearest gram. These are fatty acids with more than one unsaturated linkage. Testing procedures require that, for this specific label information, the amount of *cis, cis* methylene in-

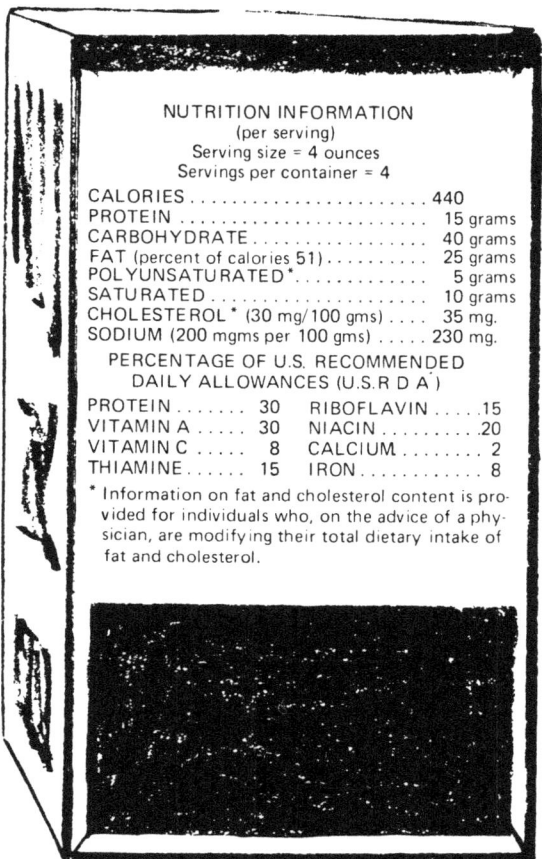

NUTRITION INFORMATION
(per serving)
Serving size = 4 ounces
Servings per container = 4

CALORIES 440
PROTEIN . . . . 15 grams
CARBOHYDRATE . . . . 40 grams
FAT (percent of calories 51) . . . . 25 grams
POLYUNSATURATED* . . . . 5 grams
SATURATED . . . . 10 grams
CHOLESTEROL* (30 mg/100 gms) . . . . 35 mg.
SODIUM (200 mgms per 100 gms) . . . . . 230 mg.

PERCENTAGE OF U.S. RECOMMENDED
DAILY ALLOWANCES (U.S.R D A)

PROTEIN 30   RIBOFLAVIN 15
VITAMIN A 30   NIACIN 20
VITAMIN C 8   CALCIUM 2
THIAMINE 15   IRON 8

* Information on fat and cholesterol content is provided for individuals who, on the advice of a phy-
sician, are modifying their total dietary intake of
fat and cholesterol.

Figure 4. Label Showing Fatty Acid and Cholesterol
Declaration by the Standard Format

terrupted fatty acids be measured. The unsaturation, then, is of the cis, cis configuration, generally thought best in maintaining low levels of cholesterol.

*The content of saturated fatty acid follows the listing of unsaturated fatty acid. This second type of fatty acid is declared to the nearest gram and includes the sum of lauric, myristic, palmitic, and stearic acids. These acids are saturated (contain no double bonds) and are twelve, fourteen, sixteen, and eighteen carbon linkages in length, respectively. They are common saturated fatty acids in foods.

*Data on cholesterol are declared either as milligrams per 100 grams or as milligrams per serving. For anyone who must restrict the intake of cholesterol, the milligrams per serving may be used in totaling the amount consumed.

*The declaratory statement appearing at the bottom of the label will be present whenever information about fatty acid or cholesterol is shown.

*In this representation of a label, the content of sodium is also shown. Sodium is a concern for heart patients, and though it need not be declared, it will appear as shown (following information on cholesterol) whenever it is declared.

*Information on labels should not be taken too literally. Regulations allow some leeway in the values that are declared. The content of fat, fatty acid, cholesterol, and calories may be understated by as much as 20 percent and still comply fully with regulations. Labeling information, like RDA's, should generally be considered as "guides" to daily nutrient intake. Guides can and should be helpful, but a literal translation of them can be misleading and perhaps even harmful.

Vitamins and Minerals

The list of nutrients on a label starts with the percentage of RDA for protein. Following protein are, as mandatory nutrients to be declared, five vitamins and two minerals. They will always appear in the order shown, i.e., vitamin A, vitamin C, thiamine, riboflavin, niacin, calcium, and iron.

Vitamins can be grouped into two useful categories relating to their source. They are either fat soluble or water soluble. Those that are soluble in fat may be expected to be found in fatty foods (unless lowfat or nonfat foods such as skim milk and nonfat dry milk are fortified with these vitamins, as is sometimes the case). Fat soluble vitamins include vitamins A, D, K, and E. Other vitamins are water soluble.

Vitamins may be considered as complex organic compounds needed by the body to hasten (catalyze) certain necessary metabolic functions. In a sense vitamins are the governors and regulators that assist other nutrients in the performance of their various functions. For the purposes of this chapter, vitamins will be discussed in the order in which they appear on the nutritional label.

Vitamin A

The two major forms of this vitamin are the preformed vitamin found only in animal products and substances with provitamin A activity. The first may be considered vitamin A as such; the others are compounds readily changed to vitamin A, though with some loss of overall efficiency. Of the latter group, carotenes are the most important provitamin A sources, and betacarotene, the yellowish pigment in carrots and squash, the most efficient of the lot. Both yellow and green vegetables are sources of carotene.

Vitamin A is poorly absorbed except in the presence of fat. It is sensitive to light, and foods containing this vitamin should be stored in the dark or in containers that screen out both sunlight and light from a refrigerator.

Preformed vitamin A, called retinol, may some day lead to a new system of rating foods for vitamin A content. Now expressed as International Units (I.U.), carotene and other provitamin A substances would be converted by the new method to retinol equivalents from which the total content of vitamin A could be determined by adding on any retinol also present. Under this system, 1 I.U. of vitamin A activity is equivalent to 0.3 mg of retinol. On the basis of absorption and actual utilization by the body as vitamin A, 1 retinol equivalent is assumed to equal 1 mg retinol. Other equivalencies become (as 1 retinol equivalent): 6 mg betacarotene, 12 mg other provitamin A carotenoids, 3.33 I.U. vitamin activity from retinol, 10 I.U. vitamin A activity from betacarotene.

The RDA for vitamin A, for labeling purposes, is 5000 I.U.; that is, the percentage of RDA values for this vitamin will be taken as a percentage of 5000 I.U.

Assuming 2500 I.U. of retinol and 2500 I.U. of provitamin A, the retinol equivalents are 750 mg retinol and 250 retinol equivalents of betacarotene, for a total of 1000 retinol equivalents. In other words, 5000 I.U. equals 1000 retinol equivalents, which is another way of expressing RDA.

Vitamin A is essential for night vision and for healthy skin (including the linings of the nose, mouth, and inner organs). A prolonged lack of this vitamin leads ultimately to blindness. Needs rise slightly during pregnancy and lactation. Vitamin A is required for growth in children, for the normal development of fetuses, and especially for the proper development of bones and teeth. Very important, considering the interest in some fad diets, is the fact that excessive intake of preformed vitamin A can be toxic (see Chapter IX). Even amounts of 7000 to 10,000 I.U. of this vitamin daily should be consumed only under medical supervision.

Sources of vitamin A. Among the richest sources of vitamin A are liver, dark green vegetables, butter, vitamin A fortified margarine, fortified skim and lowfat milk, and yellow fruit. A list of a number of sources of vitamin A follows. For the milk products it is necessary to check the label to determine if vitamin A has been added.

Sources of Vitamin A

Vegetable and Fruit Products	Animal Products
Dandelion greens	Liver
Carrots	Butter
Turnip greens	Cheese
Spinach	Eggs (whole)
Sweet potatoes	Fluid, evaporated,
Beet greens	dry whole, and
Squash	skim milk products
Broccoli	
Cantaloupe	
Tomatoes	
Apricots	
Asparagus	
Kale	
Mustard greens	
Chard	
Pumpkin	

Vitamin C

This water-soluble vitamin has been highly publicized as a preventive and cure for the common cold. Doses of vitamin C suggested for such purposes are extremely large and far exceed the RDA of the Food and Nutrition Board of the National Academy of Sciences. Even though some evidence of therapeutic value exists, it appears that the public mood in general favors exaggerated amounts of all vitamins as desirable and extra healthful. Taken to extremes, such thinking can be a serious risk to health. In moderation, assuming some small excesses, the good done by vitamins is probably more psychological than physiological. Nevertheless, these are days of high expectations of vitamins generally, and vitamin C would appear to be the prime example.

Vitamin C is essential in preventing scurvy, a rare disease in the United States which primarily affects infants and young children whose diets focus narrowly on milk. This vitamin also functions in the production of collagen which serves in the healing process of cuts and wounds. As well, it plays a role in the formation of bones and teeth.

Vitamin C is one of the nutrients readily destroyed by a variety of processes related to oxidation. Cooking in the presence of oxygen or minute amounts of copper or iron favors oxidation and loss of vitamin C. Because of its oxidizable nature, vitamin C (ascorbic acid) is sometimes added to food purely for technical reasons, to retard the oxidation of other ingredients by undergoing oxidation itself.

On food labels, the RDA for this vitamin is set at 60 milligrams. Larger amounts may be of some help in overcoming symptoms of the common cold, but regular, massive intake appears unwarranted and should be prescribed only on advice and under supervision of a medical doctor. (As more research is reported, two trends appear to be taking shape: evidence is mounting that vitamin C has little value in preventing or curing the common cold; and doses of vitamin C deemed necessary to alleviate cold symptoms are being steadily lowered, ultimately to levels we might assume are consistent with present NAS/NRC recommendations, i.e., approaching 45 mg per day.) The same could be said of all vitamins, of course, but because vitamin C has been so highly publicized, it offers a better than average opportunity for misuse. One of the latest warnings in this regard, from biochemist I. J. Wilk, points out the potential for vitamin C tablets to break down to unhealthful end-products upon prolonged storage (one year). Oxalic acid, one such compound, is a substance thought to play a role in urinary infections and to speed the formation of kidney stones. Moreover, the bulk of vitamin C tablets is filler material consisting of various sugars and other compounds. Since vitamin C degradation results in the formation of at least two sugar compounds, the collective sugar content poses a possible threat to diabetics, particularly when the vitamin is consumed at exaggerated levels.

Still another finding, by North Dakota researcher Dr. Leslie Klevay, suggests the possibility of some slight risk to the heart from massive intake of vitamin C. Theory (along with data from the study of rats) has it that vitamin C decreases trace amounts of copper ordinarily absorbed from food. A resulting imbalance of zinc and copper may then lead to increased levels of cholesterol.

Sources of vitamin C. This vitamin is present in a large number of green vegetables and in most fruits. Potatoes are also a good source. Moreover, it is a nutrient commonly added to foods, particularly fruit drinks and other breakfast beverages. Good sources of vitamin C are shown in the following list. It is necessary to consult the label to determine whether fruit drinks have been fortified with vitamin C.

Thiamine (Vitamin B$_1$)

Thiamine, also soluble in water, was the first vitamin to be discovered. It serves in the energy cycles by which the body uses food, in the normal opera-

Sources of Vitamin C

Vegetable Products	Fruit Products	Animal Products
Turnip greens	Strawberries	Liver
Green peppers	Oranges	
Broccoli	Grapefruit	
Kale	Tangerines	
Mustard greens	Limes	
Collards	Fruit juices	
Brussels sprouts	Fruit drinks	
Cauliflower		
Spinach		
Cabbage		
Rutabagas		
Beet greens		
Lima beans, green		
Radishes		
Sweet potatoes		
Potatoes		
Tomato juice		
(fortified with		
vitamin C)		

tion of the nervous system, and in growth and digestion. Thiamine encourages good appetites and helps to prevent fatigue and irritability. A severe lack of vitamin B_1 causes the disease beriberi. Symptoms of deficiency include loss of appetite, stomach pains, and constipation. Muscle tone is affected; nervous disorders may result in convulsions and even paralysis. When polished rice is a major food staple, disease symptoms are fairly common. They may also occur, as is the case with most vitamins, when alcohol becomes a major source of energy.

The need for thiamine is generally higher during the last two months of pregnancy and during lactation. The adult RDA established on food labels is 1.5 milligrams.

Sources of thiamine. Good sources of this vitamin include enriched breads and cereals, whole grains, pork, heart, liver, dry peas, kidney beans, and other dry beans. Important sources of thiamine are shown in the following list.

Sources of Thiamine (Vitamin B_1)

Vegetable Products	Cereal Grain and Bread Flour Products	Nut Products	Animal Products
Soybeans	Wheat germ	Pecans	Pork (lean)
Split peas	Soy flour	Walnuts	Heart
Beans (kidney, lima)	Oatmeal	Peanuts	Bacon
Peas	Rice flour		Fluid and dry milk products
Potatoes	Enriched wheat flour		Liver (pork, calf, beef)
	Enriched farina		
	Cornmeal		
	Whole wheat bread		
	Other enriched bread products		
	Buckwheat		
	Fortified cereal products		
	Brown rice or enriched rice		

Riboflavin (Vitamin B_2)

Without riboflavin the skin becomes greasy and scaly, cracks develop at the corners of the mouth, cataracts may form on the eyes, eyesight dims, growth is poor, birth defects occur, and muscular strength and coordination are lowered. The symptoms of its deficiency sound like a combination of general malnutrition and heavy metal poisoning, which indicate the widespread functions of this vitamin in the body. Entering metabolic pathways of oxidation, vitamin B_2, as an enzyme helpmate (a coenzyme), engineers those biochemical activities required to produce energy. If a power drill, sander, or saw is needed to convert electrical power to a meaningful use of energy, then riboflavin fills a similar function in the body. Without it, food energy is merely stored, unusable, and unuseful.

Riboflavin is quite stable in the presence of heat and acid. It survives most heat treatments needed to preserve food; it also tolerates the acid environment of certain food systems. But alkali and light both destroy vitamin B_2. Milk stored in clear glass bottles (or pitchers) and exposed to sunlight or even fluorescent light can undergo loss of active riboflavin. And milk is a major source of this vitamin, as are eggs and liver.

Riboflavin is water soluble. The RDA is set at 1.7 milligrams. Needs go up during pregnancy and lactation. In the United States riboflavin needs rarely go unfilled, although there may well be deficiency cases, many of which go unrecognized. Deficiencies are widespread in rice-consuming nations, for rice is not a good source of this vitamin. Cereal grains generally are not major sources of riboflavin except where fortification is practiced, as it is in the United States.

Sources of riboflavin. Riboflavin may be obtained from both animal and vegetable products. It is abundant in milk, liver, and heart, in enriched cereal grains and flours (bread), and in beans and dark leafy greens like turnip greens.

Sources of Riboflavin

Animal Products	Cereal and Vegetable Products
Liver	Wheat germ
Heart	Turnip and other
Kidney	dark greens
Milk (fluid, dry,	Beans
evaporated)	Spinach
Cheese	Broccoli
Eggs	Soy flour
Cocoa	Enriched bread
Bologna	Whole wheat bread
Salmon (canned)	Fortified cereals
Lamb	Enriched farina
Beef	Enriched flour
Pork (lean)	Cornmeal, whole grain
	Yeast

Niacin

Niacin, another water-soluble vitamin, is the last of the vitamins specified for mandatory listing on food labels. Other vitamins, besides the five required by regulation, are considered essential; RDA's have been established and their presence in food, when measurable amounts exist (per serving), may also be declared, but this is not mandatory *unless they are added to foods.*

Pellagra, the deficiency disease of niacin, shows up as a skin ailment resembling sunburn. Diarrhea (watery, bloody stools) occurs, and nervous disorders appear, including headaches, burning sensations in the hands and feet, and motor difficulties such as walking. So short a time ago as the first quarter of the twentieth century pellagra was a fairly common disease in the United States.

Like riboflavin, niacin enters the body in oxidation processes that produce energy. Problems usually develop when the diet is deficient in protein. A diet consisting mainly of corn will lead to a severe deficiency for reason of niacin shortages and a missing protein component, the amino acid tryptophan, which can substitute to some extent for the vitamin. About 60 milligrams of tryptophan convert to 1 milligram of niacin.

The RDA calls for 20 milligrams of niacin for adults, thus the need for tryptophan-rich protein and/or dietary niacin. The average American diet provides the equivalent of 8 to 16 milligrams of niacin from tryptophan in addition to 8 to 17 milligrams of the vitamin itself. This is important for it indicates that daily requirements are being met and can be met without resorting to vitamin supplementation (a vitamin tablet over and above food consumed). An excess of this vitamin may cause unpleasant itching or burning.

Niacin survives cooking very well as it is resistant to heat, acid, alkali, and also light. In cereals, much of the natural niacin is thought to be tied up (bound) and, therefore, unavailable to man.

Sources of niacin. Niacin is found in fish, meat, and poultry, in enriched flour, bread, and cereal products, and in nut products. A breakdown of sources follows.

Sources of Niacin

Animal and Related Products	Nut Products	Vegetable and Cereal Products
Liver	Almonds	Avocadoes
Salmon	Peanuts	Enriched bread
Tuna fish	Peanut butter	Enriched flour
Pork		Whole wheat bread
Veal		Cornmeal
Beef		Peas
Turkey		Rice
Chicken		Tomato catsup
Heart		Wheat cereals
Lamb		Wheat germ
Tongue		Enriched cereals
Sardines		Potatoes
Eggs		

The Minerals

Two minerals will always be acknowledged on labels bearing information on nutrition. They are calcium and iron. Other minerals are essential and may be listed when present naturally; it is mandatory to list them when and if they are added to foods. They will appear in the order presented in Table 26.

Calcium (Phosphorus and Vitamin D)

Without calcium man would be a shapeless blob, for, in its various forms, this mineral makes up skeletal bones. Nearly all (99.3 percent) calcium resides in the body in skeletal deposits, though in a dynamic flux that sees the mineral enter and exit the bones every day of life. And though the amount found outside the bones is small, it is no less essential as part of the vigor in muscle contractions, as the spark of nerve messages transmitted about the body, and as an agent in the clotting of blood. Calcium is a most important dietary nutrient.

To speak of the metabolism of calcium without mentioning phosphorus and vitamin D is impossible. They must be considered together, as a nutrient team, with phosphorus (as phosphate) an integral part of the chemical form in which calcium is laid down, and vitamin D the key to the efficient absorption of calcium and phosphorus into the body. To consume vitamin D along with calcium and phosphorus makes good nutritional sense; thus milk, the main food source of calcium, is almost universally fortified with vitamin D. The other major factor promoting the manufacture of vitamin D in the body is not food at all, but sunshine.

Though necessary for the utilization of calcium, phosphorus can interfere with absorption if present in excess *in absence of adequate vitamin D.* A calcium:phosphate ratio of 1:1 is considered ideal; ratios outside 2:1 and 1:2 are readily handled in the presence of ample amounts of fat-soluble vitamin D.

As discussed in the chapter on antinutritional factors in food, the oxalic and phytic acid of certain natural foodstuffs can combine with calcium to form insoluble (thus unavailable) salts. Present thinking holds this nutritional interference as unimportant as long as the intake of calcium is ample. During early infancy a calcium:phosphorus ratio of 1:5 is now recommended. Protein, too, plays a role in the utilization of calcium, a high-protein diet resulting in the greater excretion (poorer absorption) of calcium. As a result, an individual's intake of protein should be figured into estimates of his daily needs. In the United States, protein intake is thought to be somewhat too high anyway and, as one side effect, could tend to increase the need for calcium. Though not yet elucidated, high-protein, low-calcium diets may figure in the bone disease osteoporosis.

The label RDA for calcium is 1.0 gram, an ample amount for meeting most adult needs. Intakes of half that rate are considered a "practical allowance" by the Food and Agriculture Organization/World Health Organization Committee on Calcium Requirements.

Sources of calcium, phosphorus, and vitamin D. Good sources of these nutrients include the following.

Animal and Poultry Products	*Vegetable, Fruit, and Nut Products*
Fortified milk	Beans
Fortified lowfat milk	Mustard greens
Fortified skim milk	Turnip greens
Cheese	Soybeans
Ice cream	String beans
Fortified nonfat dry milk	Dates
Eggs (yolk)	Olives (green)
	Oranges
	Kale

Iron

Of the major nutrients listed on food labels, iron is the only one whose requirements may not be readily met through ordinary food sources. Though iron is present in a number of foods and is a mineral enrichment of flour and bread, the needs of infants, because of the low content of iron in milk, the needs of menstruating females, because of loss of blood, and the needs of pregnant females, because of the demands of the growing fetus, are not always adequately satisfied through dietary iron. The syndromes so common to television advertising—"tired blood," "iron deficiency anemia," "that tired, rundown feeling"—become reasonably valid expressions of the resulting state of health.

Iron is a part of the blood system. With the help of copper it combines with protein to form hemoglobin, a carrier of oxygen. Iron in combination with other protein takes the form of enzymes (cytochrome, catalase, peroxidase) that trigger oxidizing processes.

But if iron is essential and somewhat difficult to come by, there is reason to believe that an excess of iron may serve no good purpose and may in fact be detrimental to health. Besides being a health hazard to persons suffering iron storage disorders and possibly a contributing factor to Parkinson's disease in certain individuals, excess iron may mask test methods used to diagnose diseases associated with anemia, among these intestinal cancer. It may also raise the body's levels of hematocrit and hemoglobin, which are known to be associated with death in some instances. It was because of these considerations that a proposed regulatory increase in iron enrichment of flour and bread was eventually stayed.

Because of iron's unique metabolism in the body, another important consideration—with implications for other nutrients—is of more than passing interest. With iron, there is a major question of biological availability. Some professional nutritionists are concerned that fortification alone may not be an effective way of treating iron deficiency anemia solely because the form in which iron is used to fortify foods may not be readily utilized by the body. If true of iron, it is also possibly the case in foods fortified with essential amino acids to enhance protein quality. Iron and protein nutrition are the two most striking examples of dietary considerations related specifically to nutrient availability. Certainly these—and perhaps the metabolic needs for certain other nutrients—should always be kept very much in mind in all attempts to formulate foods from whatever source.

For this mineral, the label RDA is 18 milligrams. This intake is more than adequate for most males and is adequate for the dietary needs of most females, including specific requirements imposed by menstruation, pregnancy, and lactation. A fairly well balanced diet should bring in 9 to 12 milligrams of iron each day. Some concentration in iron-rich foods (meat) with a like increase in foods known to enhance the absorption of iron (citrus fruit and meat) will assure a maximum intake. During pregnancy, iron supplementation is recommended at levels of 30 to 60 milligrams daily.

It is well to note again a trend in our patterns of food consumption toward the intake of less meat (dependent upon price, availability, and commitment to world food needs). Without the additional enrichment of nonmeat foods, the overall consumption of iron from food sources can be expected to decrease, thus creating a possible increased need for adding iron to the diet through mineral supplements.

Sources of iron. With the possible exception of meat, in which iron is known to be in a form usable by the body, a list of sources cannot ensure the availability of iron in the items included. The mere presence of iron is not in itself assurance that the body can or will use it. Those iron forms that occur in eggs or vegetable products may or may not be available to the body depending upon the influence of other food ingredients. One can only assert, as for all other nutritional needs, that a varied diet is the best guarantee of meeting daily iron requirements.

Those foods known to contain reasonably high levels of iron are shown in the following list.

Sources of Iron

Animal, Fish, and Poultry Products	*Vegetable Products*	*Grain and Other Products*
Liver (pork, calf, beef)	Soybeans	Enriched bread
Heart	Spinach	Enriched flour
Beef (lean)	Dried beans	Enriched macaroni
Pork (lean)	Lima beans	products
Lamb	Peas	Fortified breakfast
Oysters and other	Chard	cereals
shellfish	Beet greens	Oat cereal
	Asparagus	Brown rice
		Almonds
		Prunes

Other Essential Nutrients

A glance at Table 26 will uncover some twelve vitamins and minerals which, known to be essential at rather precise levels, fall outside mandatory requirements for listing on the nutritional label unless the food is fortified or enriched with them. A brief analysis of each seems noteworthy. In the order in which they appear on the label, here's a view of their essential features.

Vitamin D. This vitamin is needed for the absorption of calcium (see the discussion of calcium in this chapter); thus milk, the main food source of calcium, is commonly fortified with vitamin D. However, the single most common source is not food at all, but sunshine. More accurately, 7-dehydrocholesterol found in skin converts to active vitamin D when exposed to shortwave ultraviolet sunlight rays. So important is sunlight as a source of this vitamin that rickets, the vitamin D deficiency disease of children, was formerly noted to occur rather frequently in communities blanketed by smog. That was before the fortification of milk became common practice. Now the disease is rare in the United States, though smog continues to exert an unhealthy influence in other ways.

The RDA of vitamin D is 400 I.U. This amount is known to be ample for infants during the rapid period of growth of the first six months of life. Needs differ, and the NAS/NRC Food and Nutrition Board indicates that, for most infants, 100 I.U. would suffice. The higher RDA, then, provides considerable safeguards for varying needs. Larger doses, as already mentioned, can be seriously toxic. Since individual sensitivities are different, no one can state with certainty what specific level of intake will be toxic. Simply keep in mind that toxic levels are readily attainable through a nominal overdose of supplements. Numerous cases of hypercalcemia (with serious consequences to children) are thought to have resulted from the overuse of vitamin D in England and Europe during the 1940s and 1950s. A reduction in intake suggested at that time has now made such cases rare in those countries.

Adult requirements of vitamin D are less well established than they are for children, though past experience suggests, and there is no evidence indicating otherwise, that the amount needed in infancy carries through to the adult years. Symptoms of deficiency in adults have been observed only in persons on restricted diets in which the intake of vitamin D is less than 100 I.U. For normal children, adults, and pregnant and lactating women, 400 I.U. should be adequate.

Besides milk, other food sources of vitamin D include egg yolk, salmon, sardines, herring, and tuna.

Vitamin E. This is another fat-soluble vitamin, which is a naturally occurring antioxidant necessary for healthy reproductive processes. Although essential, it appears to be receiving exaggerated acclaim as the "virility" vitamin. In food it is derived mainly from several plant sources through the tocopherols and tocotrienols. Large doses over prolonged periods may, like the fat-soluble vitamins A and D, be toxic. Such overuse is not recommended, though proof of toxicity remains undetermined. Because the toxicity of nonnutrients (poisons) is related to their chemical structure (similar structures being similarly poisonous), reasonable caution would seem well advised for vitamins of like properties or characteristics where known toxicities occur within the group.

Alpha-tocopherol, the most potent chemical form of vitamin E, is found in vegetable oils and is especially high in corn, coconut, cottonseed, peanut, and soybean oil. Margarine and shortening prepared from these oils are generally good sources of vitamin E. It may also be obtained in lesser amounts through beef steak, beef liver, chicken, and pork chops. The label RDA is 30 I.U.

Vitamin B_6. Though expressed as a single vitamin, B_6 is perhaps more properly considered a group of chemically related compounds called pyridines. Without this vitamin, infants may experience epileptic convulsions along with weight loss, stomach cramps, and vomiting. Depression and confusion are among adult symptoms.

Data on food sources of B_6 compounds are inadequate. Foods thought to provide the vitamin are meat and poultry products (especially beef liver), fish, wheat germ, and cheese.

Folic acid (folacin). Folic acic, or folacin as it is generally known, is a water-soluble vitamin functioning in the transfer of single carbon atoms and, with vitamin B_{12}, in the formation of nucleic acid. The absence of this vitamin invariably leads to a type of anemia involving large, nucleated blood cells called megaloblasts.

In foods, folic acid occurs either as "free" folacin or as polyglutamates which must be further metabolized to the active vitamin. The free form consists of about 15 percent of the total folacin activity in food.

Folacin is found in a great variety of foods: liver, asparagus, carrots, cheese, potatoes, round steak, kidney beans, peanuts, wheat and wheat germ, chicken, salmon, peaches, and other fruit. A varied diet guarantees sufficient intake of folacin. Though the label RDA is set at 0.4 milligrams, needs are known to double during pregnancy, for a special increase is required to prevent injury to the unborn infant. Needs are also greater throughout breast feeding.

Vitamin B_{12}. Also known as "cobalamin" because of its content of cobalt, this vitamin is essential for the proper functioning of all mammalian cells. Along with folacin, vitamin B_{12} serves as a coenzyme in the production of nucleic acid, the life-giving material in animal cells. Perhaps because of its need and presence in animal cells, animal products are the only source of this vitamin. Therefore, some meat, milk, cheese, and/or eggs should be included in the daily menu. Purist vegetarians ought particularly to take note.

At 6.0 micrograms, the label RDA of B_{12} will usually be met in the average American diet, which ordinarily supplies between 5.0 and 15.0 micrograms of the vitamin daily. However, the level may reach as low as 1.0 micrograms in some diets or as high as 100 micrograms in others. Recommended allowances were lowered considerably between 1968 and 1974, based primarily on updated research findings.

Phosphorus. Deficiency in this nutrient is rarely encountered in the United States and most other countries. Adequate intake is assured because phosphorus is widely distributed in foods and is also a common element of a variety of additives. Excesses, which could inhibit the utilization of calcium, are tolerable in the presence of adequate amounts of vitamin D.

Phosphorus not only contributes to the structure of bones, but it is a constituent, in the ionic state, of blood and cells. It is found in protein, carbohydrate, and lipids of the body, and forms an integral part of enzymes that transfer energy. The label RDA, at 1.0 grams (the same as calcium), is readily obtained through such diverse foods as milk, cheese, turkey, wheat, tuna fish, bread, almonds, beans, rice, and peas.

Iodine. This mineral element causes no dietary trouble where iodized salt may be had. The deficiency disease, thyroid enlargement (goiter), occurs spottily in the United States, but is more of a problem elsewhere around the world.

The main function of iodine lies in the production of thyroxin by the thyroid gland. Thyroxin governs a number of bodily activities, the most important being the rate at which energy is expended. With a sluggish thyroid gland, the body tends to store food energy as fat; with an overactive thyroid, food reserves are rapidly burned up, and the body becomes lean.

Throughout the world (though probably not in the United States to any great extent), the adequacy of a person's intake of iodine may be complicated by goitrogens, which are natural constituents of some basic foodstuffs, present as a result of either improper (or no) processing or of a lack of technical know-how concerning methods used to rid foods of these compounds. Goitrogens are found in a number of plant products of importance around the world, soybean and rapeseed being two of them. When consumed in large amounts (perhaps because of a lack of variety in the diet), goitrogens increase the need for iodine to maintain proper functioning of the thyroid.

With a label RDA at 150 micrograms, adequate intake of iodine can be assured through the use of iodized salt or from seafoods (excellent sources of iodine), dairy products, eggs, and possibly bread. Vegetable products, though providing some iodine, are not good sources. This is another nutritional concern for the strict vegetarian.

Magnesium. In the mineral content of living cells, this essential element takes second place only to potassium. It is found in many enzyme systems and seems to share a metabolic role with calcium. Lacking magnesium, persons become nervous and excitable, bones and teeth lose strength, and behavioral changes may be noted. Excess magnesium in the form of magnesium salts has a laxative effect. (Epsom salts, the common expression for these magnesium compounds, were so named because they were originally prepared from the mineral spring water of Epsom, England.)

Because of its widespread occurrence in plant products, magnesium deficiencies are thought to occur rarely. Cereal products, nuts, whole wheat bread, cocoa, peas, and beet greens are good sources. Milk and dairy products can also provide significant amounts of this element. The label RDA, at 400 milligrams, should not be difficult to come by.

Zinc. There is a temptation to be less concerned over needs of the body for nutrients required in minute quantities than for those required in large amounts. Zinc is a good example. The label RDA calls for 15 milligrams daily, a very small amount, yet in the absence of this amount dwarfish and late sexual development have been recorded in certain Middle East countries. Such a lack, though to a lesser degree, leads to poor growth and loss of appetite; it can also increase an individual's susceptibility to cadmium poisoning. Zinc is found in several enzyme systems and plays a role in the synthesis of the nuclear materials RNA and DNA. It may serve a function in the healing of wounds; it is involved in the production of certain hormones. No nutrient, however minutely required, can be taken lightly.

Some evidence exists that the intake of zinc may be borderline in small segments of American society. Generally, deficiencies appear to be rare, though possibly increasing. Phytic acid, an antinutritional factor of certain foods, may increase the need for zinc in the human diet.

Dietary needs for zinc may usually be met by an adequate intake of meat, liver, seafood, eggs, milk, and whole grain products (whole wheat bread, oatmeal, and whole corn). Beets, cabbage, and rice can also provide zinc.

Copper. Speaking of individual elements as though they existed as discrete biological forces, devoid of interaction, is overly simplistic and possibly misleading. No element performs its function in a vacuum. Putting it positively, it can be said that all elements interact with other elements and other nutrients, sometimes in complementary, sometimes in antagonistic roles. Copper aids in the absorption of iron; it reacts adversely toward molybdenum and sulfur. Similarly, molybdenum and zinc inhibit the utilization of copper, i.e., high levels of the two elements increase the body's need for copper. Copper also does battle with the ravages of heavy metal poisoning caused by cadmium. Our internal environment is a veritable battleground of ever-changing nutritional struggles. To think of it otherwise is to miss the forest for the trees.

Copper is essential to man in minute amounts (the label RDA is 2 milligrams). It exists in blood combined with protein, is found in bones, and is necessary for the proper functioning of the nervous system. Deficiencies are known, though fairly rare. A lack of copper often manifests symptoms at levels of starvation of protein calories. Oysters, liver, wheat (bread), and corn are rich depositories of dietary copper. Nuts are also good sources, as are beans and raisins.

Biotin. Although biotin is not a mineral, regulatory mandate requires that it be listed directly following copper. As a coenzyme, its main function is the synthesis of fatty acids. It also serves other purposes, one being to tie carbon dioxide into organic compounds within the body. Lacking this vitamin, an individual experiences skin diseases, nausea, and vomiting; a grayish, deathlike color invades the skin; muscle pain develops.

Symptoms of deficiency in humans generally can only be induced by consuming large amounts of the antinutritional factor avidin, which is found in raw egg white.

Because biotin is produced by intestinal bacteria, and because contribution from this source is uncertain, absolute daily needs cannot be precisely stated. The label RDA places requirements at 0.3 milligrams, which should be readily attained through a variety in the menu of fruits, vegetables, grains, meat, poultry, milk, and, for those with a sweet tooth, chocolate.

Pantothenic acid. This is the last nutrient designated for declaration on the labels of food packages. But last place should in no way imply a low priority in human nutrition. Pantothenic acid, water soluble, is widely distributed in foods and serves a critical role in the metabolism of protein, fats, and carbohydrates. As a major component of coenzyme A, this vitamin is essential to acylation reactions—the metabolic process leading to energy production—and to the synthesis of fatty acids and sterols. Though symptoms of deficiency apparently must be induced by the administration of an element antagonistic to pantothenic acid, an extreme lack can result in death.

At present, the paucity of suitable research data makes exact dietary recommendations for pantothenic acid impossible. Adult intake in the United States ranges from 5 to 20 milligrams a day and seems sufficient to meet the label RDA, which is set at 10 milligrams. Sources of pantothenic acid are almost as broad as foodstuffs in general. They include grains (wheat and bread), beef and pork, eggs, potatoes, fish, poultry, and cheese. Other sources, among them beans, cauliflower, broccoli, peanuts, and milk, make dietary deficiencies highly improbable.

Dietary Fiber

Roughage, bulk, and dietary fiber are all terms used to describe components of food that resist the normal digestive process. Rarely will such ingredients be listed on the food label. Regulations do not provide for fiber compounds as a part of nutritional information since they are nonnutrients. Yet no ingredients of food are any more essential or beneficial to human health. Indeed, nondigestibles, and their role in human nutrition, are only now beginning to receive major attention from researchers. The near future will see much said and reported about the role of fiber in the diet.

Dietary fiber should perhaps be distinguished from crude fiber, which is more often reported in dietary manuals and handbooks, but represents a smaller fraction of food components than the former. Crude fiber consists of those food constituents undigested by strong acid and alkali treatment. This is a chemical differentiation. In human nutrition, the digestive process is less severe and a number of compounds not found in crude fiber are in fact considered a part of dietary fiber, i.e., materials not digested by the human. Major components of dietary fiber include cellulose, hemicellulose, mucilages, pectins, gums, and noncarbohydrate lignin. You will recognize some of these compounds as the woody or stringy tissues of vegetables and fruit (two good sources of dietary fiber).

Since the body lacks enzymes for breaking down fibrous materials, the digestion of dietary fiber (incomplete though it may be) is left to bacteria. In the process of bacterial digestion a number of compounds, all useful in human nutrition, are formed. Major breakdown products are fatty acids, water, carbon dioxide, and methane. Together and in combination with undigestible fiber, these compounds account for the major excretory functions of the colon. And it is these varied functions that are increasingly seen as being of major significance in certain dietary-related diseases.

As in other facets of human nutrition, much remains to be learned about the role of fiber in the diet. But already evidence hints strongly that the intake of fiber may be a factor in preventing cancer of the colon and also in lowering cholesterol in the blood serum. Such seemingly unrelated ailments as hiatus hernia, varicose veins, hemorrhoids, obesity, and diabetes are all now thought to be woven into the dietary fabric influenced by fiber.

Of the changes that have taken place in eating habits in the United States during the twentieth century, none may be more significant than the reduction in the intake of fiber, which is now one-fifth that of the nineteenth century. Although it may only be circumstantial evidence, Africans on an unrefined, high-fiber diet have been found to be less predisposed to cancer of the colon and hardening of the arteries. Bacterial flora (types of bacteria), stool size, transit time in the colon, and output of bile salts are all influenced by the intake of fiber and each in turn may play a role in dietary diseases common in the United States. Under attack by certain bacteria, bile salts of the intestines may possibly break down to carcinogens. Or bile acids may be sequestered by fiber and removed as a pathway for cholesterol and the absorption of fat, thus aiding in lowering the risk of certain heart ailments. Or

Table 30. Sources and Content of Dietary Fiber

Food	Crude Fiber
Bran	9 %-12 %
Cooked beans, lentils, and soybeans	1.2%- 1.5%
Roasted nuts.	2.3%- 2.6%
Fruits and vegetables	0.5%- 1.0%

Source: Adapted from Dairy Council Digest. 1975. 46(1):1-4.

possibly, as still another postulate holds, the output of bile acid, as encouraged by the presence of fiber, may simply serve more generously its normal function of removing cholesterol manufactured inside the body. These are interesting theses, with some bases in fact.

The foregoing should not be taken to imply that Americans are stuck with a diet low in fiber. (See Table 30 for sources of fiber.) As always our chief plight is our great abundance of food and therefore the possibility of our failing to stumble by chance across foods high in fiber is more a problem than it was in the past. Diets (poor diets as most of us might gauge their quality) of the last century, consisting as they did mainly of unrefined cereal grains, vegetables, and fruit, were naturally high in fiber. Today's emphasis on meat, sweets, snacks, and alcohol can only be expected to lower our intake of fiber. That does not mean that fiber alone, or any other food constituent solely, can be considered a panacea or without risk. High amounts of fiber in the diet reduce the digestibility of other nutrients. The phytic acid of fiber, already discussed in this text, binds and thereby lowers the absorption of trace minerals. So it is a fact that those of us who do not die of violence or respiratory disease will no doubt die of something we eat—just as was the fate of our ancestors throughout history.

Summary

Labeling information now provides people with a means of determining the nutritional adequacy of their diets. Certainly a diet meeting RDA's for most nutrients—even the standard eight which must be declared—will assure a reasonable intake of all essential nutrients. Though science may yet be in the dark regarding the necessity of some nutrients and equally ignorant of the absolute daily requirements of others, no major deficiencies, and rarely, if ever, minor deficiencies, should occur where a good balance of the nutrients discussed in this chapter is met through the variety of foods such nutrients represent.

XII

Foods and Food Supplements, Their Labels and Use

A special dietary use food is one which supplies "a special dietary need that exists by reason of a physical, physiological, or other condition, including but not limited to the conditions of convalescence, pregnancy, lactation, infancy, allergic hypersensitivity to food, underweight, overweight, diabetes mellitus, or the need to control the intake of sodium . . . [or supplies] a vitamin, mineral or other dietary property for use by man to supplement his diet by increasing the total dietary intake . . . [or supplies] a special dietary need by reason of being a food for use as the sole item of the diet."

FDA definition

If you've survived comfortably, and healthily, without knowing the nutrient content of the food you eat, you probably do not need to complicate your life with such facts. Except for special dietary ailments or needs, or for tightly controlling the cost of food, there is little reason to dwell on label information. Those nutrients now arranged in formal listings on food packages have long been present without the special recognition now provided. One has always been able to live healthily (assuming there are no special dietary needs) by the temperate use of varieties of food abundant to most Americans. Our greatest nutritional faults lie in overeating and in catering to highly selective taste preferences for certain foods to the inevitable exclusion of others. A stomach packed with beer and pretzels cannot accept carrots or apples, however important they may be.

No matter what the reason—whether to avoid an additive of undetermined safety or a natural food toxin, whether to assure exposure to an adequate mix of nutrients or to cut food costs, and at the same time to fulfill the inescapable obligation of sharing our enormous food wealth—moderation and variety are still the safest bet for material and spiritual health. Spread the risk to reduce the risk. Increase variety to assure nutrient variety—including even those nutrients yet unknown to science. Reduce intake to reduce dietary diseases. These are credos of health, as valid today as they ever were in history.

If we assume an increasing variety of foods in the marketplace, and especially of plant protein substitutes for animal products, the health risks of the future will center on the potential for narrowing the diet unknowingly. It is a simple matter to walk through a supermarket and come out with a cart-load of food, seemingly a varied selection but actually foodstuffs almost entirely focused on a single nutrient (or nonnutrient) group. The coming wide-scale exploitation of plant protein substitutes with their own specialty food names can only make more hazy the natural separation of foods into nutrient sources easily recognized in the past. A hundred names of apparently different product types may in reality reflect one basic foodstuff. Beyond that, there is also a potential nutritional hazard posed by deficiencies in minor nutrients (or an improper mix of minor nutrients), particularly trace elements.

Though nutritionists will not always agree, and though fads will come and go, new approaches will be tried, regulations will be amended and promulgated anew, the near future of nutrition would seem to be best served through some attempt to categorize foods by those sources known to harbor significant amounts of essential nutrients. In that way variations in menus alone will serve the nutritional interests science has yet to understand—the interests of process technology and additives use, the interests of production efficiency and high food yields, and the interest of all of us in human health and nourishment. In this chapter, then, we will focus on basic food groups, special dietary needs, and the appropriate function of label information.

Significant Nutrient Sources

One must understand the technology of science to appreciate and use its findings for the most good. Looking at the bold-faced figures on a nutritional label, one is tempted to react to them as absolute and unvarying. But this is not the case, and it would be best to treat the figures as good estimates, but estimates only. Remember, food processors are not held to account for precise declarations of the amount of *added* nutrients: they are permitted variations above and below the amount declared that result from imprecision of different test methods; only 80 percent of the declared amount of *naturally occurring* nutrients need be found present. These are legally permissible variations from label values. Testing methods do vary in precision. Most methods have not been examined for variability over the wide range of nutrient content of foods, and levels of naturally occurring nutrients are not consistent from one field to another, from one tree to another, or even between two fruits or vegetables of the same tree or vine. Rainfall, sunshine, and regional differences all contribute to natural variations in nutrient content. Nutritional information on the food label will always reflect these several inconsistencies.

For the reasons above and also because food regulations make the distinction (thus assuring label standardization), a U.S. RDA of 10 percent appears to be an acceptable level for considering a food as a "significant" nutrient source. This is not to say that one should ignore nutrient levels below 10 percent. Even "measurable" amounts (2 percent RDA) add to daily totals and may well assure the accumulation of adequate intake of nutrients. But when foods are considered for their contribution of one or more nutrients in significant, truly meaningful, amounts, an RDA of 10 percent per serving is a suitable figure to select. Under present regulations this value will match advertising claims made by food processors that a food is a "significant" nutrient source. It will also serve as a benchmark for claims of "nutritional superiority," which must be backed by data proving that one or more nutrients are present at levels of at least 10 percent RDA per serving *higher* than a similar, competitive brand or brands.

Using Nutritional Labeling Information

For most of us, nutritional information on packages will serve, if at all, as a source for casual reference. And though more sophisticated methods of balancing nutrients and figuring menus exist (all the way from a modified abacus to computerized planning), it is the purpose here to keep the approach as simple and straightforward as possible. Since a format for nutritional labeling is available, this seems the best place to start and the best guide from which to work. So once again, let's look at the basic labeling elements but with an eye to nutrient sources and menu planning (see Figure 5).

To obtain adequate amounts of nutrients listed on labels, it is necessary

Figure 5. Nutrients Listed on Food Labels and Their Sources in Basic Food Groups

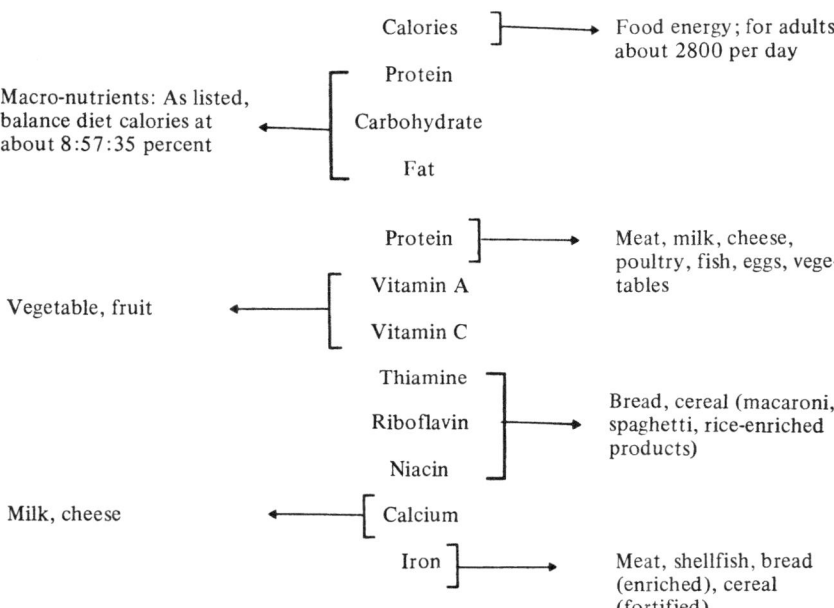

only to balance (daily) the different food sources shown. In so doing you are also essentially assured of a suitable intake of other nutrients, including the trace elements. Though you may consume milk for its calcium (or protein), you will receive, as a bonus, healthy slugs of riboflavin and vitamin D (if the milk is fortified). You may select bread or cereal for the B vitamins and, in addition, come away with a helping of calcium and iron. Or you may serve round steak for dinner because it is a source of high-quality protein, yet that same steak adds an ample sprinkling of the B vitamins (including vitamin B_6), folacin, and iron. Thus a food sought out as a source of one major nutrient may supply a mix of a number of nutrients, and small amounts added to small amounts ultimately assure a healthful intake of all essential nutrients.

Another method is to calculate, with due attention to the size of servings, the caloric and RDA values of your menu for a week or more. The intake of calories should average approximately 2800 a day (adult males) and the RDA's should approximate 100 percent a day for the basic nutrients listed. Not all foods will be labeled with nutrient information so some estimating and adjusting must be done to reveal any major shortcomings in your regular diet And you may be surprised.

Think Food Groups

There is nothing new about the concept of planning menus through a consideration of basic food groups, admittedly, and the procedure is not entirely foolproof. Still, blending a few basic foods appropriately into the daily diet can guarantee the avoidance of certain dietary pitfalls, among them the possibility of concentrating too heavily on a narrow range of foods. For the person who prepares meals for the family, a consideration of food groups can assure reasonable nutrition. For anyone who, out of preference or necessity, emphasizes convenience foods, newly proposed processor guidelines for establishing nutritional quality should provide some certainty that nutritional needs will be met. Then again, those who regularly consume meal replacements can also be assured of adequate nutrition, at least so far as the general state of such knowledge on nutrition allows. Using life insurance as an analogy, we could consider ourselves covered for most nutritional needs.

To reinforce the concept of food groups, which fits generally the main purposes of this text, a look at some actual data from labels might be helpful.

Milk group. Table 31 is an overview of representative nutritional information found on the labels of products from four basic food groups. Where you see RDA percentages of 10 or more, you may assume the food to be a significant source of that particular nutrient. In the milk group, the RDA's for protein, riboflavin, and calcium stand out, but no place else on the table is calcium found at significant levels. So as a source of calcium milk plays a major role in the diet. Calcium may be included in other dairy products such as cheese and ice cream. Since it is a component of skim milk, it is not found in significant levels in butter, nor is it present in large amounts in creams. The higher the level of fat in cream, the lower will be its content of calcium.

Overall, we can look upon milk as a good source of high-quality protein as well as calcium. Milk also consists of a number of trace elements including magnesium, chlorine, fluorine, phosphorus, sulfur, copper, cobalt, iodine, manganese, molybdenum, and zinc. Particularly does milk contribute iodine and zinc to the diet. If and when substitute milk products take over a larger share of the beverage market, it will be necessary to keep in mind the possible variation in the levels of trace minerals between the two types of products. Fabricated foods may not supply the same trace nutrients at those amounts found in milk.

Vitamin fortified milk products are good sources of vitamins A and D. Naturally present are the vitamins B_6, B_{12}, pantothenic acid, biotin, and folic acid. Milk and milk products can contribute significant amounts of vitamin B_{12} and pantothenic acid to the diet. At lower levels, but still in substantial quantities,

Table 31. Nutritional Information Found on Labels of Various Foods

Food Group	Serving Size	Calories	Protein[a]	Carbohydrate[a]	Fat[a]	Protein[b]	Vitamin A[b]	Vitamin C[b]	Thiamine[b]	Riboflavin[b]	Niacin[b]	Calcium[b]	Iron[b]
Milk group													
Milk (whole). . . .	1 cup	160	9	12	9	20	8	2	4	25	*c	30	*
Cheese (Cheddar).	1 oz.	110	7	1	9	15	4	*	*	6	*	20	*
Meat group													
Beef round (cooked, broiled)[d]	¼ lb.	290	32	0	17	70	*	*	6	15	30	*	20
Wieners.	1	140	5	1	12	10	*	*	*	4	4	*	2
Vegetable-fruit group													
Orange juice[d] . . .	6 oz.	80	1	19	0	*	8	150	10	2	4	2	2
Vegetable juice . .	6 oz.	35	1	7	0	2	40	45	4	2	6	2	2
Mixed vegetables	1 cup	90	3	14	1	4	150	20	2	4	4	4	2
Bread-cereal group													
White bread (enriched)	2 oz.	160	4	30	3	10	*	*	15	8	10	4	10
Cereal (fortified, Bran)	1 oz.	70	3	24	1	4	25	25	25	25	25	*	25

a Grams per serving.
b Percentages of U.S. RDA.
c Asterisks indicate amounts less than 2 percent U.S. RDA.
d These figures were computed from *USDA Handbook*, no. 8.

milk provides vitamin B$_6$ and biotin. Tables 32 and 33 summarize the water-soluble vitamin and mineral content of milk. These are constituents of skim milk not found in substantial amounts in butter and cream products high in fat.

Meat group. Although information on nutrients in beef round does not yet appear on the label, data shown in Table 30 are entirely adequate for illustrating the major nutrients to be expected from meat. Like poultry and fish, meat provides high-quality protein, but it also adds iron to the diet, a naturally occurring iron in readily available form. A glance at the list of the basic food groups will tell you that iron is not present in any other source in such naturally high levels. Both bread and cereal grains serve up iron in the diet, but only at significant levels in enriched and fortified products, and even then, as discussed in Chapter XI, biological availability is an important consideration. So the meat group is the place to look for iron and protein. The nutrient bonus received from this food group is generous amounts of B vita-

Table 32. Content of Water-Soluble Vitamins
in Milk

Vitamin	Content	
Choline.	123	mg./qt.
Inositol.	123	mg./qt.
*P-aminobenzoic acid	95	mg./qt.
Ascorbic acid	15	mg./qt.
Pantothenic acid	3.31	mg./qt.
Riboflavin	1.49	mg./qt.
Niacin	0.80	mg./qt.
Vitamin B_6	0.45	mg./qt.
Thiamine.	0.40	mg./qt.
Biotin.	33.10	μg./qt.
Vitamin B_{12}.	5.28	μg./qt.
Folic acid	2.17	μg./qt.

Source: The nonfat composition of milk, Dairy
Council Digest. 1963. 34(3):1-3.

mins, especially B_{12}, which can be obtained only from meat products. Meat
(and poultry) is also thought to be a major source of vitamin B_6. Certain
essential trace minerals are known to reside in meat and, like milk, the role of
meat substitutes will eventually have to be evaluated for this nutritive property. Seafoods, too, are bountiful in trace minerals, and the question of food
substitutes holds equally for them.

Looking at sources of nutrients, one is struck by the wide variety and rich
content of nutrients in liver. It is almost necessary to treat this meat product
separately because of its unique nutritional makeup. As distinguished from
meat in general, liver provides significant levels of vitamins A and C. And for
those nutrients common to meat, liver is a better than average storehouse.
For iron, particularly, this food is a veritable gold mine.

Vegetable-fruit group. Of the major nutrients other than calcium and iron,
the two most often found in short supply in nature belong to this group.
They are vitamin A and vitamin C. Both are available, naturally or as additives,
and there is no need to go without them if due attention is paid to vegetables
and fruit in the diet. A glimpse at Table 31 gives ample evidence of the nutritional value of this food group. A single serving of mixed green and yellow
vegetables can supply a day's RDA of vitamin A. Vitamin C can be as readily
obtained from citrus fruit naturally, vegetable juices, or a number of breakfast fruitlike drinks with high levels of added ascorbic acid. Bonus values of
the natural products include a low calorie/nutrient ratio plus minor amounts
of a number of vitamins and minerals. Both fruit drinks and natural fruit

Table 33. Major and Trace Mineral Content of Milk[a]

Minerals in Significant Amounts	Content (gm/qt.)	Minerals in Significant Amounts	Content (mg/qt.)	Minerals in Trace Amounts
Potassium[b]	1.31	Zinc[b]	3.60	Aluminum
Calcium[b]	1.18	Iron[b]	0.95	Barium
Chlorine[b]	0.97	Copper[b]	0.28	Chromium
Phosphorus[b]	0.91	Iodine[b]	0.20	Lead
Sodium	0.55	Bromine	0.20	Lithium
Sulfur[b]	0.28	Fluorine[b]	0.151	Rubidium
Magnesium[b]	0.11	Boron	0.151	Ruthenium
		Nickle	0.062	Silicon
		Manganese[b]	0.019	Silver
		Molybdenum[b]	0.069	Strontium
		Cobalt[b]	0.0006	Titanium
				Vanadium

Source: The nonfat composition of milk, Dairy Council Digest. 1963. 34(3):1-3.

[a] The values are average values. Certain trace minerals may vary significantly depending upon their content in feed.

[b] Considered essential.

products contain little or no fat. Like the other food groups, fruits and vegetables have much to add to good nutrition.

Proponents of natural foods would no doubt argue the benefits of natural over added vitamins. On that score one can only repeat that present evidence does not prove the case. The primary value of natural foods in this group lies in their contribution to the diet of a wide variety of nutrients—though in admittedly small amounts—over and above the major nutrients for which they are best known. Fruit drinks sometimes are empty of everything save calories and ascorbic acid. But if it's vitamin C you're after, and if the difference in cost between the substitute and pure juice is attractive to your food budget, the fruit drink may be the product to buy.

Bread-cereal group. B should stand for Bread, Bran flakes, and B vitamins, for it is in thiamine, riboflavin, and niacin that this food group shines. Although a share of that glow comes through enrichment or fortification, it is none the dimmer as a result. Fortified bread becomes a good provider of calcium. Both bread and cereal, after similar technological bracing, put nutrient muscle into the diet in the form of iron. A quick look at the last column of figures in Table 31 will indicate the necessity for shoring up the iron content of foods (at least within certain limits). Many cereal products are fortified with iron, and no better nutritional mating could be imagined than that between milk (high in calcium, short in iron) and fortified cereal (short in calcium, high in iron). It is as perfect a match as nature and man are able to come up with. In reality all nutrition is a match up of foods, like a jigsaw puzzle, with nutrient gaps of one food being filled by nutrient pieces of another. Milk and cereal, vegetable and meat, orange juice and toast—just check Table 31. The nutrient marriages are readily apparent. Good nutrition, then, is based on selection of nutritionally compatible foods.

Enriched bread, or macaroni, or noodles, consumed nutritionally for their content of B vitamins, serve that nutritional purpose admirably. Bread is also a source of protein (though of poorer quality than animal protein) and carbohydrates. This is important to note, for in many countries bread and cereal are of great importance in the diet. In the United States, because of a generally high intake of carbohydrates, these nutrients are not held in particular high regard, but it is our wealth of carbohydrates along with our natural craving for sweets that makes our perspective different. Between shortage and plenty, ours should be much the simpler of the two problems. Nonetheless, a few guides, along with a careful reading of the nutrition labels of dry cereals (both presweetened and "unsweetened") and hot cereals, will provide some basis for reaching a decision on what to buy. Here are a few thoughts ("facts") to consider:

*It is the protein of cereals, not the added vitamins, that is costly. Without fortification, cereals of corn and rice will average lower in protein than those of wheat or barley. Oat cereals usually run highest in natural protein. For the three groups of cereals, protein content ranges roughly as follows: corn/rice, 5-9 percent; wheat/barley, 9-12 percent; oat, 14-17 percent. Occasionally dry cereals are fortified with protein (milk or soy) to levels of 25 percent. Such fortification, especially with high-quality animal protein, can make cereal a significant source of protein. Good bargains in cereal purchased primarily for its protein, or for both protein and vitamin/mineral content, can be simply identified by a brief scan of the nutrition label. And remember that reasons for purchase vary. What may be the best buy for the affluent shopper could mean little or no breakfast at all for the poor—if selections were limited. It is this difference in the cost and necessity of nutrients that can make a mockery of restrictive legislation, no matter how worthwhile the benefits may seem. Benefits, like needs, depend upon one's point of reference.

*Cereal is usually consumed with milk and thus may serve to increase the intake of milk, a nutritional plus in most cases but particularly for those persons, young and old, who do not enjoy milk as a beverage.

*Cereal that is not presweetened (coated with sweetener) is usually eaten with sugar. The amount of sugar ordinarily added (about 10 grams) brings the total content of sugar of home-sweetened cereals to a level approaching (though not equal to) that of presweetened cereals.

*Sugar additives of one kind or another are found even in cereals not coated with a sweetener. Hot cereals are the major exception. Check the nutrition labels—and compare—if this is a concern.

*Presweetened cereals account for about 30 percent of the total ready-to-eat cereal consumed in the United States. One survey (sponsored by industry but conducted by an independent agency) found that presweetened cereals accounted for less than 3 percent of children's total intake of sugar. And—as cereal spokesmen will hasten to add, not without foundation—there are many foods containing sugar with far less to offer in the way of overall nutrient content.

*On the question of dental caries (tooth decay), it should be stressed that the lowering of one's intake of sugar could only prove beneficial to many Americans. But it does not necessarily follow that presweetened cereals (or other cereals) are harmful in this regard. Perhaps because milk accompanies the eating of cereal or for other reasons, several clinical studies have failed to find any relationship between the consumption of ready-to-eat cereals and the incidence of tooth decay among children. Though seemingly contradictory, the evidence strongly weighs in favor of this conclusion. Still, with no lessening of emphasis, the recommendation remains—most of us would do well to lower our overall consumption of sugar.

Shopping for bargains. Shopping for nutrient bargains is no different from shopping for bargains in clothing or household appliances, except that one should always be mindful of total dietary needs. Like the tag on an item of clothing, the label on a food product describes its quality. Comparison shopping and pricing as you would when buying an article of clothing, with an eye to the material (nutrient makeup in a food), its utility (usefulness to you or your family), and its cost is readily possible when buying foods as well. And wise purchases save on food expenses.

One way of determining which buy is best among various foods is to estimate the nutrient:calorie ratio and purchase accordingly. First decide which nutrient (or nutrients) you wish to budget. Then estimate the ratio of nutrients to calories, note the available prices of different foods, and select the least expensive source. At equivalent prices, a product with a high nutrient:calorie ratio will be the best bargain.

Suppose you are shopping for protein. At the dairy case you note several

Table 34. Selected Nutrients per 100 Calories in Whole and Skim Milk

Nutrient	U.S. RDA	Whole Milk	Skim Milk
Protein	1.61 gm	5.4 gm	10.0 gm
Vitamin A	178.0 I.U.	215.0 I.U.	314.0 I.U.[a]
Vitamin D	14.3 I.U.	63.0 I.U.[b]	112.0 I.U.[b]
Riboflavin	0.06 mg	0.26 mg	0.49 mg
Thiamine.	0.05 mg	0.04 mg	0.11 mg
Vitamin B_6	0.07 mg	0.08 mg	0.13 mg
Vitamin B_{12}	0.21 mg	0.83 mg	1.48 mg
Calcium	36.0 mg	181.0 mg	335.0 mg
Phosphorus	36.0 mg	145.0 mg	263.0 mg
Magnesium.	14.3 mg	20.0 mg	38.0 mg

Source: E. W. Speckman and R. E. Kowalski. 1975. Milk's marvelous marketing mix, American Dairy Review, 37(3):38-40.
[a] When fortified, there are 2000 I.U. per quart.
[b] When fortified, there are 400 I.U. per quart.

different kinds of cottage cheese, among them lowfat and regular creamed cottage cheese. For creamed cottage cheese (now simply called cottage cheese), the label indicates 120 calories and 14 grams of protein per serving, or a ratio of about 1:8. A lowfat cottage cheese product (½ percent fat) provides 14 grams of protein, 80 calories per serving, for a protein:calorie ratio of approximately 1:6. At comparable cost, the latter product is the better protein buy. The difference between the protein content of round steak and that of fish sticks can also be assessed. The protein:calorie ratio of round steak is 1:9, fish sticks, 1:11. At similar cost, the former is the better bargain.

The value of purchases of other nutrients can be similarly compared. Riboflavin in whole milk at 150 calories/25 percent RDA has a nutrient (riboflavin):calorie ratio of 1:6. In skim milk the ratio is 1:3. Skim milk is probably your best buy. In tomato juice, vitamin C has a nutrient:calorie ratio of approximately 1:0.8. In unsweetened orange juice the ratio is 1:0.6. In vitamin C fortified pineapple juice, you find a ratio of 1:1. If you're looking for the best buy for vitamin C and comparing these three foods, determine the cost of each per unit measure, then consider the vitamin C:calorie ratio.

The method above assumes (and is valid for most Americans) a need for minimal caloric intake. It allows you to compare values without adjusting for serving size or for differences in the content of air or water. And you only have to estimate the ratios for practical use. Similarly, whole products can be analyzed for nutritive value. On a 100-calorie basis whole and skim milk are compared in Table 34.

Protein is usually the most expensive ingredient in foods, but it is also a

Table 35. Protein:Calorie Ratios of Several
Common Foods

Food	Protein:Calorie Ratio
Cottage cheese, dry curd	1:5
Broiled chicken	1:6
Cottage cheese, creamed	1:8
Fried haddock	1:8
Round steak	1:9
Skim milk	1:10
Fish sticks	1:11
Eggs	1:13
Peas	1:13
Ham	1:14
Green beans	1:15
Cheddar cheese	1:16
Milk	1:18
Bread, whole wheat	1:23
Bread, white	1:31
Mashed potatoes	1:31
Ice cream	1:43
French-fried potatoes	1:64
Cream, light	1:72
Potato chips	1:107

fine source of a number of essential vitamins and minerals. Shopping for bargains in protein makes sense. As a guide in your quest for good protein buys, refer to the protein:calorie ratios in Table 35.

Foods for Special Dietary Needs

For certain members of the population, at birth or later, the metabolic machinery of the body malfunctions or its normal operation is somehow interrupted. A health problem exists, and it is directly related to food intake. For many of these persons special foods—foods meeting specific dietary needs —are available commercially. Often called dietetic foods, these products number in the hundreds and serve the needs, among others, of diabetics, allergy victims, overweight persons, those on low-sodium diets, and of course pregnant and lactating women and convalescents from illnesses.

It is in this, the science of food use for medical purposes, that specialists in medicine, nutrition, and food technology perform, as a team, their most miraculous and merciful work in making available preventives and cures. For often does the outcome of an inborn error of human reproduction express its evil in infancy, its chief malignancy in mental retardation, stunted growth, and an existence of pathetic dependency and lack of fulfillment, all because of an adverse reaction to food. A genetic quirk, a click in the machinery, and a whole lifetime hangs in the balance. Twelve million infants are now screened at birth for susceptibility to one such disease called *phenylketonuria*. And with computerlike efficiency the tiny, unsuspecting victims are separated from their unafflicted brothers and sisters, to be treated simply and effectively by a diet of foods of drastically reduced levels of the amino acid phenylalanine. If prior awareness of the disease is lacking, an infant can be well on his way to being nothing more than a vegetable through so common a practice as the feeding of cow's milk. That same food, so valuable to most persons in its content of essential amino acids which their "healthy" bodies cannot derive, becomes living death to these genetically wounded infants. A gap, one missing enzyme, makes it impossible for them to digest phenylalanine properly. Estimates vary, but anywhere from one in 1500 to one in 12,000 infants are so broken at birth.

A large number of modified protein (amino acid) foods are required for the treatment of at least fifteen nutritional diseases involving the metabolism of amino acids. For the person who must survive on these foods it could mean a life of eating staples with the flavor and odor of putrid milk; such are the "tastes" associated with most protein-altered or amino acid-based foods. This is the place where the chemistry and technology of flavor enter the picture. Foods of this type are used in the feeding of persons with low or no capacity of intestinal absorption. Their value lies in their near perfect assimilation (digestion) by the body, thus greatly reduced residues for the elimination process. This advantage is also of benefit to patients scheduled for major stomach or intestinal surgery.

Foul-ups in the ability to handle certain proteins or amino acids suggest similar weaknesses for other nutrients. And such ailments occur. The body's inability to cope with fats and/or fatty acids is generally well known, as are certain of the breakdowns in the human engine where carbohydrate fuels are burned. Diabetes mellitus, with its dependency on the periodic use of insulin for regulating the level of sugar in the blood, is possibly the most familiar. Another disease, hypoglycemia (or too low blood sugar), is also treated by regulating the intake of food; moreover the disease state itself is induced, for test purposes, through a hypoglycemic diet. Lactose intolerance involves the inability to digest milk sugar, and there is also a glycogen storage disease that piles up the body's store of sugar (glycogen) in the wrong places. The enzyme deficiency disease galactosemia is another dietary ailment that leads to stunted mental growth. In all, some ten carbohydrate-related nutritional diseases are known. Both infants and adults fall victims.

Elsewhere in this text mention is made of human intolerance to wheat gluten, or celiac disease as it is called. Persons with this genetic defect number about one in a thousand. Foods low in gluten are generally available for those who have this disease. Another aberration of genes, one which leads to albinism of the eyes, can be prevented by so simple a method as supplying the right amounts of manganese in the diet of pregnant women who carry faulty genes. If food has been proven teratogenic, it can also be used to prevent and cure birth defects. This science, like the victims it would treat, is in its infancy.

Almost any food that finds its way into a large number of homes may be incriminated as an allergen—a food causing allergic symptoms ranging from a rash or hives to shortness of breath and even anaphylactic shock and death. The more people exposed, the greater the certainty that someone will react adversely. Curiously enough, a food which for years may have served as a substitute for some allergenic food may itself become the offender if its use is spread over any given population. Finding his child allergic to chocolate, a parent might rightfully replace chocolate candy with one of the new sweet protein products—a good idea and perfectly logical. But should sweet proteins eventually appeal to anywhere near the number of palates now tantalized by chocolate, the probability—and that's all it is—of these alternatives reaching sensitive stomachs rises proportionately. Mainly because of its widespread use for infants, cow's milk is often classed as an allergen. Equally as often does "soy" milk serve as its replacement. Thinking ahead to the time when soy protein products become more universally accepted, one might well expect the tables to turn. Reactors to soy milk or other soy products will crop up just as frequently as, and perhaps with even greater frequency than, those with allergies to cow's milk. And ever will there be a future for nonallergenic replacements.

The future appears equally bright for low-sodium foods, for the large number of health-related problems often associated with this element. Sodium is so common that it is naturally present in ample dietary amounts in a variety of foods. If salt—sodium chloride—is added to foods, excesses become the norm. Because the body retains more water on a diet heavy in salt, its functions are sometimes strained. An illness may be initiated or the body's response to a prior ailment may be impaired.

Diets and foods low in sodium are nearly as plentiful as the "salted" variety. Assuredly, the declaration of sodium on food labels is more universally applied than is that for any other nutrient. Even nutritional labeling regulations, in provisions for mandatory nutrient listing, make exceptions for sodium in order to assure its presence on the label at a minimum of inconvenience.

Much is heard of drug dependency, little of vitamin dependency, probably because numbers of people suffering from the latter are small and present no sociological problems. In a sense, all of us are dependent upon vitamins. Some persons, though, from birth, have an exaggerated need for one or more vitamins, a need that can reach 180 times the amount normally required for good health. If it is necessary to emphasize the usefulness of synthetic vitamins, provitamins, or other body chemicals, the evidence of their metabolic value in the treatment of these nutritional "addictions" is impossible to deny. Aside from cost and availability, one might consider the practical strain on the body inherent in any attempt to satisfy such elevated vitamin requirements through conventional foods. Although sufferers of these genetic misfirings are relatively few in number, their ranks appear to be increasing. In many cases, with as simple a therapy as proper "food," these persons can lead otherwise healthy lives. But because their ranks are thin and many doctors rarely if ever have the opportunity to diagnose genetically induced nutritional malfunctions, better methods of making available information about these diseases are badly needed. Worldwide networks are under study. At least one country, Canada, has a food bank, a center for the pooling and dissemination of curative foods needed in the treatment of these diseases.

Still another disease, one that also afflicts children, is cystic fibrosis of the pancreas. Medical foods can prevent retarded growth, but often the difficulty lies not in the availability of food, for the essential food can be formulated, but in the preparation of a food product that tastes good enough to stimulate the appetite. As every mother is aware, children are prone to select foods that taste good to them. Similarly do they follow a like pattern and reject those medical foods that are ill-flavored—even when their lives depend on them.

Medical foods, or "special dietary use" foods as government regulations refer to them, serve a number of health-preserving functions, not the least of which is to assist in losing weight. And whether one is interested mainly in a good appearance or in good health, the necessity for suitable foods or suitable food ingredients continues to press science for results.

Always must industry and government be reconciled. And so there is a separate category of food regulations covering "special dietary use" foods. These paragraphs of legal responsibilities were recently modified to exclude some general-purpose foods and now are restricted to foods that serve a special dietary need, foods represented as the sole item in the diet, and dietary supplements. Most of these should be consumed only under medical supervision. Even vitamin and mineral supplements can be misused, subjecting the individual to those hazards discussed elsewhere in this text. Caution is advised in all cases.

A food labeled for special dietary use should declare those vitamins and minerals within the proximate RDA limits shown in Table 36. Upper limits for these nutrients were originally set at 150 percent of normal RDA's except for vitamins A and D and folic acid. Reduced ceilings on vitamins A and D came about because of the health risks of overuse. The limit on folic acid concerns the potential of this vitamin, in large doses, to mask the diagnosis of pernicious anemia. The RDA's of special dietary use foods must generally be based on quantities suitable for consumption within one day.

Nutrients other than those shown, among them vitamin K, choline, chlorine, potassium, sodium, sulfur, fluorine, and manganese, have not been assigned RDA's and are not considered appropriate as dietary supplements of vitamins and minerals. They may be added to other foods, such as infant formulas or foods consumed solely under medical supervision, to meet special dietary use needs.

When a food in this classification is represented for sale to more than one of the groups specified as users, the label will indicate appropriate nutrient levels for each group. Artificial sweeteners, as treated under these regulations, are defined as sweetening substances "not used in normal metabolism as a source of calories." Amino acid and related diets are restricted to levels of amino acids not in excess of that present in whole egg protein. Lastly, as a reminder of present scientific thinking, regulations covering medical foods recognize no differences between synthetic and natural vitamins. For special dietary use foods—medical foods—as for no other, one name aptly and accurately describes them: they are indeed "health" foods.

Vitamin and Mineral Supplements

About three-fourths of American households use vitamin and mineral supplements of one kind or another. Of the 20 percent of those who actually take vitamin and mineral preparations to balance (or supplement) their diets, 40 percent admit that they don't have any idea which vitamin or mineral may be deficient in their diets. Until recently, a number of supplements were being marketed with amounts of named nutrients so low that they were insig-

Table 36. Nutrients, U.S. RDA, and Order of Listing Nutrients for "Special Dietary Use" Foods

Vitamin[a] or Mineral	Unit of Measurement	Infants	Children under Age 4, with Upper Limit of Nutrient in Parentheses	Adults and Children Age 4 or Older, with Upper Limit of Nutrient in Parentheses	Pregnant or Lactating Women, with Upper Limit of Nutrient in Parentheses
Vitamin A	I.U.	1500	2500 (2500)	5000 (5000)	8000 (8000)
Vitamin D	I.U.	400	400 (400)	400 (—)	400 (400)
Vitamin E	I.U.	5	10 (15)	30 (45)	30 (60)
Vitamin C	Milligrams	35	40 (60)	60 (90)	60 (120)
Folic acid	Milligrams	0.1	0.2 (0.3)	0.4 (0.4)	0.8 (0.8)
Thiamine	Milligrams	0.5	0.7 (1.05)	1.5 (2.25)	1.7 (3.0)
Riboflavin	Milligrams	0.6	0.8 (1.2)	1.7 (2.6)	2.0 (3.4)
Niacin	Milligrams	8	9.0 (13.5)	20 (30.0)	20 (40)
Vitamin B_6	Milligrams	0.4	0.7 (1.05)	2.0 (3.0)	2.5 (4.0)
Vitamin B_{12}	Micrograms	2	3 (4.5)	6 (9.0)	8 (12.0)
Biotin	Milligrams	0.15	.15 (0.225)	0.3 (.45)	0.3 (0.6)
Pantothenic acid	Milligrams	3	5.0 (7.5)	10 (15)	10 (20)
Calcium	Grams	0.6	0.8 (1.2)	1.0 (1.5)	1.3 (2.0)
Phosphorus	Grams	0.5	0.8 (1.2)	1.0 (1.5)	1.3 (—)
Iodine	Micrograms	45	70 (105)	150 (225)	150 (300)
Iron	Milligrams	15	10 (15)	18 (27)	18 (60)
Magnesium	Milligrams	70	200 (300)	400 (600)	450 (800)
Copper	Milligrams	0.6	1.0 (1.5)	2.0 (3.0)	2.0 (4.0)
Zinc	Milligrams	5	8.0 (12)	15 (22.5)	15 (30)

Source: Federal Register. 1973. 38(148):20717.

[a] The following synonyms may be added in parentheses immediately following the name of the vitamin: Vitamin C (ascorbic acid); folic acid (folacin); riboflavin (vitamin B_2); and thiamine (vitamin B_1).

nificant in human nutrition. Other preparations contained such large quantities of nutrients that they were useful only for treating nutritional deficiencies. In overdose, some nutrients can be toxic.

These were some of the findings documented during hearings and prehearing conferences which led ultimately to an overhaul of regulations covering nutrient supplements. Clearly do they point out a need for caution and study in the purchase and use of these products. And certainly it is apparent that expectations must far exceed the benefits of the intake of supplemental nutrients for many Americans. Though vitamin and mineral supplements have their just place in meeting nutritional needs, they have not yet been proven the cure-all for mental, emotional, and physical health that many of us would make them. It is also true that nutrition is an imperfect science; much remains to be learned; much is only at best sketchily understood. Still there is a need to explain what is known, to suggest where evidence is less than complete, and to urge caution where risks are established or even suspected. I shall start by defining terms.

What is a dietary supplement? It is always necessary to define terms before embarking upon a discussion of issues. For purposes here, a legalistic definition will be used because questions regarding the use and availability of supplements will ultimately have to be resolved through legal measures.

A vitamin or mineral supplement can be defined, then, as a product marketed to supply persons with nutrients in addition to those ordinarily consumed in the daily diet as an assurance against the inadequate intake of nutrients due either to improper dietary habits or to special dietary needs.

It should be noted that this definition does not imply a daily need for vitamin pills for everyone. Nor is it at all necessary for the majority of persons to supplement their intake of food nutrients. Foods available to most American consumers, when taken in adequate balance and amounts, will supply total nutrient needs. Lacking a sound basis for making the decision, most of us take a one-a-day tablet "just to be on the safe side." Others, however, down two or more pills, or swallow extrapotent doses of a particular vitamin or mineral, in the expectation of some added health benefit—vitamin C to prevent colds, vitamin A to clear up acne, vitamin E for added sex drive or to reduce susceptibility to heart attacks, and iron to stave off that "tired, run-down feeling." Evidence of the value in these particulars is far from conclusive. Risks, perhaps as unclearly understood as benefits, may be real.

When nutrient supplements? If you're the average healthy person eating a fairly well balanced diet, you don't need to add a vitamin/mineral to your daily menu. Literally, this means most of us most of the time. A one-a-day tablet may not do you any harm—*should* not do you any harm—but neither will it provide any added health benefits. So, if you must, take one as a form of insurance, but strongly resist the urge to take two or three or more, expecting proportionately greater results from them. Better still, analyze your intake of nutrients and act accordingly. Or, if you have serious doubts, see your doctor. Be aware, though, of those situations which may make it desirable or necessary to supplement the diet. The more common are the following: extended illness; pregnancy; postsurgery recovery; use of medications that increase specific nutrient needs; and dietary restrictions which prevent the intake of a proper balance of foods.

What about risks? Certain vitamins, namely fat-soluble vitamins, are stored in the body. Since they are not readily excreted (as are most water-soluble vitamins), they build up to what can be toxic accumulations. Vitamin A in excess may cause pressure within the skull, with tumorlike symptoms. Overdoses of vitamin D cause hypercalcemia, along with kidney injury, and, in children, retarded mental and physical growth. For these reasons oral preparations of supplements of vitamins A and D are restricted in potency to 10,000 and 400 I.U., respectively, unless prescribed by a doctor. But this does not exclude the possibility of accidental or purposeful overdoses, which occur with regularity. About 4000 such cases are reported each year. Many more probably go unreported, as is generally true for most diseases.

But what are the risks of taking too much of even those vitamins normally considered readily excretable and therefore harmless at almost any level of intake? Some evidence exists, and other data are slowly accumulating, which suggest caution at least. Dr. Philip White, director of the American Medical Association's Department of Foods and Nutrition, has warned that massive doses of vitamins can "create an unnatural adaptation, a drug dependence if you will; cessation of the regimen then results in a 'withdrawal' situation—a self-induced deficiency." He states, "We know this can happen with massive doses of vitamin C . . . [and] strongly suspect it can occur with B-complex vitamins also, and vitamins D and perhaps E, too."

Charles Glen King, one of the scientists who first isolated vitamin C in 1932, says. "Vitamin C is essential to our daily diet. Ten milligrams daily are necessary to prevent scurvy. For optimum health, I would recommend 100 to 200 milligrams. Some people seem to need more than others, and there may be times when a person can use more than he ordinarily would require. But there is no proof that a dose of 100 times the normal one serves any useful purpose." And he adds, "We know that excessive vitamin C causes the body to form oxalic acid, an ingredient in some kidney stones, especially those associated with gout. . . . Also the acidifying effect of vitamin C in the urine could interfere with the treatment of diabetes [by] rendering misleading urine tests . . ." At least one other doctor, Victor Herbert, has claimed "hard evidence" of the toxicity of vitamin C. B vitamins, too, can be toxic, with evidence documented of deaths of mice upon injection of large doses. Injection, not oral feeding, was the route of administration, which makes the data vulnerable to the same kind of criticism as that leveled against other additives tested for safety in this manner.

In speaking of nutrient supplements, we tend to overlook their mineral content. Minerals are not necessarily formulated into all supplements, but they are found in some. All essential minerals are toxic when taken in overdose. The potential toxicity of iron, because it is a common mineral supplement, must be rated high. Always read the label. If you are seeking a specific nutrient, obtain it in individual dose form. Don't take multivitamin/mineral tablets to fill a single nutrient need.

Overall knowledge about nutrient toxicity is poor indeed. But enough is known to advise caution. Chronic toxicity, the toxicity that builds over a lifetime of intake, is an ever-present possibility.

It is revealing of human nature that what we will not accept as proof of safety of a food additive, we tend to accept blindly in nutrient supplements. Synthetic vitamin C in food is frowned upon as "unnatural." In tablet form we think nothing of downing a gram or two grams—huge doses. As the sage said, "Know thyself." We might add, "Know thy supplements."

Label information. Everything you need to know about a product should be on the label. Years of hearings and effort went into recently revised regulations for labeling dietary supplements. Eventually certain elements of the regulations were voided by court action, not because they were found invalid, but out of technical considerations of the law governing drugs and the FDA's authority in this area. The regulations of August 2, 1973, have since been amended, and together with the amendments may be considered about the best text yet available to prepare a person for a trip to the druggist. The essence of those rules and the reasons for them are reported here with just that thought in mind.

Table 36, though applicable to special dietary food uses, is also a governmental listing of those nutrients thought essential in the normal human diet and for which U.S. RDA's could be set for four different groups of users. The

RDA's shown are those determined by the government as appropriate. In addition, other vitamins and minerals are considered essential (or probably essential) in their biologically active forms, but to date science is unable to set specific RDA levels for them. This list includes vitamins K and choline, and the minerals chlorine, chromium, fluorine, manganese, molybdenum, nickel, potassium, selenium, silicon, sodium, tin, and vanadium. You may well find one or all of these compounds in a supplement tablet, along with a statement to the effect that RDA's have not yet been established for them. What this means of course is that you should not attempt to regulate your intake of these nutrients to any specific amount. In some cases, nickel for example, an RDA may be of scientific interest only. Nickel is so widespread among food sources that risk of deficiency is negligible.

Other nutrients are restricted in use in vitamin/mineral supplements for various reasons of safety. Vitamins A and D and folic acid, mentioned elsewhere, are among them. Minerals in this group include iodine, copper, fluorine, and potassium.

Two elements that will not be found in supplement form, though a part of body chemistry, are cobalt and sulfur. Cobalt is not an essential nutrient per se and is associated solely with vitamin B_{12} (cobalamin) in the body. This vitamin is covered by regulations, so cobalt needs can be expected to be satisfied through this channel. Sulfur is similar in this respect, having no metabolic function in the form (salt) in which it would be supplied in a dietary supplement. The body meets its sulfur needs entirely through degradation of sulfur-containing amino acids.

Table 37 is a list of nutrient makeup and the upper and lower RDA limits for three user groups. The lower limits can be considered as levels below which a supplement will not supply nutrient amounts adequate for supplemental intake. If it contains amounts less than this, the potency of the brand you have selected should be questioned. Upper limits in the table are good guides for maximum daily intake. If these limits are exceeded, especially in vitamins A and D, you run the risk of an overdose.

The different kinds of supplements. Assortments of all combinations of vitamin and mineral supplements in a wide variety of potencies are needless and confusing. Incorporation into these products of substances of unproven nutritional worth may be misleading to any but the nutritionally schooled. The use of estimated potencies, where data or human needs are unknown, offers no signficant value, may mislead, and certainly adds to confusion. For all these reasons some kind of ordered formulation seems both rational and necessary. Recently modified dietary supplement regulations were designed

Table 37. Nutrient Makeup and Upper and Lower Limits of Potency for Specific User Groups of Vitamin and Mineral Supplements

Supplement	Unit of Measurement	Children under 4 Years of Age		Adults and Children 4 or More Years of Age		Pregnant or Lactating Women	
		Lower Limit[a]	Upper Limit[a]	Lower Limit[a]	Upper Limit[a]	Lower Limit[a]	Upper Limit[a]
Vitamins							
Vitamin A	I.U.	1250	2500	2500	5000	5000	8000
Vitamin D[b]	I.U.	200	400	200	400	400	400
Vitamin E	I.U.	5	15	15	45	30	60
Vitamin C	Milligrams	20	60	30	90	60	120
Folic acid[c]	Milligrams	0.1	0.3	0.2	0.4	0.4	0.8
Thiamine	Milligrams	0.35	1.05	0.75	2.25	1.50	3.00
Riboflavin	Milligrams	0.4	1.2	0.8	2.6	1.7	3.4
Niacin	Milligrams	4.5	13.5	10.0	30.0	20.0	40.0
Vitamin B_6	Milligrams	0.35	1.05	1.00	3.00	2.00	4.00
Vitamin B_{12}	Micrograms	1.5	4.5	3.0	9.0	6.0	12.0
Biotin	Milligrams	0.075	0.225	0.150	0.450	0.300	0.600
Pantothenic acid	Milligrams	2.5	7.5	5.0	15.0	10.0	20.0
Minerals							
Calcium	Grams	0.125	1.200	0.125	1.500	0.125	2.000
Phosphorus[d]	Grams	0.125	1.200	0.125	1.500	0.125	2.000
Iodine	Milligrams	35	105	75	225	150	300
Iron	Milligrams	5	15	9	27	18	60
Magnesium	Milligrams	40	300	100	600	100	800
Copper	Milligrams	0.5	1.5	1.0	3.0	1.0	4.0
Zinc	Milligrams	4.0	12.0	7.5	22.5	7.5	30.0

Source: Federal Register. 1973. 38(148):20738.
a According to U.S. RDA.
b Optional for adults and children 4 or more years of age.
c Optional for liquid products.
d Optional for pregnant or lactating women. When present, the quantity of phosphorus may be no greater than the quantity of calcium.

toward this end. A classification system was developed. Qualitative and quantitative content was prescribed. Though some modification may yet be anticipated, the basic classification scheme remains valid. If you are at all perplexed about the vita-mania craze, the guidelines will be of help.

The FDA's classification scheme for nutrient supplements is as follows:

*Multivitamin and multimineral supplement—product containing all vitamins and minerals deemed essential in a supplement.

*Multivitamin supplement—preparation containing all the vitamins deemed essential in a supplement.

*Multimineral supplement—preparation containing all the minerals deemed essential in a supplement.

*Multivitamin and iron supplement—preparation containing all the vitamins deemed essential in a supplement plus the mineral iron.

*Single vitamin supplement—preparation containing one of the essential vitamins.

*Single mineral supplement—preparation containing one of the essential minerals.

Referring now to the specific nutrients listed in Table 37, and also to those nutrients for which RDA's are yet to be established, we can make up a minimal listing of "approved" nutrients to look for in supplement preparations; see page 243. (RDA's have not been established for the nutrients in the third column.) For a listing of approved vitamin/mineral supplements, simply combine the first two columns shown in the chart. Also, consider phosphorus as an optional mineral for pregnant or lactating women. Using this chart as a guide to the kinds of nutrients to seek out, you only have to check the product to be certain that the amounts of each are adequate. This may be determined by comparing the RDA's in tables 36 and 37. In a good supplemental preparation expect to find at least the lower nutrient limits shown in Table 37. Question your need and be aware of the risks of taking higher than the recommended doses (either as a single tablet of extra potency or as more than one tablet of an accepted dose). Also question your need for any substance (for supplemental intake) not on this list until science has established its role or level of requirement in the human diet.

Folic acid. Owing to its instability in liquid preparations, folic acid is considered an optional ingredient in liquid supplements. When a dietary supplement is prepared without folic acid, the label should declare, "This product does not contain the essential vitamin folic acid." This statement will follow the list of vitamins and minerals.

Vitamin Supplements	*Mineral Supplements*	*Other Vitamins and Minerals*
Vitamin A	Calcium	Vitamins
Vitamin D	Phosphorus	Vitamin K
Vitamin E	Iodine	Choline
Vitamin C	Iron	Minerals
Folic acid	Magnesium	Chlorine
Thiamine	Optional minerals	Chromium
Riboflavin	Copper	Fluorine
Niacin	Zinc	Manganese
Vitamin B_6		Molybdenum
Vitamin B_{12}		Nickel
Biotin		Potassium
Pantothenic acid		Selenium
		Silicon
		Sodium
		Tin
		Vanadium

Because an excess of folic acid interferes with the diagnosis of pernicious anemia (by masking evidence of vitamin B_{12} deficiency in the blood), it is restricted to 100 percent RDA in dietary supplements. Pending the evaluation of folic acid by the over-the-counter review panel, any product containing this acid either in excess of the maximum dosage allowed as a food additive or in dietary supplements will be considered a prescription drug. These maximum levels are as follows: for infants, 0.1 mg; for children under four years of age, 0.3 mg; for adults and children four or more years of age, 0.4 mg; for pregnant and lactating women, 0.8 mg. The usual therapeutic dose, oral or parenteral, of folic acid is up to 0.1 mg daily. The label of the drug will bear the statement, "Caution: Federal law prohibits dispensing without prescription." In addition, a number of warning and precautionary notes, including one on the possibilities of allergic sensitization, must appear on the label. When folic acid is added to a food for its vitamin property and no reference is made to the age or physiological state of the user, the amount added should not allow normal daily intake to exceed 0.4 mg. Or, if users are specified, the maximum daily intake mentioned above should be observed as well as the amounts appropriate for those daily intakes utilized in food products.

Supplements for infants. Dietary supplements for infants aged twelve months and under should contain not less than the lower limit of nutrients prescribed for children under four years of age or more than 100 percent of the infant RDA, except for biotin. When used, biotin must be present to a level of 0.05 mg per daily recommended quantity. Because of a commercial shortage of biotin, a lower limit of 0.05 mg is permitted in supplements until December 31, 1976.

The future. As hearings continue and new research develops, it is expected that additional vitamins and minerals will be discovered to have nutritional value and that, possibly, present standards will be found unsatisfactory. In that event, amendments to regulations for dietary supplements will be made. Amendments may be initiated either by the FDA commissioner or by interested persons using appropriate procedures as set forth in the regulations. Petitioners must be prepared to submit scientific data supporting their requests. The law requires that such data should bear directly on human nutrition or technological considerations.

The ingredients of vitamin/mineral supplements. Vitamins and minerals used in supplements may be derived from natural sources or may be chemically synthesized. If a compound is also a food additive, as defined by regulation, it must be used in conformity with regulations for food additives. As long as reasonable amounts are used, any safe and suitable compound may be used as a preservative, stabilizer, sweetener, color, seasoning, carrier, base, or vehicle. However, the utilization of such compounds is not allowed to impair the biological availability of the vitamins or minerals present.

In deriving the content of RDA's, the reference forms and molecular weights given in Table 38 are used.

A separate list of ingredients, providing the natural source or chemical form of each nutrient, will also be included on the label. This information is supplied for persons who may be allergic to minor constituents that carry over into the finished product from the various sources of vitamins and minerals used. Ordinarily, people are not allergic to vitamins per se, but rather to the contaminants that accompany them in the manufacturing process. For liquid supplements in which alcohol is present, its content should be stated as a percentage of the volume.

Expiration date. A number of factors beyond the manufacturer's control cause some nutrients in vitamin and mineral supplements to deteriorate. Among them are heat, light, oxidation, storage, and transportation. For this reason, the FDA allows moderate surpluses in the preparation of supplements, so that at the time consumed (within a reasonable limit), the nutrients will be present in the amounts declared on the label. If supplements contain one or

Table 38. Reference Forms and Molecular Weights Used in Deriving Content of
U.S. RDA's of Vitamin Supplements

Vitamin	Name	Empirical Formula	Molecular Weight
Vitamin C	L-Ascorbic acid	$C_6H_8O_6$	176.12
Folic acid	Pteroyl mono-L-glutamic acid	$C_{19}H_{19}N_7O_6$	441.41
Niacin	Nicotinic acid	$C_6H_5NO_2$	123.11
Riboflavin	Riboflavin	$C_{17}H_{20}N_4O_6$	376.37
Thiamine.	Thiamine chloride hydrochloride	$C_{12}H_{17}ClN_4OS \cdot HCl$	337.28
Vitamin B_6	Pyridoxine	$C_8H_{11}NO_3$	169.18
Vitamin B_{12}. . . .	Cyanocobalamin	$C_{63}H_{88}CoN_{14}O_{14}P$	1355.40
Biotin.	D-Biotin	$C_{10}H_{16}N_2O_3S$	244.31
Pantothenic acid	D-Pantothenic acid	$C_9H_{17}NO_5$	219.23

more nutrients susceptible to a loss in potency, and if such a loss may take place before consumption, an expiration date must appear both on the outside wrapper or container and on the container label itself. This date may be selected by the manufacturer, packer, or distributor and should be based upon appropriate information regarding expected conditions of handling.

In using vitamin and mineral supplements it might be wise to look for label information suggested in this chapter. Check to be certain that a multivitamin or mineral supplement contains the nutrients listed here and at least at the lower dose rates shown. Certain substances, such as rutin, other bioflavonoids, paraaminobenzoic acid, and inositol, are not yet considered appropriate as nutrient supplements. Also, note specifically the user group(s) designated on the label and buy for the appropriate group(s). And always keep bottles of nutrient supplements out of the reach of children; poisonings are not uncommon. Lastly, stay within the recommended upper levels of nutrient intakes, unless advised otherwise by a doctor.

One final word. Supplements are intended to assure sufficient intake of nutrients where some doubt of adequacy exists. Therefore, the first step to nutritional health must always be—eat well of a balanced diet.

XIII

Nutritional Quality Guidelines

*A nutritional guideline prescribes the minimum level or
range of nutrient composition (nutritional quality)
appropriate for a given class of food.*

FDA definition

A nutritional quality guideline (NQG) is simply a food processor's guideline for establishing the nutrient content of a class of food products. Originally proposed in 1971, NQG's were envisioned as a replacement for the traditional, more cumbersome procedure of setting standards of identity for
individual food products. It is a new approach, which offers guidelines rather
than fixed standards for, to date, six different classes of food.

What to Look for on the Label

Although the NQG is a voluntary program (as long as no claim of using an
NQG is made), any processor who meets the standard may state on the label
of the food product that "this food provides nutrients in amounts appropriate
to this class of foods as determined by the U.S. Government." However, the
converse statement will not be made if the standards are not met. Most
processors are expected to follow the guidelines, but the only assurance that
such has been done will be the presence on the label of the foregoing declaration. The guideline statement can be expected on heat and serve dinner items,
ready-to-eat cereals, hot cereal products, noncarbonated breakfast beverages,
main dish food products, and meal replacements.

Components covered by the USDA or FDA standards of identity will usually be listed by their "standardized" names as a single component, even though more than one ingredient is present. "Chicken chow mein" is an example: it contains more than one ingredient, but is called by only one name. Nonstandardized foods consisting of more than one ingredient are treated differently: in this case, each ingredient must be separately declared because no standardized name exists.

Since NQG's represent the latest scientific thinking concerning nutrition and nutrition education, a brief coverage of their use in each food class appears in order.

Heat and Serve Dinners (Frozen)

Guidelines for heat and serve dinners, the first in the list of food classes for which NQG's were proposed, reflect current regulatory philosophy concerning the nutrient fortification of food products. A few major elements are evident.

General philosophy. Nutrient composition is purposely established on a caloric basis to discourage the use of high-calorie, low-nutrient foods. No minimum caloric level is demanded, but the same purpose is served by regulating the levels of protein and certain vitamins and minerals. In this way, rationale has it, food processors are not discouraged from using good sources of protein (fish, for example) which also happen to be low in fat. Neither are processors tempted to add "empty" calories merely to satisfy a rigid caloric standard.

Protein requirements. A frozen heat and serve dinner must include three components, one of which is a significant source of protein (meat, poultry, fish, cheese, or eggs). Each component must be selected from one or more of the following: meat, poultry, fish, cheese, eggs, vegetables, fruit, potatoes, rice, or other cereal-based product (other than bread or rolls). Other servings may be present, such as soup, bread or rolls, beverage, or dessert.

Of the protein present in the dinner, 70 percent must come from the "significant" sources of protein (excluding their sauces and gravies).

Recommendations by the National Academy of Science/National Research Council, cited as the basis for adopting the foregoing protein standards, determined the Recommended Daily Allowance (RDA) on an average utilization of 70 percent of the ideal protein. Since meat and cheese proteins have biological values approximately 80 percent of this ideal, a combination of meat and cheese products with products of lesser biological value (cereal and vegetables) would appear to yield a finished heat and serve dinner of adequate protein quality.

Nutrient requirements (in total). No limitation is placed on the number of nutrients that may be added, but only the amount, which is restricted to a maximum of 150 percent of the prescribed minimum amount.

Nutrients for which no minimum level or range is established by the guideline may not be added without a "disclaimer" on the Principal Display Panel (the main panel on the package where the name of the food appears). This disclaimer will state, "The addition of ____ at the level contained in this product has been determined by the U.S. Government to be unnecessary and inappropriate and does not increase the dietary value of the food." It is not expected that many, if any, processors will add anything but "recognized" nutrients.

Added nutrients will be declared in the statement of ingredients—*but* no other claim may be made suggesting nutritional differences between products (with or without added nutrients). Table 39 shows the minimum nutrient content of heat and serve dinners as proposed by the NQG.

Other nutrients in heat and serve dinners. By their very nature food components required in heat and serve dinners will supply folic acid, magnesium, iodine, calcium, and zinc in addition to those nutrients in Table 39. At present, regulatory philosophy holds that scientific evidence is inadequate for setting minimum levels of these other nutrients. Therefore none are provided. When and if evidence warrants, specifications will be adopted. In the same vein, minimum levels for pantothenic acid, vitamin B_6, and vitamin B_{12} are tentative. Food products failing to meet the levels indicated in Table 39 will *not be deemed misbranded.* Their presence should not be counted on at the guideline levels.

Nutrient limitations. Almost as important as the amount and kind of nutrients required by the guidelines are the nutrients for which no provisions are made, i.e., those not ordinarily associated with foods found in heat and serve dinners. Meat and vegetables, for example, are poor sources of vitamins C and E and the mineral calcium. Nevertheless, they are not added to this class of foods. Reasons cited by the FDA are that the addition of nutrients ordinarily not a part of a given food system would be both irrational and, from the standpoint of nutritional education, unsound.

Calcium:phosphorus ratio. For the body to use calcium best, the ratio of calcium to phosphorus in the diet should approximate 1:1. At large amounts of phosphorus, the utilization of calcium decreases. At present, adding cal-

Table 39. Minimum Nutrient Content of Frozen Heat and Serve Dinners
That Meet NQG Standards[a]

Nutrient[b]	Content for Each 100 Calories (Kcal) of the Three Major Components	Content for Total of the Three Major Components
Protein	4.60 grams	16.0 grams
Vitamin A	150.00 I.U.	520.0 I.U.
Thiamine.	0.05 mg	0.2 mg
Riboflavin	0.06 mg	0.2 mg
Niacin	0.99 mg	3.4 mg
Pantothenic acid	0.32 mg	1.1 mg
Vitamin B_6	0.15 mg	0.5 mg
Vitamin B_{12}.	0.33 mg	1.1 mg
Iron.	0.62 mg	2.2 mg

Source: Adapted from Federal Register. 1973. 38(49):6972.
[a] Serving size is one complete dinner.
[b] If nutrients are added, they must be biologically available in the final product.

cium to foods to enhance the ratio suffers because there is a lack of suitable calcium additives, that is, derivatives of calcium that will not cause objectionable food flavors. To supply adequate amounts of this mineral from natural sources would necessitate, in heat and serve dinners, the use of milk and/or certain vegetables almost to the exclusion of other food components. The result would be a greatly decreased variety in main dishes. For this reason a specific level for calcium has not been set. Instead, the regulatory approach suggests a minimal use of phosphates (a prime source of phosphorus), i.e., those amounts needed to achieve a specific technological function.

Pantothenic acid, vitamin B_6, and vitamin B_{12}. Though we have lived by consuming nutrients throughout history, the science of their usefulness and of man's needs is yet in its infancy. Nutritionists have learned much, but much remains to be learned. At present data to support firm RDA's for pantothenic acid, vitamin B_6, and vitamin B_{12} are not conclusive. Only tentative levels have been mandated for heat and serve dinners. These nutrients may or may not be found in the amounts set by the guidelines, but if a given amount is declared on the label, this value can be expected to be an accurate reflection of the actual amount found in the product.

The question of overfortification. The United States is one of the few nations where it has become necessary to protect citizens against diseases stemming from excessive intake of nutrients, and we must guard against the dangers. If every food were fortified with every essential nutrient, our penchant for overeating would soon have many of us hospitalized with symptoms of our excess. To the extent that fortification is somewhat self-limiting (at the present state of the art), health hazards are lessened. But if extravagant nutrient fortification is coupled with a potential to supplement diets with multivitamin and mineral tablets, the risks are real. In regulatory policy for nutrient fortification in NQG's the issue is resolved by restricting the addition of nutrients at levels appropriate to the particular class of food. The addition of nutrients to a level of 50-150 percent RDA brings the food under regulatory requirements of special dietary foods (see Chapter XII). Lastly, the addition of any nutrient for which no minimum level, range, or other allowance has been established in the guideline is considered misbranding unless "a prominent and conspicuous statement appears on the . . . package label . . . indicating that such fortification as established by the U.S. Government has been found unnecessary and inappropriate and does not increase the dietary value of the food."

Nutritional labeling. When nutrients are added to a food, or when foods are advertised for their nutrient value, nutritional labeling becomes mandatory (see Chapter XI). Furthermore, use of the government declaration (as defined earlier) for heat and serve dinners legally imposes nutritional labeling requirements. Therefore, you should expect to find on all heat and serve dinners a reasonably comprehensive listing of the nutrient content.

Main Dish Products

Recommendations of the NAS/NRC Committee on Food Standards and Fortification were used as guides in setting nutritional quality guidelines for main dish dinners. Not all NAS/NRC proposals were followed to the letter, but the recommendations support most of the major provisions proposed by the guidelines.

Subclasses of main dish products. Because a wide variety of products are encompassed in the main dish classification, three subclasses were isolated and identified for separate guideline statements. These are as follows: macaroni or noodles and cheese; main dish pies and pizza; and combination recipes. Textured vegetable proteins are treated under separate regulations, as already covered in Chapter II. This is not to say that protein components or quality have not been given consideration. Each main dish has a minimum protein content per 100 calories (see Table 40). One source of protein, beans, was ruled out as a component of main dishes because of its low protein efficiency ratio (PER) rating.

General requirements for main dish products. The guidelines require the

Table 40. Minimum Nutrient Levels for Main Dish Products for Each 100 Kilocalories[a]

Nutrient	Macaroni or Noodles and Cheese		Pies and Pizza		Combination Recipes	
Protein	3.5	grams	3.5	grams	4.6	grams
Vitamin A	150.0	I.U.	200.0	I.U.	300.0	I.U.
Thiamine	.038	mg	.05	mg	.05	mg
Riboflavin	.075	mg	.06	mg	.06	mg
Niacin	.66	mg	1.0	mg	1.0	mg
Iron	.47	mg	.62	mg	.62	mg
Calcium	68.0	mg	[b]		[b]	

Source: Adapted from Federal Register. 1974. 39(116):20907.

[a] Nutrient levels include sauces, gravies, and breadings, but exclude any other food ingredient added at the time of preparation.

[b] When practical, ingredients will be selected to maintain as near as possible a calcium:phosphorus ratio of 1:1.

following: at least one source of protein derived from meat, poultry, fish (seafood), cheese, eggs, or appropriate plant protein; one or more vegetables, fruit, potatoes, rice, or other cereal-based product; conformity to specific subclass requirements.

Protein and other nutrient standards. For macaroni or noodles and cheese dishes, cheese is the major protein component. Moreover, processors are expected to use *enriched* macaroni (or noodles) in meeting suggested nutrient levels.

Protein in main dish pies and pizza stems primarily from one or more of the following sources: meat, plant protein product, poultry, fish (seafood), cheese, eggs, vegetable(s), pastry, bread dough shell, crust top.

Combination dinners contain a source of protein derived from meat, plant protein product, poultry, fish (seafood), cheese, eggs, one or more vegetables, fruit, potatoes, rice or other cereal-based product.

Other nutrient considerations. Other nutrients usually found in main dish foods are vitamin B_6, vitamin B_{12}, folic acid, magnesium, pantothenic acid, and zinc. No minimum levels will be placed on these nutrients until research validates precise levels or ranges.

Iodized salt is required by the guidelines; so is the maintenance as nearly as is technically possible of a calcium:phosphorus ratio of 1:1. And lastly, no nutrient will be found in quantities in excess of 150 percent RDA.

The food name. These foods will be identified by some descriptive name (pizza, tuna casserole, macaroni and cheese) followed by a declaration of the percentage of each major characterizing food component.

Meal Replacements (Fortified)

When an ordinary meal is replaced with a single formulated food in the form of a cooky, cake, or dairylike beverage, it is important to know what that food is intended to do in the diet. As prescribed by regulations, *formulated meal replacements* are products that serve as the sole source of nutrients for one or two meals a day, but *not as the sole item of the diet.* A formulated meal *base* is a food of similar nature and intended application, but to which another food ingredient must be added (milk, for example). These products, then, are not truly intended as the sole item in the diet. Nevertheless, an awareness that some persons might well narrow their diet to such extremes prompted the FDA to broaden the nutrient base originally suggested by the NAS/NRC. Still, it is important to note that meal replacements are not intended for use as the sole dietary food or as a food designed to meet specific medical dietary conditions. The latter products are classed as special dietary foods and are governed under an entirely separate set of regulations.

Nutrient content. Meal replacements bearing the government declaration that they have met nutritional guidelines will provide a balance of essential nutrients, to the following extent:

1. Nutrient levels—Required nutrients will be present to at least 25 percent U.S. RDA. If a nutrient is added as a discrete entity, it may not be present in excess of 50 percent U.S. RDA.

2. Caloric levels—Meal replacements will supply a minimum of 700 kilocalories (approximately 25 percent of daily needs) unless the food is marketed for reducing diets. The label will distinguish between these two functions.

3. Protein quality—As PER, protein quality will be rated at least 80 percent that of casein (a fairly high-quality protein), the amount of protein per serving being dependent upon the specific PER value. For protein with PER less than casein, the U.S. RDA is set at 65 grams; if PER equals or exceeds casein, the RDA is 45 grams. Protein quantities will be based upon these two RDA ratings.

4. Linoleic acid level—Linoleic acid, a fatty acid that must be supplied through the diet, will be present to at least 2 percent of total calories. From linoleic acid the body can manufacture any and all essential fatty acids. It is worthwhile noting that the provision for linoleic acid in meal replacements was based upon the assumption that all such products contain fat. (Fatty acids such as linoleic acid are constituents of fat.)

5. Manganese and potassium—Manganese will be present to a level between one and two milligrams per serving, potassium between one and two grams.

6. Other nutrients—Other essential nutrients to be supplied in meal replacements are shown in Table 41.

Label information. On the label the food name is specified as Formulated Meal Replacement or Formulated Meal Base. If the latter, a statement will be included describing the ingredient(s) to be added. Sodium (in salt), though not a required nutrient, will be among the nutrients listed. The declaration of sodium is mandatory in this case, primarily because meal replacements are often high in their content of salt as a direct result of processing needs. Since sodium is important in certain diseases, this information is provided.

The listing of manganese and potassium will be given in milligrams per 100 grams of food and found directly following that of sodium.

Serving size for meal replacements is considered a single meal. The label will indicate if the product is intended for persons on a reducing diet.

Fortified Ready-to-Eat Breakfast Cereals

In setting nutritional quality guidelines for ready-to-eat breakfast cereals, three factors were given consideration: the inherent nutrient makeup of whole grain cereals; nutrients included in standards of enrichment for cereal grains; and nutrients which may, according to evidence, be deficient in typical American diets. These were the major concerns of the NAS/NRC committee that made recommendations to the FDA.

Nutrients normally present in whole cereal grains. If you threshed your own grain and analyzed nutrient content of your home-grown product, you would find, in significant amounts, protein, thiamine, niacin, vitamin B_6, iron, magnesium, zinc, and copper. Were cereal products made from this grain, these would be the major nutrients contributed to the diet. But because the amounts ordinarily found in grain are not always sufficient, the diets of many persons would go wanting without some supplementation.

Nutrients usually added to enriched flour. Common enrichment nutrients of flours include thiamine, riboflavin, niacin, iron, and calcium. Vitamin D, one time a common nutrient additive to flour, has been deleted because proof exists that it is toxic to humans. In addition, vitamin D is expected to be contributed by the milk consumed with the cereal product.

Special considerations in the NQG for ready-to-eat cereals. Though copper is a common nutrient of cereal grains, it has been excluded from the guideline listing because of its ubiquitous nature. Copper is found in most common foods and will rarely, if ever, be deficient in the diet.

Magnesium and zinc are included in the guidelines for cereal products on

Table 41. NQG Nutrient Requirements for Formulated Meal Replacements

Nutrient	U.S. RDA	Nutrient	U. S. RDA
Protein	65 or 45 grams	Phosphorus	1.0 gram
Vitamin A	5000.0 I.U.	Iodine	150.0 µg
Vitamin C	60.0 mg	Magnesium	500.0 mg
Thiamine	1.5 mg	Zinc	15.0 mg
Riboflavin	1.7 mg	Copper	2.0 mg
Niacin	20.0 mg	Vitamin D	400.0 I.U.
Calcium	1.0 gram	Vitamin E	30.0 I.U.
Iron	18.0 mg	Biotin	0.3 mg
Vitamin B_6	2.0 mg	Pantothenic acid	10.0 mg
Folic acid	0.4 mg	Vitamin B_{12}	6.0 µg

Source: Adapted from Federal Register. 1974. 39(116):20906.

an *optional* basis pending more research on technical problems associated with the addition of these nutrients. Manganese was not included as a required nutrient for lack of knowledge concerning its appropriate levels of intake.

The nutrient calcium is required in ready-to-eat cereals primarily to assure the maintenance of a proper calcium:phosphorus ratio. Phosphorus is not a required nutrient because of its already high level in the diet, but it may be listed (along with biotin and other nutrients) when present naturally.

Protein, though an important nutrient, is listed as an optional ingredient in ready-to-eat cereals. The rationale in this case was that cereals are not commonly viewed as major sources of protein. Then, too, cereal protein is generally of low PER. And though protein will be listed in these foods, when the listing involves both milk and cereal, the declared amount will be the sum of the U.S. RDA percentage contributed by the cereal and the milk, a compromise approach (though not entirely precise) put forth to minimize the cost of analysis. Such costs, of course, are ultimately passed on to the consumer.

Presweetened cereals, long a subject of debate concerning their contribution to sugar in the diet, are included in this class of foods and are subject to the guideline requirements. As justification for this action, the FDA cites research studies which seem to indicate that presweetened cereals add relatively little to overall consumption of sugar. Usually presweetening simply replaces all or part of the sugar that would commonly be added by spoon before eating. However, slightly more sugar is consumed in presweetened cereals than would otherwise be the case, and this is sugar that many persons could well do without. Nevertheless there seems little to be gained (and perhaps something to be lost in the way of nutrition) by excluding sweetened cereal products from nutritional guidelines. Indeed, they will be found among the class of foods meeting the standards set for ready-to-eat cereal products.

Two exceptions. Super cereals, those with the vitamin and mineral punch of a one-a-day tablet, are on the market and touted in media advertising for their nutritional virility, as perhaps is true. Some cereals are fortified to the extent that they provide the nutritional equivalence of a multivitamin and mineral supplement. But for this reason (they *are* dietary supplements, as defined by law) such products are regulated not by nutritional quality guidelines but by standards covering dietary supplements. "High-potency" cereals and "high-potency" supplements—regulations treat them as one and the same.

The second exception is infant cereals. Because nutrient requirements for infants are different from those for children and adults, the guidelines do not apply to cereals marketed for infants.

What the guidelines require. The minimum nutrient content of fortified ready-to-eat breakfast cereals is presented in Table 42. For labeling purposes, the serving size for these products is set at one ounce of cereal, four fluid ounces of milk. Optional ingredients include protein, magnesium, and zinc. Protein may only be listed when the protein efficiency ratio is 20 percent or higher (as is required under nutritional labeling regulations). Further, the RDA for protein in cereal combined with milk will be calculated as the sum of the RDA percentages for the milk alone and the cereal alone.

Lastly, specific foods identified as cereal products for this guideline include seeds of grains and similar plants as well as oilseeds.

Fortified Hot Breakfast Cereals

Nutritional quality guidelines for hot breakfast cereals are treated distinct from ready-to-eat cereals because of major differences in patterns of consumption between the two types of cereal. Only the three B vitamins (thiamine, riboflavin, and niacin) are prescribed in hot cereals, along with calcium and, at nearly twice the amount as in cold cereals, iron. The level of iron fortification, at 45 percent U.S. RDA, makes hot cereals a better than average source of iron. At the same time, although excessive iron can be a complicating factor in certain diseases, this amount of iron is not considered a health risk.

In general, cereal products can be looked upon as good sources of the B vitamins, calcium, and iron. These foods are not particularly strong in protein, nor are they treated as such in nutritional quality guidelines. Cereal pro-

Table 42. NQG Nutrient Requirements for Ready-to-Eat Breakfast Cereals

Nutrient	Amount per Serving		Percentage of U.S. RDA
Protein	6.5	grams	10
Vitamin A	1250	I.U.	25
Thiamine	.38	mg	25
Riboflavin	.26	mg	15
Niacin	5	mg	25
Calcium	150	mg	15
Iron	4.5	mg	25
Vitamin B_6	.5	mg	25
Folic acid	.1	μg	25
Magnesium	100	mg	25
Zinc	3.8	mg	25

Source: Adapted from Federal Register. 1974. 39(116):20900.

tein tends generally to be low in PER and is further lowered in biological quality by the milling process.

The serving size of fortified hot breakfast cereals is set at one ounce of dry, *uncooked* product, with weight calculated on a 15 percent moisture basis. The guidelines do not apply to infant foods and, as is the case with ready-to-eat cereals, vitamin D has been deleted as a mandatory additive. When one or more vitamins or minerals exceed 50 percent U.S. RDA per serving, the food is regulated as a dietary supplement. Required nutrients and corresponding required amounts are given in Table 43.

Breakfast Beverage Products

Almost everyone looks to a glass of breakfast juice for a daily supply of vitamin C, and rightly so. Surveys have shown that, combined, all other foods normally consumed contribute only between 30 and 58 percent of this vitamin in the normal daily diet. But there remains a question about how much of this vitamin is actually needed. As noted earlier, massive doses show some value as a preventive or cure for the common cold. But authorities are divided on the question. And some doctors point to the possibility of ill effects to pregnant women (or the unborn child) of high doses of this vitamin. The latest recommendations would limit intake to 100-200 milligrams per day at most, with larger doses reserved for those brief periods during acute infection.

Setting standards for vitamin C in common foods is different from making recommendations for dietary supplements. The public in general must be taken into account, and special dietary risks and general nutritional well-being

Table 43. NQG Nutrient Requirements for Fortified Hot Breakfast Cereals

Nutrient	Amount per Serving	Percentage of U.S. RDA
Protein	6.5 grams	10
Thiamine	.38 mg	25
Riboflavin	.26 mg	15
Niacin	5 mg	25
Calcium	150 mg	15
Iron	8.1 mg	45

Source: Adapted from Federal Register. 1974. 39(116):20897.

must be evaluated. At the same time it is important to note that the option of individual supplementation in tablet form still remains open. For those who feel larger vitamin C intakes are desirable, use of supplements is a simple alternative.

In this light, a conservative approach to the fortification of breakfast juices with vitamin C seems appropriate. Therefore the NAS/NRC committee suggested a minimum of 50 milligrams of vitamin C per six fluid ounces and a maximum of 65. The FDA eventually settled on 60 milligrams, the present U.S. RDA for this vitamin and the amount occurring naturally in most citrus juices (the richest sources).

This nutritional quality guideline covers all full-strength juices as well as products with less than 100 percent natural juice and mixtures of juices and nectars, either in dry or in liquid form, flavored to resemble fruit or vegetable drink. But the guideline applies specifically to products *marketed as a breakfast beverage.* Those sold simply as a beverage, with neither direct nor indirect labeling or advertising as a breakfast drink, are exempt. For the present, however, the main difference would be in the content of vitamin C.

Serving size for adults is six fluid ounces, for infants and children under four years of age, four fluid ounces. When products meet the standard, the government statement will indicate only the presence of the appropriate amount of vitamin C, rather than all nutrients.

Something extra in the food name. Not all breakfast beverages will supply guideline levels of vitamin C without fortification. Certain beverages could never account for this amount of vitamin if the nutrient was not added. Therefore, an allowance is made for processors to note on the package label the presence of added vitamin C whenever the level of addition reaches 10 percent or higher. The phrase "fortified (or enriched) with vitamin C" will denote that fact. But note that some juice products may contain naturally occurring vitamin C in amounts equal in total to that found in other beverages, yet no reference to enrichment or fortification will be found on the label. In other words, absence of the words *enriched* or *fortified* should not be construed to imply nutritional inferiority.

Nutrient Addition to Food—A General Guide

Perhaps the best known example of the addition of nutrients to foods is the use of iodine in salt—iodized salt—for the prevention of goiter. It has proven a marvelously simple and effective way of preventing a disease. No wonder that advances in understanding of nutrition and general patterns of food consumption have led to nutrient fortification of other foods as a means of assuring the populace of adequate nutrition (and therefore the absence of nutritional deficiency diseases). Flour and bread are enriched with B vitamins and iron, cereals are restored and nutritionally enhanced with B vitamins and minerals, margarine has vitamin A added, fluid and dry milk products are fortified with vitamins A and D. Although some changes have taken place in the kinds and levels of enrichment, the basic purpose has remained the same— the guarantee of adequate intake of nutrients for the citizenry as a whole.

In the past, major staples have served as nutrient carriers. Indeed, there would be little need to expand on that base were it not for significant national changes in life style and eating patterns along with advances in technology which portend a future replete with dozens of new, fabricated foods. (A fabricated food is literally one which is pieced together from a variety of ingredients.) Food substitutes, imitations, meat and fish analogues, new combinations of animal and vegetable protein and fat, single cell protein are a number of the trends in food technology, which, coupled with food shortages, move inexorably to displace (to a degree) standard foods and, without sound nutrient formulation, to undermine nutritional concepts and public health. Some standard format, some rationale for the addition of nutrients, seems essential if the United States is to maintain a degree of nutritional integrity in its supermarket mix. Toward this end a regulatory proposal was issued on June 14, 1974. Hotly contested, the proposal seems fated for modification, if not retraction. But a review of its major features appears justified if only to throw light on the basic issues raised by the question of the nutrient fortification of foods.

What is a significant level of a nutrient? Assuming that science knows (or will eventually determine) what nutrients must be present in food, one must next consider what level is significant. If the loss of nutrient(s) takes place during processing, how much can be lost before the amount would be deemed significant for the diet? How much should be restored? What basic guideline should be established for fortifying with any nutrient(s) foods naturally low in nutrient content? Should such foods—snacks, for example—*be* fortified with nutrients? When a food product is pieced together from new food sources, what nutrient requirements, both in level and in kind, should be associated with it?

The above are only a few of many important questions that must be resolved in bringing about a national policy on the addition of nutrients to foods. As for significant levels of nutrients, opinions vary. Already precedence concedes 2 percent U.S. RDA per serving as a measurable (meaningful) amount. The NAS/NRC, in a statement regarding the restoration of nutrients lost in processing, regards 5 percent as a "significant level." The Canadian government requires nutrient restoration if the nutrient which is lost provides at least 10 percent of the daily requirements in a reasonable daily intake of the food. Through the FDA, the United States government now suggests 2 percent RDA per serving as a meaningful level, with restoration of lost nutrients to 10 percent RDA per serving.

Nutrient density. This term broaches the question, "How many nutrients should be packed into a given unit of food?" For a fabricated food with built-in nutrients, the question is fundamental. Since energy is the basic measure of the value of food, nutritionists apply the calorie as the standard unit. Nutrient density becomes the nutrient content per kilocalorie. One need only establish the amount of kilocalories required per capita per day and nutrient density may be expressed in terms of daily requirements. At 2800 kilocalories daily intake, a 280-kilocalorie serving would be expected to supply 10 percent of the daily needs for a given nutrient or nutrients (280 being 10 percent of 2800). Saying it differently, a nutrient, in order to supply 10 percent RDA, would have to be added in that amount which would provide 10 percent RDA for each 280-kilocalorie serving. This is the approach endorsed by the NAS/NRC and the American Medical Association Council on Foods and Nutrition. The only major unresolved issue appears to be the calorie base itself. The FDA has proposed 2800 kilocalories, a number of food companies favor 2300 kilocalories, and one company, after a detailed review of American consumer caloric intake, has suggested a 1200 kilocalorie base. There are those, too, who point out that the 2800 kilocalorie standard could result in a lower per capita nutrient intake than has been the case previously, but this premise has yet to be proven.

Nutrient balance. Natural foods (or foods as they occur in nature) contain nutrients in a balance that is probably impossible to match exactly in a fabricated food. The trace mineral balance, for example, is a delicate (though dynamic) system which could scarcely be reproduced in a synthetic food. Ratios of various nutrients, as determined by nature, in no way proportions out nutrients in a pattern which provides equal nutrient content (of all nutrients) for a given number of calories consumed. Nature's balance, whether optimal or perverse, is unique. And if there is a nutritional risk in taking present-day nutritional knowledge and applying it in a general way to a food fortification policy, it would appear to lie in science's ignorance about some known nutrients and nutrient properties and the role a balance of nutrients plays in the nutrition of man. But weighed against the risk of nutrient deficiency without some kind of fortification standard, the risk seems acceptable. It would appear wiser to fortify as best we know how than not to fortify or to leave fortification to the discretion of individual food processors.

Definition of terms. We are prone to use the words *enrichment* and *fortification* almost interchangeably, not altogether improperly. Yet, for regulatory purposes, a precise distinction is necessary and must be understood if we are to comprehend fully the meaning of information found on food labels. For words referring to the addition of nutrients to food, definitions thus far advanced are as follows:

> *Enrichment* (or *enriched*): the addition to a food of any of those nutrients found naturally associated with that food.
>
> *Fortification*: the addition to a food of nutrients *not* found naturally associated with that food.
>
> *Restoration*: the addition to conventional foods of those nutrients lost during processing.

For these words as found on most food labels of the future, the definitions above will probably apply.

Basis for adding nutrients to foods. At present four basic principles cover the addition of nutrients to foods. A nutrient may be added: (1) Whenever the intake of the nutrient is below a desirable level in the diets of a significant number of people; when the food is consumed by a significant segment of the population; when the amount contributes significantly to the diet; and when the nutrient itself is both stable and biologically available in the food. (2) When the addition is necessary to raise the nutritional quality of a food to a level appropriate for that food. (3) To balance the caloric contribution of a food. (4) To restore a nutrient lost (to a measurable amount) in processing.

Elements in (1) refer specifically to the addition of nutrients to food sources naturally poor in those nutrients.

When nutrients are used to balance calories. As now being considered, the addition of nutrients to balance calories, i.e., to yield a balanced nutrient level for a given number of calories, forecasts a whole new era, both philosophically and absolutely, of nutrient addition. Major elements of the program are, therefore, of national importance. As now proposed, these elements require (1) that the food to which the nutrient is added contribute 56 kilocalories per serving, i.e., 2 percent of the daily standard of 2800 kilocalories, and (2) that only those foods (with the exception of foods regulated under nutritional quality guidelines or standards of identity) be balanced which, if unfortified, would cause a general reduction in the nutritional quality of the total diet. These are foods, then, which are of low nutrient quality but whose bulk contributes substantially to the diet, thereby decreasing total nutrient intake. Certain snack foods would fit the mold, as might a number of textured vegetable products (if consumed in abnormally large amounts). Since foods of this kind make up as much as 25 percent of the diet of certain segments of society (adolescents particularly), a major step-up in nutrient intake is in the offing. On the label you should look for a statement to the effect that nutrients are present "in proportion to . . . caloric content."

Nutrient restoration. If current regulatory thinking prevails, nutrients lost during processing may be restored to a food when the loss amounts to at least 2 percent RDA per serving. Only those nutrients may be restored which are lost under accepted processing procedures. However, the level of addition, if the product is to bear on its label a reference to nutrient restoration, will be 10 percent RDA per serving. Conceivably, 2 percent RDA might be lost, and 10 percent restored. If less than 10 percent is restored, the only evidence of the manipulation of nutrients will be a statement indicating the percentage RDA of a given nutrient in the finished product. You might find, for example, the statement, "Contains _____ percent [less than 10 percent and more than 2 percent] of the RDA of _____ [the nutrient]." The use of the word *restored*, though, will be confined to nutrient additions of at least 10 percent.

Limitation on the use of the word fortified. The word *fortified*, which applies to the addition of nutrients not normally found in significant amounts in a particular food, will be restricted to foods fortified at a level of at least 10 percent RDA per serving. That is the present regulatory intention, anyway.

Approved nutrients for fortifying foods and balancing calories. The "essential" nutrients recognized in nutritional labeling regulations are also desig-

Table 44. Nutrients and Nutrient Levels Approved for Balancing Calories
in Certain Food Products

Nutrient	U.S. RDA[a]	Level of Nutrient Addition for Balancing Calories (per 100 kcal)[a]
Protein	65.0 grams[b]	2.32
Vitamin A	5000.0 I.U.	178
Vitamin C	60.0 mg	2.14
Thiamine	1.5 mg	.054
Riboflavin	1.7 mg	.061
Niacin	20.0 mg	.72
Calcium	1.0 grams	.036
Iron	18.0 mg	.64
Vitamin D	400.0 I.U.	14.3
Vitamin E	30.0 I.U.	1.07
Vitamin B	2.0 mg	.071
Folic acid	.4 mg	.014
Vitamin B_{12}	6.0 μg	.21
Phosphorus	1.0 grams	.036
Iodine	150.0 μg	5.36
Magnesium	400.0 mg	14.3
Zinc	15.0 mg	.54
Copper	2.0 mg	.071
Biotin	.3mg	.011
Pantothenic acid	10.0 mg	.357

Source: Adapted from Federal Register. 1974. 39(116):20904.

[a] For adults and children over four years of age.

[b] If the protein efficiency ratio of protein is equal to or better than that of casein, U.S. RDA is 45 grams and the level of nutrient addition per 100 kcal is 1.61.

nated for use in fortifying foods and balancing calories. These nutrients, along with the levels required when calories are balanced, are shown in Table 44. The column showing nutrients per 100 kilocalories (as required in caloric balancing) is based on a daily intake of 2800 kilocalories. The values shown are 1/28, i.e., 100/2800, of the U.S. RDA.

Appendixes

I

Standards of Identity for Foods

The following foods are covered by a standard of identity which sets forth, among other requirements, the presence (or limitations) of the food components indicated. Many other components or processing functions may also be spelled out, but the elements listed are those which I believe may be of most interest to consumers. Not all the food names by which the food may legally be called are mentioned, but rather the name which seems generally to describe the product. It should be remembered, too, that standards change. The standards included here were valid for the latest date available to me before the book was set in type; most (but not all) were taken from the *Code of Federal Regulations*, January 1974. Since the specific provisions reported in this appendix are major provisions (most of which have remained unchanged over the years), it is expected that the information will remain valid for some time to come. These are federal standards and may vary from those set by the individual states. Federal standards apply to products shipped in interstate commerce.

Cocoa Products

Cacao nuts (cracked cocoa): Not more than 1.75% shell content; alkali processed.

Chocolate liquor (baking chocolate): Not less than 50% and not more than 58% cacao fat.

Breakfast cocoa (high-fat cocoa): Not less than 22% fat.

Cocoa (medium-fat cocoa): Less than 22% but not less than 10% cacao fat.

Lowfat cocoa: Less than 10% cacao fat.

Sweet chocolate: Not less than 15% chocolate liquor and less than 12% milk ingredients.

Milk chocolate: Not less than 3.66% milkfat, 12% milk solids, or 10% chocolate liquor.

Skim milk chocolate: Less than 3.66% milkfat, not less than 12% skim milk solids.

Buttermilk chocolate: Less than 3.66% milkfat, not less than 12% buttermilk solids.

Mixed dairy product chocolates: Not less than 3.66% milkfat, or 12% "constituent" solids.

Sweet chocolate and vegetable fat (other than cacao fat) coating: Same as sweet chocolate except that one or any combination of two or more vegetable food oils or vegetable food fats other than cacao which has a higher melting point than cacao fat is added; emulsifying ingredients not to exceed 0.5%.

Sweet cocoa and vegetable fat (other than cacao fat) coating: Same as sweet cocoa except that cocoa is used in lieu of chocolate liquor to not less than 6.8% of finished food; one or any combination of two or more vegetable food oils, fats, or stearins other than cacao fat which has a higher melting point than cacao fat is used; milk constituent solids need not be less than 12%.

Cocoa with dioctyl sodium sulfosuccinate for manufacturing: Contains the food additive sodium dioctyl sulfosuccinate to not more than 0.4% by weight of finished food.

Wheat Flour and Related Products

Flour (white, wheat, plain): Not more than 0.75% malted barley flour, not more than 15% moisture; may contain ascorbic acid and/or bleaching agent.

Enriched flour: Provides for (on a per pound basis) added thiamine (2.9 mg), riboflavin (1.8 mg), niacin (24 mg), iron (13.0-16.5 mg), calcium (960 mg); not more than 5% wheat germ or partly defatted wheat germ.

Bromated flour: Same as flour except potassium bromate is added in amounts not to exceed 50 ppm of finished flour.

Enriched bromated flour: Same as enriched flour except potassium bromate is added in amounts not to exceed 50 ppm of finished food.

Self-rising flour: Consists of flour, sodium bicarbonate, and one or more acid-reacting substances; salt seasoning; may have bleaching agent present and label will so indicate.

Enriched self-rising flour: Same as self-rising flour, but contains added thiamine, riboflavin, niacin, iron, and calcium to levels found in enriched flour.

Phosphated flour: Same as flour except that monocalcium phosphate is added in amounts of 0.25%-0.75%.

Instantized flour: Flour particles agglomerated (clumped) by further processing.

Whole wheat flour (graham flour): Not more than 0.75% malted barley, not more than 15% moisture; may contain ascorbic acid and/or bleaching agent; proportions of natural constituents unaltered.

Bromated whole wheat flour: Same as whole wheat flour except that potassium bromate is added in amounts not to exceed 75 ppm of finished flour.

Whole durum wheat flour: Same as whole wheat flour except that cleaned durum wheat is used.

Crushed wheat, coarse ground: Larger wheat particles than normal grind; proportions of natural constituents unaltered; not more than 15% moisture.

Cracked wheat: Fragmented wheat of specified particle size; proportions of natural constituents unaltered; not more than 15% moisture.

Farina: Farina size particle fineness; freed from bran coat or bran coat and germ to 0.6% ash; not more than 15% moisture.

Enriched farina: Same as farina except for added nutrients: (per ounce) thiamine (0.38 mg), riboflavin (0.26 mg), niacin (5.0 mg), iron (8.1 mg), calcium (150 mg). (Levels were published as a proposal June 14, 1974; at that time vitamin D was dropped as an approved additive.) May contain not more than 8% wheat germ or partly defatted wheat germ, if added; not more than 15% moisture.

Semolina: Prepared from durum wheat, freed from bran coat or bran coat and germ to ash of not more than 0.92%; not more than 15% moisture.

Corn Products

White cornmeal: Prepared from cleaned white corn; not more than 15% moisture; crude fiber not less than 1.2% and not more than that of the cleaned corn; fat within 0.3% of original cleaned corn.

Bolted white cornmeal: Crude fiber not less than 1.2%; fat not less than 2.25%; not more that 15% moisture.

Bolted yellow cornmeal: Same as white cornmeal, except yellow corn is used.

Degerminated white cornmeal: Ground, cleaned white corn with bran and germ removed; crude fiber less than 1.2%; fat less than 2.25%; less than 15% moisture.

Degerminated yellow cornmeal: Same as degerminated white cornmeal except yellow corn is used.

Self-rising white cornmeal: White cornmeal, sodium bicarbonate, and acid-reacting substances may be added; seasoned with salt.

Self-rising yellow cornmeal: Same as self-rising white cornmeal except yellow corn used.

White corn flour: Not over 15% moisture; grind size must meet standard.

Yellow corn flour: Same as white corn flour except yellow corn used.

Grits (corn grits, hominy grits): Cleaned white corn, bran and germ removed so that finished product not more than 1.2% fiber, fat not more than 2.25%.

Yellow grits: Same as white grits except yellow corn used.

Quick cooking grits: Lightly steamed, slightly compressed grits.

Enriched cornmeals: Same standards as plain (unenriched) meal except for added thiamine, riboflavin, niacin, and iron; may contain added vitamin D and calcium.

Enriched corn grits: Same as plain (unenriched) grits except for added thiamine, riboflavin, niacin, and iron; may contain added vitamin D and calcium.

Rice-Related Products

Rice standards refer to various milled products. When talc and glucose are used as coating, such rice products are termed *coated rice.*

Enriched rice: Each pound contains 2-4 mg thiamine; 1.2-2.4 mg riboflavin; 16-32 mg niacin; 13-26 mg of iron; may contain added vitamin D and calcium.

Macaroni and Noodle Products

Macaroni products are dried from preformed units of dough prepared from semolina, durum flour, farina, regular flour, or any combination of two or more of these products. Optional ingredients include egg white, 0.5%-2.0%; disodium phosphate, 0.5%-1.0%; onion, celery, garlic, bay leaf seasoning; salt; gum gluten, not more than 13%; concentrated glyceryl monostearate, less than 2%. Names of products are macaroni, spaghetti, or vermicelli (depending on shape and size of units).

Milk macaroni products: Same standards as above except milk is used as sole moistening ingredient in preparing the dough and no egg white or disodium phosphate is permitted. If gum gluten is added, protein from gluten and flour must not exceed 13%.

Whole wheat macaroni: Made from whole wheat flour or whole durum wheat flour; may be seasoned with salt and/or onion, celery, garlic, and bay leaf.

Wheat and soy macaroni products: Not less than 12.5% soy flour; seasoning and gum gluten may be added, but protein not to exceed 13% from gluten and flour.

Vegetable macaroni products: Contain tomato, artichoke, beet, carrot, parsley, or spinach to not less than 3%; may contain salt and gum gluten (to not more than 13% protein from gluten and flour).

Noodle products: Prepared from semolina, durum flour, farina, flour, or any combination of two or more of these and egg products (including yolks) plus optional seasoning and gum gluten.

Wheat and soy noodle products: Same as noodle products above except soy flour added to not less than 12.5%.

Vegetable noodle products: Contain tomato, artichoke, beet, carrot, parsley, or spinach to not less than 3%.

Enriched macaroni products: Contain (per pound) 4-5 mg thiamine, 1.7-2.2 mg riboflavin, 27-34 mg niacin, 13-16.5 mg iron; may contain added vitamin D and calcium.

Enriched noodle products: Contain nutrient enrichment same as enriched macaroni products.

Enriched vegetable macaroni products: Same enrichment as for enriched macaroni products.

Enriched vegetable noodle products: Same enrichment as for enriched noodle products.

Macaroni products made with nonfat milk: Finished macaroni product contains 12%-25% nonfat (skim milk) solids; carrageenan may be used to 0.833% of nonfat milk solids.

Enriched macaroni products with nonfat milk: Finished enriched products contain 12%-25% nonfat milk solids; may contain carrageenan to 0.833% of nonfat milk solids.

Bakery Products

Bread (white), rolls, or buns: Made from flour, bromated flour, or phosphated flour, and one or more of several optional ingredients: shortening;

milk; buttermilk; eggs; sugar or sugar products; certain enzymes; yeast; lactic acid bacteria; corn flour; soybean products; sulfates or phosphates; bromates or iodates; spice; conditioners (strengtheners); L-cysteine; ascorbic acid.

Enriched bread: Provides for (on a per pound basis) added thiamine (1.8 mg), riboflavin (1.1 mg), niacin (15 mg), iron (8.0-12.5 mg), calcium (600 mg). (An order to increase iron fortification to 25 mg per pound of bread was stayed February 11, 1974. Fortification levels reverted to the earlier standard, in this case 8.0-12.5 mg.) May contain wheat germ (to not more than 5% of flour ingredient).

Milk bread: Milk used as sole moistening agent (not less than 8.2 parts milk to 100 parts flour).

Raisin bread: Not less than 50 parts raisins to 100 parts of flour.

Whole wheat bread (graham bread): Made from whole wheat flour.

Milk and Cream Products

Milk: 3.25% fat, 8.25% skim milk solids; may contain added vitamins A and D.

Lowfat milk: 0.5%-2.0% fat, 8.25% skim milk solids; added vitamin A, optional vitamin D.

Protein-fortified lowfat milk: Same as lowfat milk except 10% skim milk solids.

Skim milk: 0.5% fat, 8.25% skim milk solids; added vitamin A, optional vitamin D.

Protein-fortified skim milk: 0.5% fat, 10% skim milk solids.

Half-and-half: 10.5%-18.0% fat.

Light cream: 18%-30% fat.

Light whipping cream: 30%-60% fat.

Heavy cream: 36% fat.

Evaporated milk: 7.5% fat, 25.5% total milk solids.

Concentrated milk: 7.5% fat; 25.5% total milk solids.

Sweetened condensed milk: 8.5 % fat; 28.0% total milk solids.

Nonfat dry milk: Not over 1½% fat and 5.0% moisture; or, if over 1½% fat, must declare amount of fat present.

Vitamin-fortified nonfat dry milk: Same as nonfat dry milk except fortified with vitamins A and D.

Sour cream: Cultured (soured by bacteria) product with not less than 18% fat; optional ingredients include (among others) sweeteners, salt, flavoring (either artificial or natural).

Acidified sour cream: Same as sour cream except souring is produced by addition of suitable acids with or without bacterial cultures.

Sour half-and-half: Cultured product; not less than 10.5% but less than 18% fat; optional ingredients same as sour cream.

Acidified half-and-half: Same as acidified sour cream except fat content must be not less than 10.5% or more than 18%.

Sour cream dressing: Same as sour cream (and resembles sour cream) except that dairy ingredients other than cream are allowed; not less than 18% fat.

Sour half-and-half dressing: Resembles sour half-and-half except that dairy ingredients other than milk and cream are allowed; not less than 10.5% or more than 18% fat.

Cheeses, Processed Cheeses, Cheese Foods,
Cheese Spreads, and Related Products

Cheddar: 50% fat (dry weight), 39% moisture.

Cheddar cheese for manufacturing: Same as Cheddar cheese except that pasteurization of the milk and cheese during curing is not required.

Low-sodium Cheddar: Same as Cheddar, contains not more than 96 mg sodium per pound of cheese.

Washed curd cheese: 50% fat (dry weight), 42% moisture.

Washed curd cheese for manufacturing: Same as washed curd except that milk is not pasteurized and cheese need not be cured.

Colby cheese: 50% fat (dry weight), 40% moisture.

Colby cheese for manufacturing: Same as Colby cheese except that milk is not pasteurized and cheese need not be cured.

Low-sodium Colby: Contains not more than 96 mg sodium per pound of cheese.

Cream cheese: 33% fat in finished cheese, 55% moisture.

Neufchatel: 20%-33% fat in finished cheese, 65% moisture.

Cottage cheese (creamed): 4% fat in finished cheese, 80% moisture.

Lowfat cottage cheese: 0.5%-2.0% fat; 82.5% moisture.

Dry curd cottage cheese: Less than 0.5% fat; 80% moisture.

Granular cheese (stirred curd): 50% fat (dry weight); 39% moisture.

Granular cheese for manufacturing: Same as granular cheese except that milk is not pasteurized and cheese need not be cured.

Swiss cheese (Emmentaler): 43% fat (dry weight); 41% moisture.

Swiss cheese for manufacturing: Same as Swiss cheese except that milk is not pasteurized and cheese need not be cured.

Gruyère cheese: 45% fat (dry weight); 39% moisture.

Samsoe cheese: 45% fat (dry weight); 41% moisture.

Brick cheese: 50% fat (dry weight); 44% moisture.

Brick cheese for manufacturing: Same as brick cheese except that milk is not pasteurized and cheese need not be cured.

Muenster cheese: 50% fat (dry weight); 46% moisture.

Muenster cheese for manufacturing: Same as Muenster cheese except that milk is not pasteurized and cheese need not be cured.

Edam: 40% fat (dry weight); 45% moisture.

Gouda: 46% fat (dry weight); 45% moisture.

Blue cheese: 50% fat (dry weight); 46% moisture.

Gorgonzola cheese: 50% fat (dry weight); 42% moisture.

Nuworld cheese: 50% fat (dry weight); 46% moisture.

Roquefort cheese: 50% fat (dry weight); 45% moisture.

Limburger cheese: 50% fat (dry weight); 50% moisture.

Monterey cheese (jack): 50% fat (dry weight); 44% moisture.

High-moisture jack cheese: 50% fat (dry weight); more than 44% moisture.

Provolone cheese: 45% fat (dry weight); 45% moisture.

Caciocavallo Siciliano cheese: 42% fat (dry weight); 40% moisture.

Parmesan cheese: 32% fat (dry weight); 32% moisture.

Mozzarella cheese (scamorza): 45% fat (dry weight); 52%-60% moisture.

Part skim Mozzarella: 30%-45% fat (dry weight); 52%-60% moisture.

Low-moisture Mozzarella: 45% fat (dry weight); 45%-52% moisture.

Low-moisture part skim Mozzarella: 30%-45% fat (dry weight); 45%-52% moisture.

Romano cheese: 38% fat (dry weight); 34% moisture.

Asiago fresh (soft) cheese: 50% fat (dry weight); 45% moisture.

Asiago medium cheese: 45% fat (dry weight); 35% moisture; cured not less than 6 months.

Asiago old cheese: 42% fat (dry weight); 32% moisture; cured not less than 1 year.

Cook cheese (Koch): Made from skim milk; 80% moisture.

Sap sago cheese: Made from skim milk; 38% moisture.

Gammelost: Made from skim milk; 52% moisture.

Hard cheeses: 50% fat (dry weight); 39% moisture.

Semisoft cheeses: 50% fat (dry weight); 39%-50% moisture.

Semisoft part skim cheeses: 45%-50% fat (dry weight); 50% moisture.

Soft ripened cheeses: 50% fat (dry weight).

Spiced cheeses: 50% fat (dry weight); spices added.

Part skim spiced cheeses: 20%-50% fat (dry weight); spices added.

Hard grating cheeses: 32% fat (dry weight); 34% moisture.

Skim milk cheese for manufacturing: Not more than 50% moisture; made from skim milk.

Pasteurized process cheese: Fat and moisture may vary according to cheese varieties from which the product is made; 43%-47% fat (dry weight); moisture 40%-44%.

Pasteurized blended cheese: Blended cheeses; fat and moisture required to be arithmetical averages of original cheeses.

Pasteurized process cheese with fruits, vegetables, or meats: Fat 1% lower, moisture 1% higher than pasteurized process cheese; contains fruit, vegetables, or meats (or mixture of these).

Pasteurized process pimento cheese: 41% fat (dry weight); 49% moisture.

Pasteurized blended cheese with fruit, vegetables, or meats: Fat 1% lower, moisture 1% higher than that of pasteurized blended cheese.

Pasteurized process cheese food: 23% fat (finished food); 44% moisture.

Pasteurized process cheese food with fruits, vegetables, or meats: Not less than 22% fat (finished food).

Pasteurized process cheese spread: Not less than 20% fat (finished food); 44%-60% moisture.

Pasteurized cheese spread: Same as pasteurized process cheese spread except no emulsifying agent used.

Pasteurized process cheese spread with fruits, vegetables, or meats: Same as pasteurized process cheese spread except for added fruit, vegetables, or meats (or mixtures of these).

Pasteurized cheese spread with fruits, vegetables, or meats: Same as pasteurized cheese spread except for added fruit, vegetables, or meats (or mixtures of these).

Cream cheese with other foods: 27%-33% fat (finished cheese); 60% moisture; other foods added.

Pasteurized Neufchatel cheese spread: Not less than 20% fat (finished food); 65% moisture.

Cold-pack cheese: 43%-47% fat (dry weight); 39%-42% moisture.

Cold-pack cheese food: 23% fat (finished food); 44% moisture.

Cold-pack cheese food with fruits, vegetables, or meats: 22% fat (finished food); added fruit, vegetables, or meat (or mixtures of these).

Grated American cheese food: Not less than 23% fat (finished food).

Grated cheeses: 31% fat minimum.

Frozen Desserts

Ice cream: Plain and bulky (nut and fruit) ice creams vary in requirements. Plain, 10% fat, 2.7% protein: Bulky, 8% fat, 2.2% protein. Weight, 4.5 pounds per gallon. Total food solids, 1.6 pounds per gallon.

Frozen custard: Same as ice cream except must have 1.4% egg yolk solids.

Ice milk: 2%-7% fat for plain ice milk; for bulky ice milk fat reductions do not apply; 2.7% protein for both plain and bulky ice milk; weight of 4.5 pounds per gallon; 1.3 pounds per gallon total food solids.

Fruit sherbets: 1%-2% fat; 2%-5% total milk solids; weight of 6 pounds per gallon.

Water ices: Prepared from sweetened fruit or fruit juices.

Nonfruit sherbets: 1%-2% fat; 2%-5% total milk solids; any natural or artificial flavoring except fruit or fruitlike flavoring.

Nonfruit water ices: Weigh not less than 6 pounds per gallon.

Mellorine: Plain flavors, 6% fat; 2.7% protein. Bulky flavors, 4.8% fat; 2.2% protein. Weight of 4.5 pounds per gallon; 1.6 pounds total food solids per gallon; minimum of 40 I.U. of vitamin A per gram.

Food Flavoring

Vanilla extract: 35% ethyl alcohol; at least one "unit" of vanilla constituent per gallon.

Concentrated vanilla extract: Not less than 35% alcohol; two or more "units" of vanilla constituent per gallon.

Vanilla flavoring: Less than 35% ethyl alcohol.

Concentrated vanilla flavoring: Part of the solvent (alcohol) is removed; two or more "units" of vanilla constituent per gallon.

Vanilla-vanillin extract: 35% ethyl alcohol; for each "unit" of vanilla, not less than one ounce of added vanillin.

Vanilla-vanillin flavoring: Less than 35% ethyl alcohol; vanillin added as above.

Vanilla powder: One "unit" of vanilla constituent per 8 pounds of product; mixture of vanilla beans (or oleoresin) and sugar, dextrose, lactose, starch, dried corn syrup, and gum acacia, with anticaking compounds.

Vanilla-vanillin powder: Same as vanilla powder except vanillin added and content declared on the label.

Dressings for Food

Mayonnaise: Not less than 65% vegetable oil, 2½% vinegar, 2½% lemon or lime juice (or both), spices, sweeteners, egg products, and EDTA preservative.

French dressing: Vegetable oil, tomato product mixture; not less than 35% vegetable oil; contains vinegar, spices, lemon or lime juice (or both), stabilizers, emulsifiers, EDTA preservative, safe and suitable color additives.

Salad dressing: Not less than 30% vegetable oil; contains starch paste, spices, sweeteners, egg products, vinegar, lemon or lime juice (or both), and EDTA preservative.

Nutritive Sweeteners

Sweetener content (chemical form) and total solids content prescribed. Dextrose monohydrate must have not less than 90% total solids, anhydrose dextrose 98%, glucose syrup 70%, and dried glucose syrup 90%. These are the only nutritive sweeteners for which identity standards have been promulgated.

Canned Fruits and Fruit Juices

Canned fruit includes peaches, apricots, prunes, pears, seedless grapes, cherries, plums, pineapples, figs, applesauce, and grapefruit.

Standards of identity for canned fruit specify the following:

1. Fruit ingredients: The fruit itself, whether peeled or unpeeled, and the size and shape of the fruit (whole, half, quarter, slice, diced, etc.).

2. Optional ingredients: Spice, flavoring (whether natural or artificial—note the label), vinegar, ascorbic acid, pits, kernels, etc.

3. Optional packing media: Packing media may vary from one fruit to another, but may consist of one of the following: slightly sweetened water (extra light syrup); light, heavy, or extra heavy syrup; slightly, lightly, heavily, or extra heavily sweetened fruit juice; slightly, lightly, heavily, or extra heavily sweetened fruit juice and water. An additional category has been proposed of over 50% fruit juice in water. Each of the categories of packing medium is regulated according to specific sweetener amounts, usually measured by Brix hydrometer after 15 or more days of storage.

4. A variety of quality factors are specified: (a) weight of fruit, whether whole, half, quarter; (b) texture of fruit; (c) amount of peel per unit weight; (d) percentage of units allowed with scab, hail injury, discoloration, scar tissue, or other blemish; (e) amount of allowable broken or crushed units; (f) in case of pitted fruits, the allowable number of pits per unit.

5. A standard of fill, usually the maximum quantity that can be sealed in the container and heat processed (which causes expansion and, upon cooling, contraction and the appearance of underfill), is also specified for each fruit.

Fruit with rum: Conforms to the standard of identity for the specific fruit and contains added rum to an alcohol content of 3%-5%.

Canned berries: Blackberries, blueberries, boysenberries, dewberries, gooseberries, huckleberries, loganberries, red raspberries, black raspberries, strawberries, and youngberries. Packing media same as for fruit, only the berry juice is used instead of fruit juice. Density is controlled to specific Brix hydrometer readings.

Canned fruit cocktail: Packing media same as for fruit, controlled at specific Brix hydrometer levels; peaches (30%-50%), pears (25%-45%), whole grapes (6%-20%), pineapple (6%-16%), cherry ingredients (2%-6%).

Canned applesauce: Comminuted or chopped apple; optional ingredients: water, apple juice, salt, organic acid, nutritive carbohydrate sweetener, spices, flavoring (natural or artificial), antioxidant, color additives. Container filled to not less than 90% of total capacity; glass containers not less than 85% of total capacity.

A number of standards for canned fruit juice and juice drink products have been stayed pending establishment of new or amended standards. The following list includes both stayed and active standards, with the former designated by asterisks. Stayed standards are acknowledged in order to lend insight into the role standards play in attempting to control the makeup of products (to prevent consumer deception) and into the necessity and complexity of amending or introducing standards to cover new product lines. Fruit *drinks* are typical of the ever-changing nature of food standards. These products may contain little or no actual fruit juice or they may contain sizable amounts of juice. The only way of determining the amounts of real juice is to read the label. If none is present the label will so indicate; if some is present, information on the label will tell approximately how much.

Lemonade*: Lemon ingredients to an acidity of 0.70 grams/100 ml, preservatives, buffering salts.

Colored lemonade*: Same as above, but with safe and suitable color allowed (artificial or natural, with appropriate label declaration).

Frozen concentrate for lemonade: Not less than 48% soluble solids (as sucrose—sugar—value); when reconstituted with water, the concentrate yields the same acidity as lemonade, with soluble solids of not less than 10.5%.

Frozen concentrate for colored lemonade: Same as above except with safe and suitable color added (either artificial or natural and appropriately labeled).

Frozen concentrate for artificially sweetened lemonade: Same as frozen concentrate for lemonade, except artificial sweetener allowed; solids specifications do not apply; reconstitutes to same level of acidity as lemonade.

Orange juice: Unfermented juice of mature oranges of species *Citrus sinensis*.

Frozen orange juice: Same as above, and frozen.

Pasteurized orange juice: Unfermented juice of mature oranges; not more than 10% (by volume) of unfermented mature oranges of *Citrus reticulata* species (or hybrids); not less than 10.5% orange juice soluble solids; pasteurized.

Canned orange juice: Unfermented juice of mature oranges; not more than 10% (by volume) of unfermented mature oranges of *Citrus reticulata* species; concentration specified; canned.

Frozen concentrated orange juice: Mix of (1) *Citrus sinensis*, (2) *Citrus reticulata*, and (3) *Citrus aurantium*; unconcentrated product may contain not more than 10% of (2) and 5% of (3). Diluted (reconstituted) juice contains not less than 11.8% solids (by weight).

Canned concentrated orange juice: Same as above, only canned and not frozen.

Orange juice from concentrate: Made from frozen concentrate or concentrate for manufacture; finished product contains at least 11.8% solids (by weight).

Orange juice for manufacturing: Made for further processing; 10% *Citrus reticulata* juice allowed.

Orange juice with preservative: Same as above (for manufacturing) except preservative is allowed to retard spoilage.

Concentrated orange juice for manufacturing: A concentrate similar in makeup, but not identical to frozen concentrate orange juice.

Concentrated orange juice with preservative: Similar to concentrated orange juice for manufacturing, only preservatives allowed.

Orange juice-drink*: Not less than 50% orange juice; ascorbic acid (vitamin C) added to 30-60 mg/4 fluid ounces; preservatives allowed.

Orangeade*: Same as above except consists of 25%-50% orange juice.

Orange drink*: 10%-30% orange juice.

Orange flavored drink*: Less than 10% but more than 0% orange juice.

Canned pineapple-grapefruit juice-drink*: Consistency specified by flow rate measurements; juices not less than 50% of the finished product weight; ascorbic acid added to 20 mg/4 fluid ounces.

Canned fruit nectars*: Consistency specified by flow rate; percentage of single fruit nectar or combined fruit nectars is specified and ranges from 15% to 40% in the finished product.

Cranberry juice cocktail*: 25% cranberry juice; may contain added vitamin C; acid content specified.

Artificially sweetened cranberry juice cocktail*: Same as above, except artificial sweeteners allowed.

Limeade*: Same as lemonade except lime juice ingredients used.

Fruit Pies

Frozen cherry pie: 25% cherries in finished product; not more than 15% blemished cherries.

Fruit Butters, Fruit Jellies, Fruit Preserves, and Related Products

Fruit butter: One or more specified fruit ingredients; 43% soluble solids content in finished product; preservatives allowed; saccharin allowed.

Fruit jelly: One or more specified fruit ingredients; 65% soluble solids in finished product; preservatives allowed; contains specified amounts of saccharin.

Fruit preserves (jams): One or more specified fruit ingredients; 65%-68% soluble solids in finished product; preservatives allowed; contains specified amount of saccharin.

Artificially sweetened fruit jelly: 55% fruit juice in finished product; preservatives allowed; some nutritive sweetener present, but limited; majority of sweetness provided by artificial sweeteners.

Artificially sweetened fruit preserves: Same as above, except in preserve (jam) form.

Table Syrups

Table syrup: At least 65% soluble sweetener solids; butter, if present, not less than 2%; maple and/or honey, if present, not less than 10%; no added vitamins, minerals, protein, or artificial sweeteners; a number of safe and suitable ingredients permitted, including preservatives and color additives.

Maple syrup: At least 66% soluble solids from maple sap; salt, chemical preservatives, and defoaming agents allowed as optional ingredients.

Cane syrup: At least 74% soluble solids from sugar cane juice; same provisions for optional ingredients as maple syrup.

Sorghum syrup: At least 74% soluble solids from sorghum cane juice; optional ingredients include salt, chemical preservatives, defoaming agents, enzymes, anticrystallizing and antisolidifying agents.

Nonalcoholic beverages

Soda water: A class of beverages made from carbonated water; contain no alcohol or only the alcohol (not in excess of 0.5% by weight) contributed by the flavoring; artificial flavoring, natural or artificial coloring; contain caffeine, but content limited to 0.02% by weight; may contain safe and suitable optional ingredients, except vitamins, minerals, protein, and artificial sweetener.

Frozen Fruit

Frozen strawberries: Must have soluble solids content of not less than 18% or more than 35%. Extraneous matter limited, also amount of uncolored berries in the pack.

Shellfish

Canned shrimp (wet and dry pack): Wet pack: cut-out weight of shrimp must equal 64% of water capacity of container; dry pack cut-out weight must equal 60% of container water capacity.

Canned oysters: Drained weight of oysters cannot be less than 59% of container water capacity. Oyster species identified.

Raw oysters: Oyster species specified; amount of liquid in oyster pack specified; washing procedure specified.

Extra large oysters: Same as raw oysters except one gallon contains not more than 160 oysters; one quart of smallest oysters contains not more than 44 oysters.

Large oysters: Same as raw oysters except one gallon contains more than 160 but less than 210 oysters; a quart of the smallest contains not more than 58 oysters, of the largest more than 36 oysters.

Medium oysters: Same as raw oysters except one gallon contains more than 210, but not more than 300 oysters; a quart of the smallest contains not more than 83 oysters, of the largest more than 46 oysters.

Small oysters: Same as raw oysters except one gallon contains more than 300 but not more than 500 oysters; one quart of the smallest contains not more than 138 oysters, of the largest more than 68 oysters.

Very small oysters: Same as raw oysters except one gallon contains more than 500 oysters; a quart of the largest oysters contains more than 112 oysters.

Olympia oysters: Species is defined as *Ostrea lurida.*

Large Pacific oysters: Species of oyster specified; one gallon contains not more than 64 oysters; largest oyster cannot be more than twice the weight of the smallest.

Medium Pacific oysters: Same as large Pacific oysters except one gallon contains more than 64 oysters and not more than 96 oysters; largest oyster cannot be more than twice the weight of the smallest.

Small Pacific oysters: Same as large Pacific oysters except one gallon contains more than 96 oysters but not more than 144; largest oyster cannot be more than twice the weight of the smallest.

Extra small Pacific oysters: Same as large Pacific oysters except one gallon contains more than 144 oysters; the largest oyster cannot be more than twice the weight of the smallest.

Frozen raw breaded shrimp: Breaded product containing not less than 50% shrimp; antioxidants allowed.

Frozen raw lightly breaded shrimp: Same as above except that shrimp must be 65% of the product.

Canned tuna: Species specified; not more than 18% free flakes of tuna in solid pack form; chunk size in chunk tuna is specified; flakes and grated tuna defined by piece size; fill of container specified.

Fish

Canned tuna: Class of fish (twelve in all) specified; style (solid, chunk, flakes, or grated) and color (white, light, dark, or blended) also specified; optional packing media include vegetable oil (without olive oil), olive oil, or water; seasoning agents: salt, MSG, hydrolyzed protein with reduced MSG, spices, vegetable broth, garlic, and lemon flavoring.

Canned salmon: Species specified; forms include regular or skinless (with backbone removed); optional ingredients are salt and edible salmon oil.

Eggs and Egg Products

Liquid eggs: Must be rendered free of salmonella organisms; no chemical preservatives allowed.

Frozen eggs: Same as liquid eggs; phosphate additives allowed.

Dried eggs: Same as liquid eggs only dried and with the optional allowance of the removal of glucose.

Egg yolks (liquid): Yolks separated from whites to contain not less than 43% total egg solids; salmonella free.

Frozen yolks: Same as egg yolks, only frozen.

Dried yolks: 95% egg solids; anticaking compound allowed; optional removal of glucose.

Egg whites (liquid): Material separated from yolks; rendered salmonella free; optional use of added whipping agents.

Frozen egg whites: Same as egg whites, only frozen.

Dried egg whites: Must be rendered salmonella free; optional removal of glucose.

Oleomargarine

Oleomargarine: Not less than 80% fat; can be mixture of animal and vegetable fat; preservatives allowed; vitamin fortified.

Liquid oleomargarine: Same as oleomargarine, only liquid form.

Nut Products

Peanut butter: Fat content cannot exceed 55%; no preservatives or artificial flavors; no color additives; no added vitamins.

Mixed nuts: Four or more shelled tree nut products except in transparent containers of 2 ounces or less, where three nut products permissible; each nut product must be present to 2% but not more than 80% by weight of the finished food; single nut species present at 50%-80% levels must be declared (by percentage) on the label; antioxidants allowed.

Shelled nuts in rigid containers: Same as mixed nuts; fill of container must average 85% of volume.

Frozen Vegetables

Frozen peas: Species specified; natural and artificial flavors allowed; not more than 4% by weight bland peas, 10% blemished peas, 2% seriously blemished peas, 15% peas fragments, 0.5% extraneous vegetable matter.

Canned Vegetables

Canned peas: Species of peas specified; no artificial flavoring; artificial color optional; not more than 4% spotted peas, 0.5% extraneous vegetable matter, 10% pea fragments, 25% of skins ruptured to 1/10 of inch or more; texture requirements specified; alcohol insoluble components limited; fill of container specified.

Canned dry peas: Same as canned peas, except the dry form is used and optional ingredients are not allowed.

Canned green and wax beans: Species specified; style and length of cut specified; texture requirements; not more than 10% blemished units; not more than 8 unstemmed units/340 grams (12 oz); fill of container specified (not less than 90% of total capacity).

Canned corn: Styles (whole kernel, cream, etc.) specified; no artificial flavoring; specifies maximum allowable amounts of discolored corn, pieces of cob, husk, and corn silk; container fill specified (not less than 90% of total container capacity).

Canned field corn: Same as canned corn except that succulent field corn is used.

Canned mushrooms: Weight of mushrooms per container specified; vitamin C optional.

Canned Vegetables Other Than Those Specifically Regulated

In this group the source of the vegetable and optional forms is designated. As well, optional ingredients are specified. Natural flavoring may be added to all except mushrooms. Citric acid and vinegar are optional ingredients for all except artichokes (where it is required) and mushrooms. Vegetables in this group (and certain specific requirements, where pertinent) are as follows.

Artichokes: Vinegar added to pH 4.5; ascorbic acid optional when packed in glass containers.

Asparagus: Stannous chloride optional when packed in glass containers.

Bean sprouts: Calcium lactate (texturizer) optional; vinegar and citric acid optional.

Lima or butter beans: Calcium salts allowed.

Carrots: Calcium salts allowed as firming agents.

Black-eye (or black-eyed) peas: Calcium disodium EDTA allowed to preserve color.

Green sweet peppers: Calcium salts allowed as firming agents.

Potatoes: Calcium salts allowed as firming agents.

Other vegetables regulated under this standard include shelled beans, beets, beet greens, broccoli, Brussels sprouts, cabbage, cauliflower, celery, collards, dandelion greens, kale, mushrooms, okra, onions, parsnips, field peas, red sweet peppers, pimentos, salsify, spinach, sweet potatoes, Swiss chard, truffles, turnip greens, and turnips.

Tomato Products

Tomato juice: Unconcentrated liquid strained from red tomatoes; may contain added ascorbic acid (vitamin C) to a level of 10 mg per fluid ounce.

Yellow tomato juice: Same as tomato juice, only yellow tomatoes used.

Catsup (ketchup): Strained, concentrated liquid from tomatoes, seasoned with salt, vinegar, spices and/or flavoring, onions and/or garlic, and sweetener.

Tomato puree: Liquid from tomatoes; concentrated to 8% but not more than 24% natural tomato soluble solids; salt.

Tomato paste: Liquid of tomatoes; concentrated to at least 24% natural tomato solids; flavoring allowed and baking soda.

Canned tomatoes: Tomato product; optional ingredients include calcium salts, organic acid, salt, spice, flavoring; drained weight must not be less than 50% of the weight of water required to fill the container; blemishes regulated; fill must be not less than 90% of the total capacity of the container.

Standards for Meat and Poultry Products

The following are USDA standards for meat and poultry products as of November 1973. (The information was taken from the publication *Standards for Meat and Poultry Products—A Consumer Reference List*, USDA Animal and Health Inspection Service, November 1973.)

Meat Products

Unless otherwise stated, all percentages are figured on the basis of fresh can cooked weight.

Baby food

High meat dinner: At least 30% meat.

Meat and broth: At least 65% meat.

Vegetable and meat: At least 8% meat.

Bacon (cooked): Weight of cooked bacon cannot exceed 40% of cured, smoked bacon.

Bacon and tomato spread: At least 20% cooked bacon.

Bacon dressing: At least 8% cured, smoked bacon.

Barbecued meats: Weight of meat when barbecued cannot exceed 70% of the fresh uncooked meat; must have barbecued (crusted) appearance and be prepared over burning or smoldering hardwood or its sawdust; if cooked by other dry heat means, product name must mention the type of cookery.

Barbecue sauce with meat: At least 35% meat (cooked basis).

Beans and meat in sauce: At least 20% meat.

Beans in sauce with meat: At least 20% cooked or cooked and smoked meat.

Beans with bacon in sauce: At least 12% bacon.

Beans with frankfurters in sauce: At least 20% frankfurters.

Beans with meatballs in sauce: At least 20% meatballs.

Beans and dumplings with gravy or beef and gravy with dumplings: At least 25% beef.

Beef and pasta in tomato sauce: At least 17½% beef.

Beef carbonade: At least 50% tenderloin.

Beef burger sandwich: At least 35% hamburger (cooked basis).

Beef burgundy: At least 50% beef; enough wine to characterize the sauce.

Beef sauce with beef and mushrooms: At least 25% beef and 7% mushrooms.

Beef sausage (raw): No more than 30% fat; no by-products, no extenders.

Beef stroganoff: At least 45% fresh uncooked beef or 30% cooked beef, and at least 10% sour cream or a "gourmet" combination of at least 7½% sour cream and 5% wine.

Beef with barbecue sauce: At least 50% beef (cooked basis).

Beef with gravy: At least 50% beef (cooked basis).

Gravy with beef: At least 35% beef (cooked basis).

Breaded steaks, chops, etc.: Breading cannot exceed 30% of finished product weight.

Breakfast (frozen product containing meat): At least 15% meat (cooked basis).

Breakfast sausage: No more than 50% fat.

Brown and serve sausage: No more than 35% fat and no more than 10% added water.

Brunswick stew: At least 25% of at least two kinds of meat and/or poultry. Must contain corn as one of the vegetables.

Burgundy sauce with beef and noodles: At least 25% beef (cooked basis); enough wine to characterize the sauce.

Burritos: At least 15% meat.

Cabbage rolls with meat: At least 12% meat.

Cannelloni with meat and sauce: At least 10% meat.

Cappelletti with meat in sauce: At least 12% meat.

Cheesefurter: At least 15% cheese.

Chili con carne: At least 40% meat.

Chili con carne with beans: At least 25% meat.

Chili hot dog with meat: At least 40% meat in chili.

Chili macaroni: At least 16% meat.

Chili pie: At least 20% meat; filling must be at least 50% of the product.

Chili sauce with meat or chili hot dog sauce with meat: At least 6% meat.

Chop suey (American style) with macaroni and meat: At least 25% meat.

Chop suey vegetables with meat: At least 12% meat.

Chopped ham: Must be prepared from fresh, cured, or smoked ham plus certain kinds of curing agents and seasonings. May contain dehydrated onions, dehydrated garlic, corn syrup, and not more than 3% water to dissolve the curing agents.

Chorizos empanadillos: At least 25% fresh chorizos or 17% dry chorizos.

Chow mein vegetables with meat: At least 12% meat.

Chow mein vegetables with meat and noodles: At least 8% meat and the chow mein must equal 2/3 of the product.

Condensed, creamed dried beef or chipped beef: At least 18% dried or chipped beef (figured on reconstituted total content).

Corned beef and cabbage: At least 25% corned beef (cooked basis).

Corned beef hash: At least 35% beef (cooked basis); must contain potatoes, curing agents, and seasonings; may contain onions, garlic, beef broth, beef fat, or others; no more than 15% fat; no more than 72% moisture.

Corn dog: Must meet standards for frankfurters; batter cannot exceed the weight of the frank.

Country ham: Dry-cured product frequently coated with spices.

Crackling corn bread: At least 10% cracklings (cooked basis).

Cream cheese with chipped beef (sandwich spread): At least 12% chipped beef.

Crepes: At least 20% meat (cooked basis), or 10% meat (cooked basis) if the filling has other major characterizing ingredient, such as cheese.

Croquettes: At least 35% meat.

Curried sauce with meat and rice (casserole): At least 35% meat (cooked basis) in the sauce and meat part; no more than 50% cooked rice.

Deviled ham: No more than 35% fat.

Dinners (frozen product containing meat): At least 25% meat or meat food product (cooked basis) figured on total meal minus appetizer, bread, and dessert; minimum weight of a consumer package, 10 ounces.

Dumplings and meat in sauce: At least 18% meat.

Egg foo yong with meat: At least 12% meat.

Egg rolls with meat: At least 10% meat.

Enchilada with meat: At least 15% meat.

Entrees

    Meat or meat food product and one vegetable: At least 50% meat or meat food product (cooked basis).

    Meat or meat food product, gravy or sauce, and one vegetable: At least 30% meat or meat food product (cooked basis).

Frankfurter, bologna, and similar cooked sausage: May contain only skeletal meat; no more than 30% fat, 10% added water, and 2% corn syrup; no more than 15% poultry meat (exclusive of water in formula).

Frankfurter, bologna, and similar cooked sausage with by-products or variety meats: Same limitations as above on fat, added water, and corn syrup.

Must contain at least 15% skeletal meat. Each by-product or variety meat must be specifically named in the list of ingredients. These include heart, tongue, spleen, tripe, stomach.

Frankfurter, bologna, and similar cooked sausage with by-products or variety meats and which also contain nonmeat binders: Product made with the formulas above and also containing up to 3½% nonmeat binders (or 2% isolated soy protein). These products must be distinctively labeled, such as "frankfurters with by-products, nonfat dry milk added." The binders must be named in their proper order in the list of ingredients.

Fried rice with meat: At least 10% meat.

Fritters: At least 35% meat; a breaded product.

German style potato salad with bacon: At least 14% bacon (cooked basis).

Goulash: At least 25% meat.

Gravies: At least 25% meat stock or broth, or at least 6% meat.

Ham (canned): Limited to 8% total weight gain after processing.

Ham (cooked or cooked and smoked), not canned: Must not weigh more after processing than the fresh ham weighs before curing and smoking; if contains up to 10% added weight, must be labeled "Ham, Water Added"; if more than 10%, must be labeled "Imitation Ham."

Ham a la king: At least 20% ham (cooked basis).

Ham and cheese spread: At least 25% ham (cooked basis).

Hamburger, hamburg, burger, ground beef, or chopped beef: No more than 30% fat; no extenders.

Ham chowder: At least 10% ham (cooked basis).

Ham croquettes: At least 35% ham (cooked basis).

Ham salad: At least 35% ham (cooked basis).

Ham spread: At least 50% ham.

Hash: At least 35% meat (cooked basis).

Hors d'oeuvre: At least 15% meat (cooked basis) or 10% bacon (cooked basis).

Jambalaya with meat: At least 25% meat (cooked basis).

Knishes: At least 15% meat (cooked basis) or 10% bacon (cooked basis).

Kreplach: At least 20% meat.

Lasagna with meat and sauce: At least 12% meat.

Lasagna with sauce, cheese, and dry sausage: At least 8% dry sausage.

Lima beans with ham or bacon in sauce: At least 12% ham or bacon.

Liver products such as liver loaf, liver paste, liver paté, liver cheese, liver spread, and liver sausage: At least 30% liver.

Macaroni and beef in tomato sauce: At least 12% beef.

Macaroni and meat: At least 25% meat.

Macaroni salad with ham or beef: At least 12% meat (cooked basis).

Manicotti (containing meat filling): At least 10% meat.

Meat and dumplings in sauce: At least 25% meat.

Meat and seafood egg roll: At least 5% meat.

Meat shortcake: At least 25% meat (cooked basis).

Meat and vegetables: At least 50% meat.

Meatballs: No more than 12% extenders (cereal, etc., including textured vegetable protein); at least 65% meat.

Meatballs in sauce: At least 50% meatballs (cooked basis).

Meat casseroles: At least 25% fresh uncooked meat or 18% cooked meat.

Meat curry: At least 50% meat.

Meat loaf (baked or oven-ready): At least 65% meat and no more than 12% extenders including textured vegetable protein.

Meat pastry: At least 25% meat.

Meat pies: At least 25% meat.

Meat ravioli: At least 10% meat in ravioli.

Meat ravioli in sauce: At least 10% meat.

Meat salads: At least 35% meat (cooked basis).

Meat soups

   Ready-to-eat: At least 5% meat.

   Condensed: At least 10% meat.

Meat spreads: At least 50% meat.

Meat taco filling: At least 40% meat.

Meat tacos: At least 15% meat.

Meat turnovers: At least 25% meat.

Meat Wellington: At least 50% cooked tenderloin spread with a liver paté or similar coating and covered with not more than 30% pastry.

Mince meat: At least 12% meat.

Oleomargarine or margarine: If product is entirely of animal fat, or contains some animal fat, processed under federal inspection. Must contain, individually or in combination, pasteurized cream, cow's milk, skim milk, a combination of nonfat dry milk and water or finely ground soybeans and water. May contain butter, salt, artificial coloring, vitamins A and D, and permitted functional substances. Finished product must contain at least 80% fat. Labels must clearly state which types of fat are used.

Omelet with bacon: At least 9% bacon (cooked basis).

Omelet with dry sausage or with liver: At least 12% dry sausage or liver (cooked basis).

Omelet with ham: At least 18% ham (cooked basis).

Pan haus: At least 10% meat.

Paté de foie: At least 30% liver.

Pepper steaks: At least 30% beef (cooked basis).

Peppers and Italian-brand sausage in sauce: At least 20% sausage (cooked basis).

Petcha: At least 50% calves feet.

Pizza sauce with sausage: At least 6% sausage.

Pizza with meat: At least 15% meat.

Pizza with sausage: At least 12% sausage (cooked basis) or 10% dry sausage, such as pepperoni.

Pork sausage: Not more than 50% fat; may contain no by-products or extenders.

Pork with barbecue sauce: At least 50% pork (cooked basis).

Pork and dressing: At least 50% pork (cooked basis).

Pork with dressing and gravy: At least 30% pork (cooked basis).

Proscrutti: A flat, dry-cured ham coated with spices.

Salisbury steak: At least 65% meat and no more than 12% extenders including textured vegetable protein.

Sandwiches (containing meat): At least 35% meat in total sandwich; filling must be at least 50% of sandwich.

Sauce with chipped beef: At least 18% chipped beef.

Sauce with meat, or meat sauce: At least 6% meat.

Sauerbraten: At least 50% meat (cooked basis).

Sauerkraut balls with meat: At least 30% meat.

Sauerkraut with wieners and juice: At least 20% wieners.

Scalloped potatoes and ham: At least 20% ham (cooked basis).

Scallopine: At least 35% meat (cooked basis).

Scrambled eggs with ham in a pancake: At least 9% cooked ham.

Scrapple: At least 40% meat and/or meat products.

Shepherd's pie: At least 25% meat; no more than 50% mashed potatoes.

Sloppy Joe (sauce with meat): At least 35% meat (cooked basis).

Snacks: At least 15% meat (cooked basis) or 10% bacon (cooked basis).

Spaghetti with sliced franks and sauce: At least 12% franks.

Spanish rice with beef or ham: At least 20% beef or ham (cooked basis).

Stews (beef, lamb, and the like): At least 25% meat.

Stuffed cabbage with meat in sauce: At least 12% meat.

Stuffed peppers with meat in sauce: At least 12% meat.

Sukiyaki: At least 30% meat.

Sweet and sour pork or beef: At least 25% meat and at least 16% fruit.

Sweet and sour spareribs: At least 50% bone-in spareribs (cooked basis).

Swiss steak with gravy: At least 50% meat (cooked basis).

Gravy and swiss steak: At least 35% meat (cooked basis).

Tamale pie: At least 20% meat; filling must be at least 40% of total product.

Tamales: At least 25% meat.

Tamales with sauce (or with gravy): At least 20% meat.

Taquitos: At least 15% meat.

Tongue spread: At least 50% tongue.

Tortellini with meat: At least 10% meat.

Veal birds: At least 60% meat and no more than 40% stuffing.

Veal cordon bleu: At least 60% veal, 5% ham, and containing Swiss, Gruyère, or Mozzarella cheese.

Veal fricassee: At least 40% meat.

Veal parmagiana: At least 40% breaded meat product in sauce.

Veal steaks: Can be chopped, shaped, cubed, frozen. Beef can be added with product name shown as "Veal Steaks, Beef Added, Chopped, Shaped and Cubed" if no more than 20% beef, or must be labeled "Veal and Beef Steak, Chopped, Shaped and Cubed." No more than 30% fat.

Vegetable and meat casserole: At least 25% meat.

Vegetable and meat pie: At least 25% meat.

Vegetable stew and meat balls: At least 12% meat in total product.

Won ton soup: At least 5% meat.

Poultry Products

All percentages of poultry—chicken, turkey, or other kinds of poultry—are on cooked deboned basis unless otherwise indicated. When standard indicates poultry meat, skin, and fat, the skin and fat are in proportions normal to poultry.

Baby food

High poultry dinner: At least 18¾% poultry meat, skin, fat, and giblets.

Poultry with broth: At least 43% poultry meat, skin, fat, and giblets.

Beans and rice with poultry: At least 6% poultry meat.

Breaded poultry: No more than 30% breading.

Cabbage stuffed with poultry: At least 8% poultry meat.

Canned boned poultry

Boned (kind), solid pack: At least 95% poultry meat, skin, and fat.

Boned (kind): At least 90% poultry meat, skin, and fat.

Boned (kind), with broth: At least 80% poultry meat, skin, and fat.

Boned (kind), with specified percentage of broth: At least 50% poultry meat, skin, and fat.

Cannelloni with poultry: At least 7% poultry meat.

Chicken cordon bleu: At least 60% boneless chicken breast (raw basis), 5% ham, and either Swiss, Gruyère, or Mozzarella cheese; if breaded, no more than 30% breading.

Creamed poultry: At least 20% poultry meat; product must contain some cream.

Eggplant parmagiana with poultry: At least 8% poultry meat.

Egg roll with poultry: At least 2% poultry meat.

Entree

Poultry or poultry food product and one vegetable: At least 37½% poultry meat or poultry food product.

Poultry or poultry food product with gravy or sauce and one vegetable: At least 22% poultry meat.

Poultry à la Kiev: Must be breast meat (may have attached skin) stuffed with butter and chives.

Poultry a la king: At least 20% poultry meat.

Poultry almondine: At least 50% poultry meat.

Poultry barbecue: At least 40% poultry meat.

Poultry blintz filling: At least 40% poultry meat.

Poultry Brunswick stew: At least 12% poultry meat; must contain corn.

Poultry burgers: 100% poultry meat, with skin and fat.

Poultry burgundy: At least 50% poultry; enough wine to characterize the product.

Poultry cacciatore: At least 20% poultry meat, or 40% with bone.

Poultry casserole: At least 18% poultry meat.

Poultry chili: At least 28% poultry meat.

Poultry chili with beans: At least 17% poultry meat.

Poultry chop suey: At least 4% poultry meat.

Chop suey with poultry: At least 2% poultry meat.

Poultry chow mein, without noodles: At least 4% poultry meat.

Poultry croquettes: At least 25% poultry meat.

Poultry croquettes with macaroni and cheese: At least 29% croquettes.

Poultry dinners (frozen product): At least 18% poultry meat, figured on total meal minus appetizer, bread, and dessert.

Poultry empanadillo: At least 25% poultry meat including skin and fat (raw basis).

Poultry fricassee: At least 20% poultry meat.

Poultry fricassee of wings: At least 40% poultry wings (cooked basis, with bone).

Poultry hash: At least 30% poultry meat.

Poultry lasagna: At least 8% poultry meat (raw basis).

Poultry livers with rice and gravy: At least 30% livers in poultry and gravy portion or 17½% in total product.

Poultry paella: At least 35% poultry meat or 35% poultry meat and other meat (cooked basis); no more than 35% cooked rice; must contain seafood.

Poultry pies: At least 14% poultry meat.

Poultry ravioli: At least 2% poultry meat.

Poultry roll: No more than 3% binding agents, such as gelatin, in the cooked product; no more than 2% natural cooked-out juices.

Poultry roll with natural juices: Contains more than 2% natural cooked-out juices.

Poultry roll with broth: Contains more than 2% poultry broth in addition to natural cooked-out juices.

Poultry roll with gelatin: Gelatin exceeds 3% of cooked product.

Poultry salad: At least 25% poultry meat (with normal amounts of skin and fat).

Poultry scallopini: At least 35% poultry meat.

Poultry soup

    Ready-to-eat: At least 2% poultry meat.

    Condensed: At least 4% poultry meat.

Poultry stew: At least 12% poultry meat.

Poultry stroganoff: At least 30% poultry meat and at least 10% sour cream or a "gourmet" combination of at least 7½% sour cream and 5% wine.

Poultry tamales: At least 6% poultry meat.

Poultry tetrazzini: At least 15% poultry meat.

Poultry Wellington: At least 50% boneless poultry breast, spread with a liver or similar paté coating and covered with not more than 30% pastry.

Poultry with gravy: At least 35% poultry meat.

Gravy with poultry: At least 15% poultry meat.

Poultry with gravy and dressing: At least 25% poultry meat.

Poultry with noodles or dumplings: At least 15% poultry meat, or 30% with bone.

Noodles or dumplings with poultry: At least 6% poultry meat.

Poultry with noodles au gratin: At least 18% poultry meat.

Poultry with vegetables: At least 15% poultry meat.

Stuffed cabbage with poultry: At least 8% poultry meat.

Sauce with poultry or poultry sauce: At least 6% poultry meat.

II

Categories of Additive Functions as Used in the Regulation of Food Ingredients

1. *Anticaking agents and free-flow agents:* Substances added to finely powdered or crystalline food products to prevent caking, lumping, or agglomeration.

2. *Antimicrobial agents:* Substances used to preserve food by preventing growth of microorganisms and subsequent spoilage, including fungistats, mold and rope inhibitors, and the effects listed by the National Academy of Sciences/National Research Council under "preservatives."

3. *Antioxidants:* Substances used to preserve food by retarding deterioration, rancidity, or discoloration due to oxidation.

4. *Colors and coloring adjuncts:* Substances used to impart, preserve, or enhance the color or shading of a food, including color stabilizers, color fixatives, color-retention agents, etc.

5. *Curing and pickling agents:* Substances imparting a unique flavor and/or color to a food, usually producing an increase in shelf life, stability.

6. *Dough strengtheners:* Substances used to modify starch and gluten, thereby producing a more stable dough, including the applicable effects listed by the National Academy of Sciences/National Research Council under "dough conditioners."

7. *Drying agents:* Substances with moisture-absorbing ability, used to maintain an environment of low moisture.

8. *Emulsifiers and emulsifier salts:* Substances that modify surface tension in the component phase of an emulsion to establish a uniform dispersion or emulsion.

9. *Enzymes:* Enzymes used to improve food processing and the quality of the finished food.

10. *Firming agents:* Substances added to precipitate residual pectin, thus strengthening the supporting tissue and preventing its collapse during processing.

11. *Flavor enhancers:* Substances added to supplement, enhance, or modify the original taste and/or aroma of a food, without imparting a characteristic taste or aroma of its own.

12. *Flavoring agents and adjuvants:* Substances added to impart or help impart a taste or aroma in food.

13. *Flour treating agents:* Substances added to milled flour, at the mill, to improve its color and/or baking qualities, including bleaching and maturing agents.

14. *Formulation aids:* Substances used to promote or produce a desired physical state or texture in food, including carriers, binders, fillers, plasticizers, film-formers, and tableting aids, etc.

15. *Fumigants:* Volatile substances used for controlling insects or pests.

16. *Humectants:* Hygroscopic substances incorporated in food to promote retention of moisture, including moisture-retention agents and antidusting agents.

17. *Leavening agents:* Substances used to produce or stimulate production of carbon dioxide in baked goods to impart a light texture, including yeast, yeast foods, and calcium salts listed by the National Academy of Sciences/ National Research Council under "dough conditioners."

18. *Lubricants and release agents:* Substances added to food contact surfaces to prevent ingredients and finished products from sticking to them.

19. *Nonnutritive sweeteners:* Substances having less than 2% of the caloric value of sucrose per equivalent unit of sweetening capacity.

20. *Nutrient supplements:* Substances necessary for the body's nutritional and metabolic processes.

21. *Nutritive sweeteners:* Substances having greater than 2% of the caloric value of sucrose per equivalent unit of sweetening capacity.

22. *Oxidizing and reducing agents:* Substances that chemically oxidize or reduce another food ingredient, thereby producing a more stable product, including the applicable effect listed by the National Academy of Sciences/ National Research Council under "dough conditioners."

23. *pH control agents:* Substances added to change or maintain active acidity or basicity, including buffers, acids, alkalies, and neutralizing agents.

24. *Processing aids:* Substances used as manufacturing aids to enhance the appeal or utility of a food or food component, including clarifying agents, clouding agents, catalysts, flocculents, filter aids, and crystallization inhibitors, etc.

25. *Propellants, aerating agents, and gases:* Gases used to supply force to expel a product or used to reduce the amount of oxygen in contact with the food in packaging.

26. *Sequestrants:* Substances that combine with polyvalent metal ions to form a soluble metal complex, to improve the quality and stability of products.

27. *Solvents and vehicles:* Substances used to extract or dissolve another substance.

28. *Stabilizers and thickeners:* Substances used to produce viscous solutions or dispersions, to impart body, improve consistency, or stabilize emulsions, including suspending and bodying agents, setting agents, jellying agents, bulking agents, etc.

29. *Surface-active agents:* Substances used to modify surface properties of liquid food components for a variety of effects, other than emulsifiers, but including solubilizing agents, dispersants, detergents, wetting agents, rehydration enhancers, whipping agents, foaming agents, and defoaming agents.

30. *Surface-finishing agents:* Substances used to increase palatability, preserve gloss, and inhibit discoloration of foods, including glazes, polishes, waxes, and protective coatings.

31. *Synergists:* Substances used to act or react with another food ingredient to produce a total effect different from or greater than the sum of the effects produced by the individual ingredients.

32. *Texturizers:* Substances that affect the appearance or feel of the food.

Source: Federal Register, September 23, 1974, pp. 34173-34176.

III

101 Varieties of Food Additives

In the following list of the functions and sources of additives no distinction is made between intentional and incidental additives. Some examples are provided, but are not necessarily inclusive of all compounds in a given category, especially since certain categories comprise dozens if not hundreds of different additives (flavor compounds, for instance).

The applications selected were especially chosen to lend specificity to the breadth of technological functions additives serve and also to offer insight into the many external sources from which "incidental" (contaminating) additives arise.

Some categories overlap; some additives have more than one function. But as a group, the following examples yield palpable evidence of the breadth of essential uses additives serve in present-day society. Equally obvious is evidence of the potential for food contamination from a host of environmental sources.

1. Acidulants: Compounds used to make products more acid (sour).

Fumaric acid: fruit juice drinks, gelatin desserts

Malic acid, citric acid: wine

2. Adhesives: Used to glue containers together, to bind inserts in bottle-caps and jar lids, etc. Such compounds may come in direct contact with food and must be cleared through food regulatory agencies for this kind of use. Adhesive components form a very large category of additives, with well over 500 compounds.

bis-(sodium 2-sulfoethyl) maleate, butyraldehyde: components of adhesives

3. Aerating agents: For foamed foods such as whipped toppings. Some common aerating compounds include *chloropentafluorethane, carbon dioxide, nitrous oxide, propane,* and *octafluorocyclobutane.*

4. Air: Compressed air, used for processing sherry wine.

5. Air exclusion agent: To keep air out of products. An example is the use of *mineral oil* in wine casks. No mineral oil is allowed in the finished wine.

6. Antibrowning agents (enzymatic browning): Compounds to prevent brown discoloration caused by natural food enzymes (reaction hasteners). This is a problem primarily in fruits and vegetables.

> *Ascorbic acid (vitamin C):* fruits and vegetables
>
> *Erythorbic acid:* fruits and vegetables

7. Anticaking compounds: To prevent food products or ingredients from balling up or forming chunks.

> *Calcium silicate:* baking powder
>
> *Aluminum calcium silicate, magnesium silicate:* table salt

8. Antifogging compounds: Used in food packaging materials to maintain transparency.

9. Antimycotics: To retard or prevent mold growth.

> *Calcium propionate:* cheese, processed cheese, baked goods
>
> *Sodium benzoate:* fruit drinks, fruit juices, purees, margarine, pie fillings, salad dressing

10. Antioxidants: To retard or prevent rancidity of food fats and oils; also used in plastics manufacture for food containers.

> *Butylated hydroxyanisole:* fresh pork, fresh pork sausage, dry yeast, beverages and desserts from dry mixes, dry breakfast cereals, dry diced glacéed fruit, dry mixes for beverages and desserts, emulsifiers and stabilizers for shortening, potato flakes and granules, sweet potato flakes, potato shreds, chewing gum base, beet sugar, yeast, flavoring substances, dry food, frozen dried fruits and vegetables, dry sausage, margarine, rendered animal fat
>
> *Citric acid:* lard, shortening, dry sausage, chili con carne, margarine

11. Antisegregating agents: To keep trace mineral nutrients dispersed in animal feed. *Mineral oil* is an example.

12. Antistatic compounds: Used in food packaging materials to prevent buildup of static electricity, the same kind of problem that causes clothing to crackle and cling to the body or an electric discharge when a metal object is touched.

> *N-acyl sarcosines* and *N, N-Bis (2-hydroxyethyl) alkyl (C_{12}-C_{18}) amine:* antistatic agents

13. Antisticking agent: To prevent food from being sticky in texture, or from adhering to packaging material. *Lecithin*, a common ingredient of milk and cream, is added to process cheese food and process cheese spread as an antisticking agent.

14. Bactericides: Chemical compounds with the ability to kill or retard bacteria that cause food spoilage.

> *Sodium benzoate* (a salt of benzoic acid, the acid itself found naturally occurring in cranberries, prunes, and other food products): carbonated and noncarbonated beverages, fruit drinks, fruit juices, salted margarine, pie fillings, jams, and jellies
>
> *Acetic acid* (vinegar acid): catsup, mayonnaise, pickles

15. Beef cut softener: Organic compounds with the ability to break down tough, fibrous material in beef.

> *Papain:* beef cut softener derived from the juice and leaves of a plant (*Carica papaya L., Carricacae*). Papain contains enzymes similar to pepsin which are found in the human digestive system and perform similar digestive action on meat products we eat.
>
> *Bromelin:* a protein-digesting enzyme found in pineapple
>
> *Aspergillus flavus*, oryzae group: a mold with protein-digesting ability

16. Binder (emulsifier): A food additive used to hold ingredients together, to prevent separation. *Aluminum caprate* is an example.

17. Bleaching agents: Compounds used to whiten food products.

> *Calcium sulfate, dicalcium phosphate:* flour
>
> *Ammonium persulfate:* starch

18. Blood-clotting inhibitors: Used on fresh beef blood. *Citric acid*, a blood-clotting inhibitor, is found in lemon juice.

19. Buffer: A chemical agent that resists changes in acidity or alkalinity. *Hydrochloric acid*, while not considered a buffer in the usual sense, is defined as a buffer when used for modifying food starch. Other buffers include *ammonium phosphate* and *calcium lactate*.

20. Bulking agents: Used to provide a foundation for foods, much as flour is the foundation for pie crust. Many bulking agents are highly nutritious; most are sources of protein. They should not be considered nonnutritive fillers.

> *Caseinates* (salts of milk protein)
>
> *Isolated soy protein* (soybean protein)

21. Carriers (diluents): Compounds in which a product is dispersed. Water might be considered the most common carrier of all. Many products, though,

are not readily dispersible in water, or are not as active, chemically, in water as in other compounds. Thus, carriers are needed. Among other uses, carriers may be required for flavor compounds and pesticide formulations (insect killers). In the latter regard carriers should be distinguished from the "active ingredient(s)" though when and if a carrier ends up in food it is considered an additive the same as the active ingredient itself. These would be termed *incidental*—unintentional—*additives.*

> *Monoglycerides* and *diglycerides* of fat-forming fatty acids: flavor for bakery products and spice oil carrier
>
> *Sorbitan monostearate:* flavor and spice oil carrier in cookies, cream fillings, bread, and rolls

22. Catalyst: Any agent that speeds up a reaction. Catalysts are used frequently in resin and adhesive manufacture for food containers. They are also used in many food processes. When added intentionally, a desired end—a finished product—is reached more quickly, thereby lowering cost of production (cost of food). Some catalysts, though, must be scrupulously avoided in order to prevent bad changes. Copper in amounts as low as one ppm can cause fat oxidation—rancidity—within hours, if present in some foods.

> *Alpha amylase:* breaks down starch to dextrins and maltose
>
> *Pectinase:* breaks down pectin to produce clear (uncloudy) apple juice, grape juice, wines, etc.

23. Chewing gum base: The stuff that makes chewing gum chewy. You may not swallow your gum, but anything in the gum base that is extractable in mouth juices will find its way to the stomach. *Natural plant gums, paraffin, waxes,* and *resins* are all used in chewing gum.

24. Clarifying agents: Compounds for clearing away suspended fruit solids in juices and wines. Not always is clarity desired, but in apple or grape juice and in most wines, crystal-clear products have the most market appeal. *Pectinase,* an enzyme that can break down pectin, which occurs naturally in fruit and which holds fruit ingredients in suspension, is an additive that causes suspended solids to clump together and drop out of solution. This processing treatment is also needed in the manufacture of many "clear" jams and jellies. It is a perfectly harmless food preparation method.

25. Cloud inhibitor: Chemicals that prevent foods from becoming cloudy (opaque). Two compounds used in vegetable and salad oils are *coconut oil acid polyglycerol esters* and *oleic acid polyglycerol ester.*

26. Coatings: Used to make containers, closures, gaskets, etc., more resistant to erosion, leakage, damage, and so on. Nobody likes leaky milk cartons.

Coatings are also used on fresh fruits and vegetables both to protect and to enhance appearance.

Acrylamide: closures and gaskets

Oleic acid, potassium persulfate: fresh fruit and vegetables

27. Color fixative: For maintaining the natural color of a foodstuff. In many cases color changes in foods take place at a much faster rate than spoilage. A food will retain its flavor and nutritive value, while "going bad" in appearance. *Potassium nitrite* is a color fixative in cured meat products.

28. Coloring: Color means different things to different people. Some of us will eat only brown-shelled eggs, others only white-shelled eggs, even though we never eat the shell as such and what's inside is exactly the same in either case. Just as color is associated with spoilage (whether rightly or wrongly), so is it associated with quality. Some markets demand highly colored (yellow to orange) Cheddar cheese; other markets the lighter (or uncolored) Cheddar. The difference? A little added color.

Annatto, an extract of annatto seeds (*Bina orellana L.*): food coloring in general

Alkanet, an extract of the root of *Alkanna tinctoria*, a plant native to the Mediterranean region, Hungary, and western Asia: sausage casings, margarine, shortening

29. Conditioning agent: Used to achieve the kind of consistency required to process a food.

Sodium stearyl fumarate: dough, dehydrated potatoes, processed cereals, starch, flour-thickened foods

Calcium stearoyl-2-lactate: potato flakes, dehydrated potatoes

30. Cook-out juice preventive: Chemical used to prevent excessive loss of meat juices during cooking.

Sodium acid pyrophosphate, monosodium phosphate: meats

31. Corrosion inhibitor: Steel, tinplate, and other metals used in processing equipment and packaging material for transporting steel and tinplate articles. Corrosion inhibitors add life to processing equipment and also help to prevent erosion of metal components into foods, some metals of which hasten off-flavor development in food.

Sodium silicate: a corrosion inhibitor

Dicylcohexylamine nitrite: a corrosion inhibitor for use in metal packaging equipment

32. Cross-linking agents: Used in the manufacture of plastic (epoxy resin). Cross-linking agents can be likened to a stitch in a knitting pattern. They

form "linkages"—a matrix—that binds the plastic together. *Butyl acrylate* and *styrene* are two cross-linking agents.

33. Crystallization control agents: To prevent crystal formation which, like ice crystals, may cause texture changes (crunchiness where you don't want it), or appearance changes (cloudiness), or undesirable thickening.

> *Sorbitol:* soft candies and confectionaries
>
> *Sodium acid pyrophosphate:* tuna fish products

34. Curing agent: A substance used to bring about or maintain desired characteristics in a food, either flavor, texture, or color characteristics.

> *Potassium nitrite:* meat products
>
> *Salt:* meat products

35. Decolorizers: Products used to remove unwanted color in foods.

> *Dimethyl dialkyl ammonium chloride:* sugar
>
> *Activated charcoal:* wines

36. Defoaming agents: Compounds used to prevent foam formation. If you've ever had a pan of food boil over on a stove, you've experienced the need for defoaming compounds. In bulk food manufacture, defoamer use falls into several categories: (a) Food processing (cooking purposes): examples are *aluminum stearate* and *mineral oil* in beet sugar and yeast processing. (b) Wash water: for washing certain raw foods. (c) Coating manufacture: for food cartons. (d) Paper and paperboard manufacture: an example is *methyl alcohol*. (e) Components of formulations applied to growing crops: to allow better spray and coverage characteristics. (f) Filling operations: to prevent foam from interfering with the level of fill of containers.

37. Denuding agent: To remove the mucous membrane on tripe. A compound used for this purpose is *sodium metasilicate*.

38. Drugs (veterinary): Used to treat animal and poultry diseases, by either infusion or injection or in the feed. No matter which application is used, the drugs per se and any associated chemicals (carriers, diluents, etc.) may find their way into food. Specific withdrawal times are required for milk, meat, swine, and poultry products when drugs are used. A number of *antibiotics* typify this class of food additives.

39. Drying oils: Products which, upon exposure to air, dry to resistant film. Drying oils are often used as components of finished resins. In this category would fall *candlenut oil, dehydrated castor oil, hempseed oil, linseed oil*.

40. Dust inhibitors: To hold down dust in animal and poultry feeds. Concentrated feeds can be very dusty and dirty, and create an unhealthy environment for both man and animal. *Mineral oil* is often used as a dust inhibitor.

41. Dyes: Coloring agents used in some pesticide formulations. One such dye is *methyl violet 2 B.*

42. Electron beam radiation: Used in a limited way, as a method of food preservation. It may also be used in treating wheat to control insect infestation.

43. Emulsifying agents: A sizable class of food additives whose major function is to prevent ingredient separation, or—to put it positively—to maintain homogeneity. Without emulsifiers, many fatty foods would separate into an oily top layer and a watery underlayer. Appearance is affected, also pourability. Think of vinegar and oil salad dressings (where separation of oil and vinegar is actually desired). Without emulsifiers, most foods would go onto the market shelf similarly separated into component parts. Gravies, sauces, dressings, mixes would all be greasy messes far different from what we now know them to be.

> *Stearyl monoglyceride:* shortening
>
> *Polysorbate 60:* cakes and mixes, whipped topping, shortening, vegetable oil, nonstandardized dressings, chocolates and chocolate syrup, ice cream, etc.

44. Enrichment compounds: (See dietary nutrients and supplements.) Nutrients added to food to increase the nutritional value. *Enrichment* is the term often used to designate nutrient fortification of flour and bread, although similar "enrichment" is processed into many other foods. Chemical derivatives of *thiamine, riboflavin, niacin,* and *iron* are common nutrients used to enrich flour and bread.

45. Enzymes: Enzymes are organic catalysts. That is, they are products of living cells and are able to speed up certain specific biological reactions. Enzymes are one of the means by which human and animal life carry on life processes. Enzymes can cause off-flavors or brown discoloration in food. They may also be used to bring about desired changes in flavor or physical characteristics of foods. *Rennet* extract, an enzyme base, causes milk to coagulate in the cheese-making process. *Lipase,* another enzyme, breaks down fat to produce rancid flavors which blend with and enhance the flavor of such diverse foods as blue cheese and chocolate.

46. Extractants (solvents): In food processing it may be necessary to separate or purify ingredients (as in soybean processing for protein and fat recovery) or to remove a specific unwanted component in order to meet given market demands. Some compounds are leached out. Others are left as residues, while the desired component is removed. Solvents and extractants are used for this separation process.

Methylene chloride: for extracting caffein in the preparation of decaffeinated coffee

Hexane: an extractant for the color additive turmeric oleoresin

47. Feed additives: Compounds added to animal and poultry feed to promote growth or prevent disease.

Bacitracin methylene disalicylate: for chicken feed. Bacitracin is an antibiotic produced by *Bacillus subtilus* and is an effective killer of certain diseases of poultry.

Coccidiostats: for poultry. These germ killers reduce poultry losses due to a disease that strikes in the early weeks following hatching.

48. Fermentation adjunct: Compounds used in conjunction with fermenting processes. *Starch* products obtained from corn and rice are added to the mix from which beer is brewed.

49. Fermentation food: Yeast, the fermenting agent in wines and other foods, requires food for growth and multiplication. *Phosphates* are additives that are food for yeast used in the manufacture of champagne and sparkling wine. Specifically, phosphates are added to initiate secondary fermentation.

50. Filters: To separate or purify food products or ingredients. Most of us have used cheesecloth at one time or another to filter juices or cheese curd or other food. Filter materials vary in composition and makeup, and, because food contact occurs, filter materials are regulated as a potential contaminant (incidental additive) of foods.

51. Firming agents: To maintain and improve the texture of processed fruits and vegetables. Firming agents are responsible for the crunchy characteristics of pickles and apples. They also preserve texture and prevent foods from becoming mushy during and after processing.

Alum (aluminum sulfate and other aluminum compounds): pickles

Calcium chloride: canned potatoes, canned sliced apples

52. Flavor enhancer: Certain additives have the ability to "bring out" natural or added flavors of food; the food becomes more tasty. *Monosodium glutamate* (a salt of a naturally occurring amino acid found in protein) is a meat flavor enhancer. Other flavor enhancers include 5' *inosinate* and *disodium guanylate* (salt of guanylic acid, a yeast or pancreatic extract).

53. Flavoring: One of the largest categories of food additives is flavoring. Flavoring may be natural (derived from natural sources) or artificial (chemically synthesized). Most foods have added flavor.

Natural vanilla (an extract of the vanilla bean): a natural flavor

Vanillin acetate: an artificial vanilla flavor

54. Floating agent: Used in the manufacture of vinegar where exposure to

air (oxygen) is necessary to allow growth of organisms which convert alcohol to acetic acid.

> *Petroleum hydrocarbons:* additives in the production of wine and
> vinegar

55. Fumigants: Literally, compounds whose "fumes" are used for disinfection of food handling and storage areas; for ridding areas of germs and insect pests which might otherwise contaminate food.

> *Methyl formate:* bulk and package raisins
>
> *Propylene oxide:* dried prunes, glacéed fruit, bulk cocoa

56. Fungicides: Chemical agents that can destroy fungus, a cause of food spoilage and food-borne diseases. Various *phosphates* have antifungal activity in cherries, fruits, cheese, wine, meats, and poultry.

57. Growth promoter: Compounds to stimulate growth of poultry or animals.

> *Progesterone:* lambs and steers; no residue is allowed in finished,
> slaughtered animals
>
> *Diethylstilbesterol (DES):* cattle; status as an additive is currently under
> study

58. Herbicides: Weed-killing compounds for application to preemergent and growing crops. Tolerances (amounts allowed on or in finished products) are established when deemed necessary.

> *Benefin:* preplanting soil treatment
>
> *Dalapon:* almonds, corn grain, bananas, cranberries, fresh corn, sugar
> cane, grapefruit, oranges, lemons, limes, tangerines, walnuts, pecans

59. Humectants: For preventing certain foods from drying out. Without humectants, shredded coconut would be dry and crumbly, marshmallows hard and gummy. Most humectants are found naturally occurring in fruits and vegetables, and in chemical combination in meat fats.

> *Sorbitol:* marshmallows, peppermint patties, sugarless chewing gum
>
> *Glycerine:* coconut

60. Ink transfer inhibitors: For inks used on packaging material. An example is *carbon tetrachloride.*

61. Iodizing compounds: For providing iodine in iodized salt, iodine being necessary to prevent goiter. An example is *potassium iodide.*

62. Leavening agents: Ingredients that make bread rise and bakery goods light and fluffy in texture. Leavening agents used in commercial bakeries are essentially the same as those used in the home, but to meet the demands of eight-hour workdays, they are designed to leaven at, in some cases, a slower rate of speed.

Yeast: bread making

Baking powder: bread, cakes, and other bakery goods

63. Lubricants: Used in a number of ways to provide lubricating effect.

Aluminum stearoyl benzoyl hydroxide: for metallic articles in contact with food

Mineral oil: animal feeds in manufacture of pellet, cube, or block forms of feed

64. Maturing agent: To hasten otherwise slower, often natural processes. Storage of foods for maturing (curing) purposes is very costly. Where maturing time can be reduced, food cost can be lowered.

Hydrogen peroxide: some cheeses

Calcium sulfate: flour

65. Molded plastic: Containers for food.

66. Neutralizing agent: To neutralize acidity or alkalinity.

Hydrochloric acid: food starch

Monocalcium phosphate monohydrate: bread dough

67. Nutrients: Food and feed (animal and poultry) supplementation.

Isolated soy protein: meat extender

Trace minerals: human food, animal and poultry feed

Vitamins and minerals: various fortified foods

68. Nonnutritive sweeteners: For sweetening low-calorie products. A common nonnutritive sweetener is *saccharin.*

69. Oiling-off preventer: An agent to prevent separation of an oily layer in foods.

Monoglycerides and *diglycerides* of certain edible fats and oils: peanut butter

Monoglycerides and *diglycerides* from the glycerolysis of certain edible fats and oils: margarine

70. Paperboard components: Any material that is a part of those food packages made of paperboard. Like food flavors, this is one of the largest categories of food additives listed in FDA files. *Itaconic acid* and *nitrocellulose* are examples of paperboard component compounds.

71. Peeling agent enhancer: Chemical compounds to assist mechanized fruit and vegetable peeling operations. *Pelargonic acid* (from oil of pelargonium) is used in the lye-peeling of certain fruits.

72. Pesticides, insect killers: Although much controversy has raged over the use of pesticides, it is generally understood that insect control, by whatever means, is absolutely essential to an adequate food supply.

Omite: sulfite miticide (mite killer) used on commercial fruit crops

Dimethoate: insecticide spray for dairy barns

73. Plastic colorant: Coloring agents for plastic food containers. *Aluminum* is a plastic colorant compound.

74. Plasticizer: To make a product more pliable.

Mineral oil: plasticizer for rubber

Lactylic esters of fatty acids: food plasticizer (or emulsifier)

75. Preservatives: Compounds used to retard growth of spoilage microorganisms (bacteria, yeast, mold, fungus). Antioxidants are preservatives also, but a distinction is made here between microbial spoilage and chemical (oxidative) spoilage.

Sodium benzoate: antimycotic (mold inhibitor) for meat, margarine

Sodium propionate: bread and bakery products

Stannous chloride: fruit drinks and other foods

76. Refining agent: For purification of food. *Diatomaceous earth* is used in refining rendered fats.

77. Rehydration aids: Agents used to aid in reliquefying dried food products. Drying is an excellent way of preserving foods, but getting dried food back into liquid form for consumption can pose problems. *Polyhydric alcohols* are added to some foods to assist in rehydration.

78. Release agents: For manufacture of food-packaging materials. Examples are *linoleamide* (a derivative of a food fatty acid) and *polyethylene glycol 400.*

79. Rodenticides (mice and rat killers): Used to keep food-processing and warehouse facilities free of these carriers of human disease; necessary to maintain adequate supplies of food. In some countries of the world rodents consume large quantities of food fit for human needs.

80. Salt substitute: Provides a saltlike flavor for foods in which table salt (sodium chloride) is unwanted; for example, foods for persons on low-sodium diets. *Glutamic acid* (an amino acid of food protein) is used as a salt substitute.

81. Sanitizing solutions: Chemicals used to kill microorganisms on the surface of food-processing equipment. Food contact surface of processing equipment is a major source of microbial contamination of food, microbes which can cause disease and spoilage. Residual sanitizer remaining on equipment is a potential contaminant of food and is therefore considered an incidental food additive.

Sodium hypochlorite (a chemical used in clothes bleach): sanitizer for most processing equipment

Quaternary ammonium compounds: large group of chemical agents of similar (though not identical) molecular structure; applicable to most food contact surfaces

82. Seasoning, spices: Any spice used to season food is a food additive (for example, sage, mustard, horseradish, dill, nutmeg). Some common and some not-so-common spices fall in this category of additives.

83. Sequestrants: As the word implies, these are compounds that have the ability to "sequester," to seize and hold other compounds. Sequestrants may be added in food to hold ingredients in solution (to prevent them from sedimenting out as a separate layer) or to hold desired ingredients in suspension (during processing) while other ingredients or contaminants are removed.

Calcium chloride: evaporated milk

Sodium thiosulfate: table salt

84. Setting agents: For use in manufacture of paper and paperboard containers, to harden or firm protein components, casein (milk protein) for example. *Hexamethylene tetramine* is one setting agent.

85. Slimicides: Compounds used to inhibit growth of slime-producing molds during paper and paperboard manufacture. This is another large category of incidental food additives. Examples are *acetone, capric nitrate,* and *potassium pentachlorophenate.*

86. Solvents: For dissolving materials, either to rid a product of unwanted components or to carry desired components in solution. Water is the oldest known solvent, but water will not dissolve all food materials, nor will it handle the tough solvent jobs required in the manufacture of food-packaging material. Solvents are used in foods as carriers for flavors, colors, stabilizers, antioxidants, emulsifiers, etc.

Propylene glycol: solvent for flavor compounds

Ethyl alcohol: solvent for some coloring compounds

87. Sprays and dusts (preharvest): Incidental additives that function as herbicides or insecticides are classified solely as preharvest sprays or dusts. One such compound is *metaldehyde,* an insecticide applied to strawberry plants.

88. Stabilizers: Thickening agents. *Gelatin,* the ingredient that causes gelatin desserts to gel (thicken), is a stabilizer. If added specifically to a food to cause thickening, gelatin is an additive. There are a fairly large number of food stabilizers, many of which, like gelatin, are natural foods in and of themselves. Some examples found in many foods, from ice cream to jellies and jams, are *dextrins* (starch gum), chemically treated *starches, hydroxymethyl-*

cellulose, pectin (from apple peel), *guar gum,* and *tragacanth* (an Asiatic plant derivative).

89. Supplements, dietary: Vitamins or minerals added to food or combined in tablets (as "vitamin" pills) to standardize or supplement the intake of these nutrients. Vitamin C found in fruit juice, when added to fruit juice as ascorbic acid (a chemical name for vitamin C), is an additive. Natural and synthetic vitamins are considered nutritionally identical; they are identical in molecular structure. But added vitamins (and vitamin supplements) must be carried in solvents or diluents, and these latter compounds are also additives. In vitamin supplements it is the carrier compounds to which some people may be allergic. Read the statement of ingredients to identify such substances.

Ascorbic acid: fruit juices and drinks

Vitamin A, vitamin A acetate, alpha tocopherolacetate, vitamin A palmitate: skim milk, nonfat dry milk

Calcium oxide: a source of the mineral calcium

Ferric phosphate (or *pyrophosphate* or *sulfate*): sources of the mineral iron

90. Surfactants: Compounds with "surface-active" properties. They are used to provide penetrating power (spreading and wetting power) in detergents or in foods to (1) aid the reconstitution (rewetting) of dry foods to their liquid form; (2) improve texture of cake, cake mixes, and doughnuts; (3) hold the butterfat in suspension in ice cream (also to improve dryness and stiffness); (4) prevent oily layers from forming in coffee when coffee whiteners are added; (5) increase the volume of cake icing; (6) emulsify (hold in suspension) fat in meat products; (7) emulsify fluid shortenings.

Dioctyl sodium sulfo succinate: dry gelatin dessert, dry beverage base, fruit and juice drinks (where standard of identity permits)

Propylene glycol monostearate: fluid shortening

91. Sweeteners: The additive that far outweighs all others in amount consumed is *sugar,* either cane or beet sugar. The average annual consumption in the United States is about 100 pounds per person.

92. Tenderizers: For making food products (meat primarily) more tender in texture (see beef cut softener).

Proteases: meat, baked goods

Ficin: meat

93. Thickeners: Used primarily to thicken jams and jellies, to provide consistency in gel formation. Without thickeners, some fruit preserves would be liquid or semiliquid in body. *Pectin* and *gelatin* are two common thickeners. (See stabilizers for other examples.)

94. Vitamins: (See supplements.)

95. Washing compounds: Cleaning compounds for food-processing equipment. Formulations consist of detergents, caustics, phosphates, chelating agents, wetting agents, and organic and mineral acids in various combinations. As in the home, cleaning of food and food-handling equipment is essential to food purity. Since small residuals of cleaning compounds may be swept into foods, they must be considered additives.

Caustic soda: basic alkali found in some cleaning compounds

Phosphoric acid: mineral acid found in some cleaning compounds

96. Water binding agents: Used in canned hams to prevent moisture loss (leakiness) which in turn influences the texture of ham. Examples are *monosodium phosphate* and *sodium acid pryophosphate.*

97. Water repellant: For animal and poultry feeds. *Mineral oil* is an additive that performs this function.

98. Water treatment (softening): Methods of treating water to reduce "hardness" compounds, calcium and magnesium salts, that cause cleaning problems. In ion exchange—a common water-softening method—chemical agents are, as the name implies, exchanged. Hardness compounds are removed, exchanged for, other chemicals. It is these "other" chemicals that are then classed as additives when water is used in food processing.

99. Weed control (see herbicides, also): For drinking water. *Dichlorobenzil* is one compound used for this purpose.

100. Whipping agents: Compounds that enhance incorporation of air, the process by which a stable foam (whipping) is formed.

Sodium lauryl sulfate: marshmallows

Calcium stearyl-2-lactate: liquid and frozen egg whites, dried egg whites, vegetable oil topping

101. Wood preservatives: To retard rotting of wood products in which foods are handled or stored.

Pentachlorophenol: plywood components of containers

Mineral spirits, paraffin waxes: wood for handling raw agricultural products

IV

Food Categories
Used in the Regulation
of Food Ingredients

1. Baked goods and baking mixes, including all ready-to-eat and ready-to-bake products, flours, and mixes requiring preparation before serving.

2. Beverages, alcoholic, including malt beverages, wines, distilled liquors, and cocktail mixes.

3. Beverages and beverage bases, nonalcoholic, including only special or spiced teas, soft drinks, coffee substitutes, and fruit- and vegetable-flavored gelatin drinks.

4. Breakfast cereals, including ready-to-eat and instant and regular hot cereals.

5. Cheeses, including curd and whey cheeses, cream, natural, grating, processed, spread, dip, and miscellaneous cheeses.

6. Chewing gum, including all forms.

7. Coffee and tea, including regular, decaffeinated, and instant types.

8. Condiments and relishes, including plain seasoning sauces and spreads, olives, pickles, and relishes, but not spices or herbs.

9. Confections and frostings, including candy and flavored frostings, marshmallows, baking chocolate, and brown, lump, rock, maple, powdered, and raw sugars.

10. Dairy product analogues, including nondairy milk, frozen or liquid creamers, coffee whiteners, toppings, and other nondairy products.

11. Egg products, including liquid, frozen, or dried eggs, and egg dishes made therefrom, i.e., egg roll, egg foo yong, egg salad, and frozen multicourse egg meals, but not fresh eggs.

12. Fats and oils, including margarine, dressings for salads, butter, salad oils, shortenings, and cooking oils.

13. Fish products, including all prepared main dishes, salads, appetizers, frozen multicourse meals, and spreads containing fish, shellfish, and other aquatic animals, but not fresh fish.

14. Fresh eggs, including cooked eggs and egg dishes made only from fresh shell eggs.

15. Fresh fish, including only fresh and frozen fish, shellfish, and other aquatic animals.

16. Fresh fruits and fruit juices, including only raw fruits, citrus, melons, and berries, and home-prepared ades and punches made therefrom.

17. Fresh meats, including only fresh or home-frozen beef or veal, pork, lamb or mutton and home-prepared fresh meat-containing dishes, salads, appetizers, or sandwich spreads made therefrom.

18. Fresh poultry, including only fresh or home-frozen poultry and game birds and home-prepared fresh poultry-containing dishes, salads, appetizers, or sandwich spreads made therefrom.

19. Fresh vegetables, tomatoes, and potatoes, including only fresh and home-prepared vegetables.

20. Frozen dairy desserts and mixes, including ice cream, ice milks, sherbets, and other frozen dairy desserts and specialties.

21. Fruit and water ices, including all frozen fruit and water ices.

22. Gelatins, puddings, and fillings, including flavored gelatin desserts, puddings, custards, parfaits, pie fillings, and gelatin base salads.

23. Grain products and pastas, including macaroni and noodle products, rice dishes, and frozen multicourse meals, without meat or vegetables.

24. Gravies and sauces, including all meat sauces and gravies, and tomato, milk, buttery, and specialty sauces.

25. Hard candy and cough drops, including hard-type candies.

26. Herbs, seeds, spices, seasonings, blends, extracts, and flavorings, including all natural and artificial spices, blends, and flavors.

27. Jams and jellies, home-prepared, including only home-prepared jams, jellies, fruit butters, preserves, and sweet spreads.

28. Jams and jellies, commercial, including only commercially processed jams, jellies, fruit butters, preserves, and sweet spreads.

29. Meat products, including all meats and meat-containing dishes, salads, appetizers, frozen multicourse meat meals, and sandwich ingredients prepared by commercial processing or using commercially processed meats with home preparation.

30. Milk, whole and skim, including only whole, lowfat, and skim fluid milks.

31. Milk products, including flavored milks and milk drinks, dry milks, toppings, snack dips, spreads, weight-control milk beverages, and other milk-origin products.

32. Nuts and nut products, including whole or shelled tree nuts, peanuts, coconut, and nut and peanut spreads.

33. Plant protein products, including the National Academy of Sciences/National Research Council "reconstituted vegetable protein" category, and meat, poultry, and fish substitutes, analogues, and extender products made from plant proteins.

34. Poultry products, including all poultry and poultry-containing dishes, salads, appetizers, frozen multicourse poultry meals, and sandwich ingredients prepared by commercial processing or using commercially processed poultry with home preparation.

35. Processed fruits and fruit juices, including all commercially processed fruits, citrus, berries, and mixtures; salads, juices and juice punches, concentrates, dilutions, ades, and drink substitutes made therefrom.

36. Processed vegetables and vegetable juices, including all commercially processed vegetables, vegetable dishes, frozen multicourse vegetable meals, and vegetable juices and blends.

37. Snack foods, including chips, pretzels, and other novelty snacks.

38. Soft candy, including candy bars, chocolates, fudge mints, and other chewy or nougat candies.

39. Soups, home-prepared, including meat, fish, poultry, vegetable, and combination home-prepared soups.

40. Soups and soup mixes, including commercially prepared meat, fish, poultry, vegetable, and combination soups and soup mixes.

41. Sugar, white, granulated, including only white granulated sugar.

42. Sugar substitutes, including granulated, liquid, and tablet sugar substitutes.

43. Sweet sauces, toppings, and syrups, including chocolate, berry, fruit, corn syrup, and maple sweet sauces and toppings.

Source: Federal Register, September 23, 1974, pp. 34173-34176.

V

A Listing of Some
Natural and Artificial
Flavoring Compounds

The use of natural and synthetic flavoring compounds is limited to those minimum amounts required to produce their intended effect in accordance with good manufacturing practices. The antioxidant BHA (butylated hydroxyanisole) may be used in flavoring substances at levels not to exceed 0.5 percent of the essential (volatile) oil content of the flavoring substances.

Source for the following information on "Natural Flavoring Substances" and "Synthetic Flavoring Substances" is *Code of Federal Regulations*, revised, April 1, 1974, pts. 120-129, pp. 450-459.

Common name	Scientific name	Limitations
Aloe	*Aloe perryi* Baker, *A. barbadensis* Mill., *A. ferox* Mill., and hybrids of this sp. with *A. africana* Mill and *A. spicata* Baker.	
Althea root and flowers . .	*Althea officinalis* L.	
Amyris (West Indian sandalwood).	*Amyris balsamifera* L.	
Angola weed.	*Roccella fuciformis* Ach.	In alcoholic beverages only.
Arnica flowers.	*Arnica montana* L., *A. fulgens* Pursh, *A. sororia* Greene, or *A. cordifolia* Hooker.	Do.

Common name	Scientific name	Limitations
Artemisia (wormwood). . .	*Artemisia* spp.	Finished food thujone free.
Artichoke leaves.	*Cynara scolymus* L.	In alcoholic beverages only.
Beeswax, white (Cire d'abeille).	*Apis mellifera* L.	
Benzoin resin	*Styrax benzoin* Dryander, *S. paralleloneurus* Perkins, *S. tonkinensis* (Pierre) Craib ex Hartwich, or other spp. of the Section *Anthostyrax* of the genus *Styrax*.	
Blackberry bark	*Rubus*, Section *Eubatus*.	
Boldus (boldo) leaves. . . .	*Peumus boldus* Mol.	Do.
Boronia flowers	*Boronia megastigma* Nees.	
Bryonia root.	*Bryonia alba* L., or *B. dioica* Jacq.	Do.
Buchu leaves.	*Barosma betulina* Bartl. et Wendl., *B. crenulata* (L.) Hook. or *B. serratifolia* Willd.	
Buckbean leaves.	*Menyanthes trifoliata* L.	Do.
Cajeput.	*Melaleuca leucadendron* L. and other *Melaleuca* spp.	
Calumba root	*Jateorhiza palmata* (Lam.) Miers.	Do.
Camphor tree	*Cinnamomum camphora* (L.) Nees et Eberm.	Safrole free.
Cascara sagrada	*Rhamnus purshiana* DC.	
Cassie flowers	*Acacia farnesiana* (L.) Willd.	
Castor oil.	*Ricinus communis* L.	
Catechu, black.	*Acacia catechu* Willd.	
Cedar, white (arborvitae), leaves and twigs	*Thuja occidentalis* L.	Finished food thajone free.
Centaury	*Centaurium umbellatum* Gilib.	In alcoholic beverages only.
Cherry pits.	*Prunus avium* L. or *P. cerasus* L.	Not to exceed 25 p.p.m. prussic acid.
Cherry-laurel leaves	*Prunus laurocerasus* L.	Do.
Chestnut leaves	*Castanea dentata* (Marsh.) Borkh.	
Chirata	*Swertia chirata* Buch.-Ham.	In alcoholic beverages only.
Cinchona, red, bark.	*Cinchona succirubra* Pav. or its hybrids.	In beverages only; not more than 83 p.p.m. total cinchona alkaloids in finished beverage.

Common name	Scientific name	Limitations
Cinchona, yellow, bark. . .	*Cinchona ledgeriana* Moens, *C. calisaya* Wedd., or hybrids of these with other spp. of *Cinchona.*	Do.
Copaiba	South American spp. of *Copaifera* L.	
Cork, oak	*Quercus suber* L., or *Q. occidentalis* F. Gay.	In alcoholic beverages only.
Costmary.	*Chrysanthemum balsamita* L.	Do.
Costus root	*Saussurea lappa* Clarke.	
Cubeb	*Piper cubeba* L. f.	
Currant, black, buds, and leaves.	*Ribes nigrum* L.	
Damiana leaves	*Turnera diffusa* Willd.	
Davana	*Artemisia pallens* Wall.	
Dill, Indian.	*Anethum sowa* Roxb. (*Peucedanum graveolens* Benth. et Hook., *Anethum graveolens* L.).	
Dittany (fraxinella) roots	*Dictamnus albus* L.	In alcoholic beverages only.
Dittany of Crete.	*Origanum dictamnus* L.	
Dragon's blood (dracorubin).	*Daemonorops* spp.	
Elder tree leaves.	*Sambucus nigra* L.	In alcoholic beverages only; not to exceed 25 p.p.m. prussic acid in the flavor.
Elecampane rhizome and roots	*Inula helenium* L.	In alcoholic beverages only.
Elemi.	*Canarium commune* L. or *C. luzonicum* Miq.	
Erigeron	*Erigeron canadensis* L.	
Eucalyptus globulus leaves.	*Eucalyptus globulus* Labill.	
Fir ("pine") needles and twigs	*Abies sibirica* Ledeb., *A. alba* Mill., *A. sachalinesis* Masters or *A. mayriana* Miyabe et Kudo.	
Fir, balsam, needles and twigs	*Abies balsamea* (L.) Mill.	
Galanga, greater	*Alpinia galanga* Willd.	Do.
Galbanum	*Ferula galbaniflua* Boiss. et Buhse and other *Ferula* spp.	
Gambir (catechu, pale). . .	*Uncaria gambir* Roxb.	
Genet flowers . .˙.	*Spartium junceum* L.	

Common name	Scientific name	Limitations
Gentian rhizome and roots	*Gentiana lutea* L.	
Gentian, stemless	*Gentiana acaulis* L.	Do.
Germander, chamaedrys . .	*Teucrium chamaedrys* L.	Do.
Germander, golden	*Teucrium polium* L.	In alcoholic beverages only.
Guaiac	*Guaiacum officinale* L., *G. santum* L., *Bulnesia sarmienti* Lor.	
Guarana	*Paullinia cupana* HBK.	
Haw, black, bark	*Viburnum prunifolium* L.	
Hemlock needles and twigs	*Tsuga canadensis* (L.) Carr. or *T. heterophylla* (Raf.) Sarg.	
Hyacinth flowers	*Hyacinthus orientalis* L.	
Iceland moss	*Cetraria islandica* Ach.	Do.
Imperatoria	*Peucedanum ostruthium* (L.) Koch (*Imperatoria ostruthium* L.).	Do.
Iva	*Achillea moschata* Jacq.	Do.
Labdanum	*Cistus* spp.	
Lemon-verbena	*Lippia citriodora* HBK.	Do.
Linaloe wood	*Bursera delpechiana* Poiss. and other *Bursera* spp.	
Linden leaves	*Tillia* spp.	Do.
Lovage	*Levisticum officinale* Koch.	
Lungmoss (lungwort). . . .	*Sticta pulmonacea* Ach.	
Maidenhair fern	*Adiantum capillus-veneris* L.	Do.
Maple, mountain	*Acer spicatum* Lam.	
Mimosa (black wattle) flowers	*Acacia decurrens* Willd. var. *dealbata.*	
Mullein flowers	*Verbascum phlomoides* L. or *V. thapsiforme* Schrad.	Do.
Myrrh.	*Commiphora molmol* Engl., *C. abyssinica* (Berg) Engl., or other *Commiphora* spp.	
Myrtle leaves.	*Myrtus communis* L.	Do.
Oak, English, wood	*Quercus robur* L.	Do.
Oak, white, chips	*Quercus alba* L.	
Oak moss.	*Evernia prunastri* (L.) Ach., *E. furfuracea* (L.) Mann, and other lichens.	Finished food thujone.
Olibanum	*Boswellia carteri* Birdw. and other *Boswellia* spp.	

Common name	Scientific name	Limitations
Opopanax (bisabol- myrrh)	*Opopanax chironium* Koch (true opopanax) or *Com- miphora erythraea* Engl. var. *labrescens.*	
Orris root	*Iris germanica* L. (includ- ing its variety *florentina* Dykes) and *I. pallida* Lam.	
Pansy.	*Viola tricolor* L.	In alcoholic beverages only.
Passion flower.	*Passiflora incarnata* L.	
Patchouly	*Pogostemon cablin* Benth. and *P. heyneanus* Benth.	
Peach leaves	*Prunus persica* (L.) Batsch.	In alcoholic beverages only; not to exceed 25 p.p.m. prussic acid in the flavor.
Pennyroyal, American . . .	*Hedeoma pulegioides* (L.) Pers.	
Pennyroyal, European . . .	*Mentha pulegium* L.	
Pine, dwarf, needles and twigs	*Pinus mugo* Turra var. *pumilio* (Haenke) Zenari.	
Pine, Scotch, needles and twigs.	*Pinus sylvestris* L.	
Pine, white, bark	*Pinus strobus* L.	In alcoholic beverages only.
Pine, white oil	*Pinus palustris* Mill., and other *Pinus* spp.	
Poplar buds	*Populus balsamifera* L. (*P. tacamahacca* Mill.), *P. candicans* Ait., or *P. nigra* L.	In alcoholic beverages only.
Quassia.	*Picrasma excelsa* (Sw.) Planch. or *Quassia amara* L.	
Quebracho bark.	*Aspidosperma quebracho- blanco* Schlecht, or (*Quebrachia lorentzii* Griseb).	*Schinopsis lorentzii* (Griseb.) Engl.
Quillaia (soapbark)	*Quillaja saponaria* Mol.	
Red saunders (red sandalwood)	*Pterocarpus san alinus* L.	In alcoholic beverages only.
Rhatany root	*Krameria triandra* Ruiz et Pav. or *K. argentea* Mart.	
Rhubarb, garden root. . . .	*Rheum rhaponticum* L.	In alcoholic beverages only.
Rhubarb root	*Rheum officinale* Baill., *R. palmatum* L., or other spp. (excepting *R. rhapon-*	

Common name	Scientific name	Limitations
	ticum L.) or hybrids of *Rheum* grown in China.	
Roselle	*Hibiscus sabdariffa* L.	Do.
Rosin (colophony)	*Pinus palustris* Mill., and other *Pinus* spp.	Do.
St. Johnswort leaves, flowers, and caulis	*Hypericum perforatum* L.	Hypericin-free alcohol distillate form only; in alcoholic beverages only.
Sandalwood, white (yellow, or East Indian). . . .	*Santalum album* L.	
Sandarac	*Tetraclinis articulata* (Vahl.), Mast.	In alcoholic beverages only.
Sarsaparilla.	*Smilax aristolochiaefolia* Mill. (Mexican sarsaparilla), *S. regelii* Killip et Morton (Honduras sarsaparilla), *S. febrifuga* Kunth (Ecuadorean sarsaparilla), or undetermined *Smilax* spp. (Ecuadorean or Central American sarsaparilla).	
Sassafras leaves	*Sassafras albidum* (Nutt.) Nees.	Safrole free.
Senna, Alexandria.	*Cassia acutifolia* Delile	
Serpentaria (Virginia snakeroot)	*Aristolochia serpentaria* L.	In alcoholic beverages only.
Simaruba bark.	*Simaruba amara* Aubl.	Do.
Snakeroot, Canadian (wild ginger)	*Asarum canadense* L.	
Spruce needles and twigs. .	*Picea glauca* (Moench) Voss or *P. mariana* (Mill.) BSP.	
Storax (styrax)	*Liquidambar orientalis* Mill. or *L. styraciflua* L.	
Tagetes (marigold)	*Tagetes patula* L., *T. erecta* L., or *T. minuta* L. (*T. glandulifera* Schrank).	As oil only.
Tansy.	*Tanacetum vulgare* L.	In alcoholic beverages only; finished alcoholic beverage thujone free.
Thistle, blessed (holy thistle).	*Onicus benedictus* L.	In alcoholic beverages only.
Thymus capitatus (Spanish "origanum") . .	*Thymus capitatus* Hoffmg. et Link.	
Tolu	*Myroxylon balsamum* (L.) Harms.	

Common name	Scientific name	Limitations
Turpentine.	*Pinus palustris* Mill. and other *Pinus* spp. which yield terpene oils exclusively.	
Valerian rhizome and roots.	*Valeriana officinalis* L.	
Veronica	*Veronica officinalis* L.	In alcoholic beverages only.
Vervain, European	*Verbena officinalis* L.	Do.
Vetiver	*Vetiveria zizanioides* Stapf.	
Violet, Swiss.	*Viola calcarata* L.	In alcoholic beverages only.
Walnut husks (hulls), leaves, and green nuts. . .	*Juglans nigra* L., or *J. regia* L.	
Woodruff, sweet.	*Asperula odorata* L.	Do.
Yarrow	*Achillea millefolium* L.	In beverages only; finished beverage thujone free.
Yerba santa	*Eriodictyon californicum* (Hook. et Arn.) Torr.	
Yucca, Joshua-tree	*Yucca brevifolia* Engelm.	
Yucca, Mohave	*Yucca schidigera* Roezl ex Ortgies (*Y. mohavensis* Sarg.).	

Synthetic Flavoring Substances

Acetal; acetaldehyde diethyl acetal.
Acetaldehyde phenethyl propyl acetal.
Acetanisole; 4'-methoxyacetophenone.
Acetophenone; methyl phenyl ketone.
Adipic acid; 1,4-butanedicarboxylic acid.
Allyl anthranilate.
Allyl butyrate.
Allyl cinnamate.
Allyl cyclohexaneacetate.
Allyl cyclohexanebutyrate.
Allyl cyclohexanehexanoate.
Allyl cyclohexanepropionate.
Allyl cyclohexanevalerate.
Allyl disulfide.
Allyl 2-ethylbutyrate.
Allyl hexanoate; allyl caproate.
Allyl *a*-ionone; 1-(2,6,6-trimethyl-2-cyclo-hexen-1-yl) -1,6-heptadiene-3-one.
Allyl isothiocyanate; mustard oil.
Allyl isovalerate.
Allyl mercaptan; 2-propene-1-thiol.
Allyl nonanoate.

Allyl octanoate.
Allyl phenoxyacetate.
Allyl phenylacetate.
Allyl propionate.
Allyl sorbate; allyl 2,4-hexadienoate.
Allyl sulfide.
Allyl tiglate; allyl *trans*-2-methyl-2-butenoate.
Allyl 10-undecenoate.
Ammonium isovalerate.
Ammonium sulfide.
Amyl alcohol; pentyl alcohol.
Amyl butyrate.
a-Amylcinnamaldehyde.
a-Amylcinnamaldehyde dimethyl acetal.
a-Amylcinnamyl acetate.
a-Amylcinnamyl alcohol.
a-Amylcinnamyl formate.
a-Amylcinnamyl isovalerate.
Amyl formate.
Amyl heptanoate.
Amyl hexanoate.

Amyl octanoate.
Anisole; methoxybenzene.
Anisyl acetate.
Anisyl alcohol; *p*-methoxybenzyl alcohol.
Anisyl butyrate.
Anisyl formate.
Anisyl phenylacetate.
Anisyl propionate.
Beechwood creosote.
Benzaldehyde dimethyl acetal.
Benzaldehyde glyceryl acetal; 2-phenyl-*m*-dioxan-5-ol.
Benzaldehyde propylene glycol acetal; 4-methyl-2-phenyl-*m*-dioxolane.
Benzenethiol; thiophenol.
Benzoin; 2-hydroxy-2-phenylacetophenone.
Benzophenone; diphenylketone.
Benzyl acetate.
Benzyl acetoacetate.
Benzyl alcohol.
Benzyl benzoate.
Benzyl butyl ether.
Benzyl butyrate.
Benzyl cinnamate.
Benzyl 2,3-dimethylcrotonate; benzyl methyl tiglate.
Benzyl disulfide; dibenzyl disulfide.
Benzyl ethyl ether.
Benzyl formate.
3-Benzyl-4-heptanone; benzyl dipropyl ketone.
Benzyl isobutyrate.
Benzyl isovalerate.
Benzyl mercaptan; *a*-toluenethiol.
Benzyl methoxyethyl acetal; acetaldehyde benzyl β-methoxyethyl acetal.
Benzyl phenylacetate.
Benzyl propionate.
Benzyl salicylate.
Birch tar oil.
Borneol; *d*-camphanol.
Bornyl acetate.
Bornyl formate.
Bornyl isovalerate.
Bornyl valerate.
β-Bourbonene; 1,2,3,3a,3bβ,4,5,6,6aβ, 6bα-decahydro-1α-isopropyl-3aα-methyl-6-methylene-cyclobuta [1,2:3,4] dicyclopentene.

2-Butanone; methyl ethyl ketone.
Butter acids.
Butter esters.
Butter starter distillate.
Butyl acetate.
Butyl acetoacetate.
Butyl alcohol; 1-butanol.
Butyl anthranilate.
Butyl butyrate.
Butyl butyryllactate; lactic acid, butyl ester, butyrate.
a-Butylcinnamaldehyde.
Butyl cinnamate.
Butyl 2-decenoate.
Butyl ethyl malonate.
Butyl formate.
Butyl heptanoate.
Butyl hexanoate.
Butyl *p*-hydroxybenzoate.
Butyl isobutyrate.
Butyl isovalerate.
Butyl lactate.
Butyl laurate.
Butyl levulinate.
Butyl phenylacetate.
Butyl propionate.
Butyl stearate.
Butyl sulfide.
Butyl 10-undecenoate.
Butyl valerate.
Butyraldehyde.
Cadinene.
Camphene; 2,2-dimethyl-3-methylene-norbonane.
d-Camphor.
Carvacrol; 2-*p*-cymenol.
Carvacryl ethyl ether; 2-ethoxy-*p*-cymene.
Carveol; *p*-mentha-6,8-dien-2-ol.
4-Carvomenthenol; 1-*p*-menthen-4-ol; 4-terpinenol.
cis Carvone oxide; 1,6-epoxy-*p*-menth-8-en-2-one.
Carvyl acetate.
Carvyl propionate.
β-Caryophyllene.
Caryophyllene alcohol.
Caryophyllene alcohol acetate.
β-Caryophyllene oxide; 4,12,12,-tri-

methyl-9-methylene-5-oxatricyclo [8.2.0.0^{4,6}] dodecane.

Cedarwood oil alcohols.

Cedarwood oil terpenes.

1,4-Cineole.

Cinnamaldehyde ethylene glycol acetal.

Cinnamic acid.

Cinnamyl acetate.

Cinnamyl alcohol; 3-phenyl-2-propen-1-ol.

Cinnamyl anthranilate.

Cinnamyl benzoate.

Cinnamyl butyrate.

Cinnamyl cinnamate.

Cinnamyl formate.

Cinnamyl isobutyrate.

Cinnamyl isovalerate.

Cinnamyl phenylacetate.

Cinnamyl propionate.

Citral diethyl acetal; 3,7-dimethyl-2,6-octadienal diethyl acetal.

Citral dimethyl acetal; 3,7-dimethyl-2,6-octadienal dimethyl acetal.

Citral propylene glycol acetal.

Citronellal; 3,7-dimethyl-6-octenal; rhodinal.

Citronellol; 3,7-dimethyl-6-octen-1-ol; *d*-citronellol.

Citronelloxyacetaldehyde.

Citronellyl acetate.

Citronellyl butyrate.

Citronellyl formate.

Citronellyl isobutyrate.

Citronellyl phenylacetate.

Citronellyl propionate.

Citronellyl valerate.

p-Cresol.

Cuminaldehyde; cuminal; *p*-isopropyl benzaldehyde.

Cyclohexaneacetic acid.

Cyclohexaneethyl acetate.

Cyclohexyl acetate.

Cyclohexyl anthranilate.

Cyclohexyl butyrate.

Cyclohexyl cinnamate.

Cyclohexyl formate.

Cyclohexyl isovalerate.

Cyclohexyl propionate.

p-Cymene.

γ-Decalactone; 4-hydroxy-decanoic acid, γ-lactone.

δ-Decalactone; 5-hydroxy-decanoic acid, δ-lactone.

Decanal dimethyl acetal.

1-Decanol; decylic alcohol.

2-Decenal.

3-Decen-2-one; heptylidene acetone.

Decyl acetate.

Decyl butyrate.

Decyl propionate.

Dibenzyl ether.

4,4-Dibutyl-γ-butyrolactone; 4,4-dibutyl-4-hydroxy-butyric acid, γ-lactone.

Dibutyl sebacate.

Diethyl malate.

Diethyl malonate; ethyl malonate.

Diethyl sebacate.

Diethyl succinate.

Diethyl tartrate.

2,5-Diethyltetrahydrofuran.

Dihydrocarveol; 8-*p*-menthen-2-ol; 6-methyl-3-isopropenylcyclohexanol.

Dihydrocarvone.

Dihydrocarvyl acetate.

m-Dimethoxybenzene.

p-Dimethoxybenzene; dimethyl hydroquinone.

2,4-Dimethylacetophenone.

a,a-Dimethylbenzyl isobutyrate; phenyldimethylcarbinyl isobutyrate.

2,6-Dimethyl-5-heptenal.

2,6-Dimethyl octanal; isodecylaldehyde.

3,7-Dimethyl-1-octanol; tetrahydrogeraniol.

a,a-Dimethylphenethyl acetate; benzylpropyl acetate; benzyldimethylcarbinyl acetate.

a,a-Dimethylphenethyl alcohol; dimethylbenzyl carbinol.

a,a-Dimethylphenethyl butyrate; benzyldimethylcarbinyl butyrate.

a,a-Dimethylphenethyl formate; benzyldimethylcarbinyl formate.

Dimethyl succinate.

1,3-Diphenyl-2-propanone; dibenzyl ketone.

delta-Dodecalactone; 5-hydroxydodecanoi acid, deltalactone.

γ-Dodecalactone; 4-hydroxydodecanoic acid γ-lactone.

2-Dodecenal.

Estragole.
p-Ethoxybenzaldehyde.
Ethyl acetoacetate.
Ethyl 2-acetyl-3-phenylpropionate; ethyl-
  benzyl acetoacetate.
Ethyl aconitate, mixed esters.
Ethyl acrylate.
Ethyl-p-anisate.
Ethyl anthranilate.
Ethyl benzoate.
Ethyl benzoylacetate.
a-Ethylbenzyl butyrate; a-phenylpropyl
  butyrate.
Ethyl brassylate; tridecanedioic acid cyclic
  ethylene glycol diester; cyclo 1,13-ethyl-
  enedioxytridecan-1,13-dione.
2-Ethylbutyl actate.
2-Ethylbutyraldehyde.
2-Ethylbutyric acid.
Ethyl cinnamate.
Ethyl crotonate; trans-2-butenoic acid ethyl-
  ester.
Ethyl cyclohexanepropionate.
Ethyl decanoate.
Ethyl formate.
2-Ethylfuran.
Ethyl 2-furanpropionate.
4-Ethylguaiacol; 4-ethyl-2-methoxy-
  phenol.
Ethyl heptanoate.
2-Ethyl-2-heptenal; 2-ethyl-3-butyl-
  acrolein .
Ethyl hexanoate.
Ethyl isobutyrate.
Ethyl isovalerate.
Ethyl lactate.
Ethyl laurate.
Ethyl levulinate.
Ethyl maltol; 2-ethyl-3-hydroxy-4H-
  pyran-4-one.
Ethyl 2-methylbutyrate.
Ethyl myristate.
Ethyl nitrite.
Ethyl nonanoate.
Ethyl 2-nonynoate; ethyl octyne
  carbonate.
Ethyl octanoate.
Ethyl oleate.
Ethyl phenylacetate.
Ethyl 4-phenylbutyrate.

Ethyl 3-phenylglycidate.
Ethyl 3-phenylpropionate; ethyl hydro-
  cinnamate.
Ethyl propionate.
Ethyl pyruvate.
Ethyl salicylate.
Ethyl sorbate; ethyl 2,4-hexadienoate.
Ethyl tiglate; ethyl trans-2-methyl-2-
  butenoate.
Ethyl undecanoate.
Ethyl 10-undecenoate.
Ethyl valerate.
Eucalyptol; 1,8-epoxy-p-menthane;
  cineole.
Eugenyl acetate.
Eugenyl benzoate.
Eugenyl formate.
Eugenyl methyl ether; 4-allylveratrole;
  methyl eugenol.
Farnesol; 3,7,11-trimethyl-2,6,10-
  dodecatrien-1-ol.
d-Fenchone; d-1,3,3-trimethyl-2-
  norbornanone.
Fenchyl alcohol; 1,3,3-trimethyl-2-
  norbornanol.
Formic acid.
(2-Furyl)-2-propanone; furyl acetone.
1-Furyl-2-propanone; furyl acetone.
Fusel oil,-refined (mixed amyl alcohols).
Geranyl acetoacetate; trans-3,7-dimethyl-
  2,6-octadien-1-yl acetoacetate.
Geranyl acetone; 6,10-dimethyl-5,9-
  undecadien-2-one.
Geranyl benzoate.
Geranyl butyrate.
Geranyl formate.
Geranyl hexanoate.
Geranyl isobutyrate.
Geranyl isovalerate.
Geranyl phenylacetate.
Geranyl propionate.
Glucose pentaacetate.
Glyceryl monooleate.
Guaiacol; o-methoxyphenol.
Guaiacyl acetate; o-methoxyphenyl
  acetate.
Guiaicyl phenylacetate.
Guaiene; 1,4-dimethyl-7-isopropenyl-
  Δ9, 10-octahydroazulene.
Guaiol acetate; 1,4-dimethyl-7-

(α-hydroxyisopropyl)-Δ9,10-octahydro-
azulene acetate.
γ-Heptalactone; 4-hydroxyheptanoic
acid, γ-lactone.
Heptanal; enanthaldehyde.
Heptanal dimethyl acetal.
Heptanal 1,2-glyceryl acetal.
2,3-Heptanedione; acetyl valeryl.
3-Heptanol.
2-Heptanone; methyl amyl ketone.
3-Heptanone; ethyl butyl ketone.
4-Heptanone; dipropyl ketone.
cis-4-Heptenal; cis-4-hepten-1-al.
Heptyl acetate.
Heptyl alcohol; enanthic alcohol.
Heptyl butyrate.
Heptyl cinnamate.
Heptyl formate.
Heptyl isobutyrate.
Heptyl octanoate.
1-Hexadecanol; cetyl alcohol.
ω-6-Hexadecenlactone; 16-hydoxy-6-
hexadecenoic acid, ω-lactone; ambret-
tolide.
γ-Hexalactone; 4-hydroxyhexanoic acid,
γ-lactone; tonkalide.
Hexanal; caproic aldehyde.
2,3-Hexanedione; acetyl butyryl.
Hexanoic acid; caproic acid.
2-Hexenal.
2-Hexen-1-ol.
3-Hexen-1-ol; leaf alcohol.
2-Hexen-1-yl acetate.
3-Hexenyl isovalerate.
3-Hexenyl 2-methylbutyrate.
3-Hexenyl phenylacetate; cis-3-hexenyl
phenylacetate.
Hexyl acetate.
2-Hexyl-4-acetoxytetrahydrofuran.
Hexyl alcohol.
Hexyl butyrate.
α-Hexylcinnamaldehyde.
Hexyl formate.
Hexyl hexanoate.
2-Hexylidene cyclopentanone.
Hexyl isovaleratel.
Hexyl 2-methylbutyrate.
Hexyl octanoate.
Hexyl phenylacetate; n-hexyl phenylacetate.

Hexyl propionate.
Hydroxycitronellal; 3,7-dimethyl-7-
hydroxyoctanal.
Hydroxycitronellal diethyl acetal.
Hydroxycitronellal dimethyl acetal.
Hydroxycitronellol; 3,7-dimethyl-1,7-
octanediol.
N-(4-Hydroxy-3-methoxybenzyl)-non-
anamide; pelargonyl vanillylamide.
5-Hydroxy-4-octanone; butyroin.
4-(p-Hydroxyphenyl-2-butanone; p-
hydroxybenzyl acetone.
Indole.
α-Ionone; 4-(2,6,6-trimethyl-2-cyclohex-
en-1-yl)-3-buten-2-one.
β-Ionone; 4-(2,6,6-trimethyl-1-cyclohex-
en-1-yl)-3-buten-2-one.
α-Irone; 4-(2,5,6,6-tetramethyl-2-cyclo-
hexene-1-yl)-3-buten-2-one; 6-methyl-
ionone.
Isoamyl acetate.
Isoamyl acetoacetate.
Isoamyl alcohol; isopentyl alcohol; 3-
methyl-1-butanol.
Isoamyl benzoate.
Isoamyl butyrate.
Isoamyl cinnamate.
Isoamyl formate.
Isoamyl 2-furanbutyrate; α-isoamyl
furfurylpropionate.
Isoamyl 2-furanpropionate; α-isoamyl
furfurylacetate.
Isoamyl hexanoate.
Isoamyl isobutyrate.
Isoamyl isovalerate.
Isoamyl laurate.
Isoamyl-2-methylbutyrate; isopentyl-2-
methylbutyrate.
Isoamyl nonanoate.
Isoamyl octanoate.
Isoamyl phenylacetate.
Isoamyl propionate.
Isoamyl pyruvate.
Isoamyl salicylate.
Isoborneol.
Isobornyl acetate.
Isobornyl formate.
Isobornyl isovalerate.
Isobornyl propionate.
Isobutyl acetate.

Isobutyl acetoacetate.
Isobutyl alcohol.
Isobutyl angelate; isobutyl *cis*-2-methyl-2-butenoate.
Isobutyl anthranilate.
Isobutyl benzoate.
Isobutyl butyrate.
Isobutyl cinnamate.
Isobutyl formate.
Isobutyl 2-furanpropionate.
Isobutyl heptanoate.
Isobutyl hexanoate.
Isobutyl isobutyrate.
a-Isobutylphenethyl alcohol; isobutyl benzyl carbinol; 4-methyl-1-phenyl-2-pentanol.
Isobutyl phenylacetate.
Isobutyl propionate.
Isobutyl salicylate.
Isobutyraldehyde.
Isobutyric acid.
Isoeugenol; 2-methoxy-4-propenyl-phenol.
Isoeugenyl acetate.
Isoeugenyl benzyl ether; benzyl iso-eugenol.
Isoeugenyl ethyl ether; 2-ethoxy-5-propenylanisole; ethyl isoeugenol.
Isoeugenyl formate.
Isoeugenyl methyl ether; 4-propenylvera-trole; methyl isoeugenol.
Isoeugenyl phenylacetate.
Isojasmone; mixture of 2-hexylidene-cyclopentanone and 2-hexyl-2-cyclo-penten-1-one.
a-Isomethylionone; 4-(2,6,6-trimethyl-2-cyclohexen-1-yl)-3-methyl-3-buten-2-one; methyl γ-ionone.
Isopropyl acetate.
p-Isopropylacetophenone.
Isopropyl alcohol; isopropanol.
Isopropyl benzoate.
p-Isopropylbenzyl alcohol; cuminic alcohol; *p*-cymen-7-ol.
Isopropyl butyrate.
Isopropyl cinnamate.
Isopropyl formate.
Isopropyl hexanoate.
Isopropyl isobutyrate.
Isopropyl isovalerate.

p-Isopropylphenylacetaldehyde; *p*-cymen-7-carboxaldehyde.
Isopropyl phenylacetate.
3-(*p*-Isopropylphenyl)-propionaldehyde; *p*-isopropylhydrocinnamaldehyde; cuminyl acetaldehyde.
Isopropyl propionate.
Isopulegol; *p*-menth-8-en-3-ol.
Isopulegone; *p*-menth-8-en-3-one.
Isopulegyl acetate.
Isoquinoline.
Isovaleric acid.
cis-Jasmone; 3-methyl-2-(2-pentenyl)-2-cyclopenten-1-one.
Lauric aldehyde; dodecanal.
Lauryl acetate.
Lauryl alcohol; 1-dodecanol.
Lepidine; 4-methylquinoline.
Levulinic acid.
Linalool oxide; *cis*- and *trans*-2-vinyl-2-methyl-5-(1′-hydroxy-1′-methylethyl) tetrahydrofuran.
Linalyl anthranilate; 3,7-dimethyl-1,6-octadien-3-yl anthranilate.
Linalyl benzoate.
Linalyl butyrate.
Linalyl cinnamate.
Linalyl formate.
Linalyl hexanoate.
Linalyl isobutyrate.
Linalyl isovalerate.
Linalyl octanoate.
Linalyl propionate.
Maltol; 3-hydroxy-2-methyl-4H-pyran-4-one.
Menthadienol; *p*-mentha-1,8(10)-dien-9-ol.
p-Mentha-1,8-dien-7-ol; perillyl alcohol.
Menthadienyl acetate; *p*-mentha-1,8(10)-dien-9-yl acetate.
p-Menth-3-en-1-ol.
1-*p*-Menthen-9-yl acetate; *p*-menth-2-en-9-yl acetate.
Menthol; 2-isopropyl-5-methylcyclohex-anol.
Menthone; *p*-menthan-3-one.
Menthyl acetate; *p*-menth-3-yl acetate.
Menthyl isovalerate; *p*-menth-3-yl isoval-erate.
o-Methoxybenzaldehyde.

p-Methoxybenzaldehyde; *p*-anisalde-
hyde.

o-Methoxycinnamaldehyde.

2-Methoxy-4-methylphenol; 4-methyl-
guaiacol; 2-methoxy-*p*-cresol.

4-(*p*-Methoxyphenyl)-2-butanone; anisyl
acetone.

1-(4-Methoxyphenyl)-4-methyl-1-
penten-3-one; methoxystyryl isopropyl
ketone.

1-(*p*-Methoxyphenyl)-1-penten-3-one;
a-methylanisylidene acetone; ethone.

1-(*p*-Methoxyphenyl)-2-propanone;
anisyl methyl ketone; anisic ketone.

2-Methoxy-4-vinylphenol; *p*-vinyl-
guaiacol.

Methyl acetate.

4'-Methylacetophenone; *p*-methyl-
acetophenone; methyl *p*-tolyl ketone.

2-Methoxy-4-vinylphenol; *p*-vinyl-
guaiacol. 1-yl butyrate.

Methyl anisate.

o-Methylanisole; *o*-cresyl methyl ether.

p-Methylanisole; *p*-cresyl methyl ether;
p-methoxytoluene.

Methyl benzoate.

Methylbenzyl acetate, mixed *o*-,*m*-,*p*-.

a-Methylbenzyl acetate; styralyl acetate.

a-Methylbenzyl alcohol; styralyl alcohol.

a-Methylbenzyl butyrate; styralyl
butyrate.

a-Methylbenzyl isobutyrate; styralyl isobutyrate.

a-Methylbenzyl formate; styralyl
formate.

a-Methylbenzyl propionate; styralyl
propionate.

2-Methylbutyl isovalerate.

Methyl *p-tert*-butylphenylacetate.

2-Methylbutyraldehyde; methyl ethyl
acetaldehyde.

3-Methylbutyraldehyde; isovaleralde-
hyde.

Methyl butyrate.

2-Methylbutyric acid.

a-Methylcinnamaldehyde.

p-Methylcinnamaldehyde.

Methyl cinnamate.

2-Methyl-1,3-cyclohexadiene.

Methylcyclopentenolone; 3-methylcyclo-
pentane-1,2-dione.

Methyl disulfide; dimethyl disulfide.

Methyl ester of rosin, partially
hydrogenated (as defined in
121.1059); methyl dihydroabietate.

Methyl heptanoate.

2-Methylheptanoic acid.

6-Methyl-5-hepten-2-one.

Methyl hexanoate.

Methyl 2-hexenoate.

Methyl *p*-hydroxybenzoate; methylparaben.

Methyl *a*-ionone; 5-(2,6,6-trimethyl-2-
cyclohexen-1-yl)-4-penten-3-one.

Methyl *β*-ionone; 5-(2,6,6-trimethyl-1-
cyclohexen-1-yl)-4-penten-3-one.

Methyl *δ*-ionone; 5-(2,6,6-trimethyl-3-
cyclohexen-1-yl)-4-penten-3-one.

Methyl isobutyrate.

2-Methyl-3-(*p*-isopropylphenyl)-
propionaldehyde; *a*-methyl-*p*-iso-
propylhydrocinnamaldehyde; cycla-
men aldehyde.

Methyl isovalerate.

Methyl laurate.

Methyl mercaptan; methanethiol.

Methyl *o*-methoxybenzoate.

Methyl *N*-methylanthranilate; dimethyl
anthranilate.

Methyl 2-methylbutyrate.

Methyl 2-methylthiopropionate.

Methyl 4-methylvalerate.

Methyl myristate.

Methyl *β*-naphthyl ketone; 2'-acetonaph-
thone.

Methyl nonanoate.

Methyl 2-nonenoate.

Methyl 2-nonynoate; methyloctyne
carbonate.

2-Methyloctanal; methyl hexyl acetaldehyde.

Methyl octanoate.

Methyl 2-octynoate; methyl heptine
carbonate.

4-Methyl-2,3-pentanedione; acetyl iso-
butyryl.

4-Methyl-2-pentanone; methyl isobutyl
ketone.

β-Methylphenethyl alcohol; hydratropyl alcohol.

Methyl phenylacetate.

3-Methyl-4-phenyl-3-butene-2-one.

2-Methyl-4-phenyl-2-butyl acetate; dimethylphenylethyl carbinyl acetate.

2-Methyl-4-phenyl-2-butyl isobutyrate; dimethylphenyl-ethylcarbinyl isobutyrate.

3-Methyl-2-phenylbutyraldehyde; a-isopropyl phenylacetaldehyde.

Methyl 4-phenylbutyrate.

4-Methyl-1-phenyl-2-pentanone; benzyl isobutyl ketone.

Methyl 3-phenylpropionate; methyl hydrocinnamate.

Methyl propionate.

3-Methyl-5-propyl-2-cyclohexen-1-one.

Methyl sulfide.

3-Methylthiopropionaldehyde; methional.

2-Methyl-3-tolylpropionaldehyde, mixed *o-, m-, p-*.

2-Methylundecanal; methyl nonyl acetaldehyde.

Methyl 9-undecenoate.

Methyl 2-undecynoate; methyl decyne carbonate.

Methyl valerate.

2-Methylvaleric acid.

Myrcene; 7-methyl-3-methylene-1,6-octadiene.

Myristaldehyde; tetradecanal.

d-Neomenthol; 2-isobutyl-5-methyl-cyclohexanol.

Nerol; *cis*-3,7-dimethyl-2,6-octadien-1-ol.

Nerolidol; 3,7,11-trimethyl-1,6,10-dodecatrien-3-ol.

Neryl acetate.

Neryl butyrate.

Neryl formate.

Neryl isobutyrate.

Neryl isovalerate.

Neryl propionate.

2,6-Nonadien-1-ol.

γ-Nonalactone; 4-hydroxynonanoic acid, γ-lactone; aldehyde C-18.

Nonanal; pelargonic aldehyde.

1,3-Nonanediol acetate, mixed esters.

Nonanoic acid; pelargonic acid.

2-Nonanone; methylheptyl ketone.

3-Nonanon-1-yl acetate; 1-hydroxy-3-nonanone acetate.

Nonyl acetate.

Nonyl alcohol; 1-nonanol.

Nonyl octanoate.

Nonyl isovalerate.

Nootkatone; 5,6-dimethyl-8-isopropenyl-bicyclo [4,4,0] -dec-1-en-3-one.

Ocimene; *trans*-β-ocimene; 3,7-dimethyl-1,3,6-octatriene.

γ-Octalactone; 4-hydroxyoctanoic acid, γ-lactone.

Octanal; caprylaldehyde.

Octanal dimethyl acetal.

Octanoic acid; caprylic acid.

1-Octanol; octyl alcohol.

2-Octanol.

3-Octanol.

2-Octanone; methyl hexyl ketone.

3-Octanone; ethyl amyl ketone.

3-Octanon-1-ol.

1-Octen-3-ol; amyl vinyl carbinol.

1-Octen-3-yl acetate.

Octyl acetate.

3-Octyl acetate.

Octyl butyrate.

Octyl formate.

Octyl heptanoate.

Octyl isobutyrate.

Octyl isovalerate.

Octyl octanoate.

Octyl phenylacetate.

Octyl propionate.

ω-Pentadecalactone; 15-hydroxypentadecanoic acid, ω-lactone; pentadecanolide; angelica lactone.

2,3-Pentanedione; acetyl propionyl.

2-Pentanone; methyl propyl ketone.

4-Pentenoic acid.

1-Penten-3-ol.

Perillaldehyde; 4-isopropenyl-1-cyclohexene-1-carboxaldehyde; *p*-mentha-1,8-dien-7-al.

Perillyl acetate; *p*-mentha-1,8-dien-7-yl acetate.

a-Phellandrene; *p*-mentha-1,5-diene.

Phenethyl acetate.

Phenethyl alcohol; β-phenylethyl alcohol.
Phenethyl anthranilate.
Phenethyl benzoate.
Phenethyl butyrate.
Phenethyl cinnamate.
Phenethyl formate.
Phenethyl isobutyrate.
Phenethyl isovalerate.
Phenethyl 2-methylbutyrate.
Phenethyl phenylacetate.
Phenethyl propionate.
Phenethyl salicylate.
Phenethyl senecioate; phenethyl 3,3-dimethylacrylate.
Phenethyl tiglate.
Phenoxyacetic acid.
2-Phenoxyethyl isobutyrate.
Phenylacetaldehyde; α-toluic aldehyde.
Phenylacetaldehyde 2,3-butylene glycol acetal.
Phenylacetaldehyde dimethyl acetal.
Phenylacetaldehyde glyceryl acetal.
Phenylacetic acid; α-toluic acid.
4-Phenyl-2-butanol; phenylethyl methyl carbinol.
4-Phenyl-3-buten-2-ol; methyl styryl carbinol.
4-Phenyl-3-buten-2-one.
4-Phenyl-2-butyl acetate; phenylethyl methyl carbinyl acetate.
1-Phenyl-3-methyl-3-pentanol; phenylethyl methyl ethyl carbinol.
1-Phenyl-1-propanol; phenylethyl carbinol.
3-Phenyl-1-propanol; hydrocinnamyl alcohol.
2-Phenylpropionaldehyde; hydratropaldehyde.
3-Phenylpropionaldehyde; hydrocinnamaldehyde.
2-Phenylpropionaldehyde dimethyl acetal; hydratropic aldehyde dimethyl acetal.
3-Phenylpropionic acid; hydrocinnamic acid.
3-Phenylpropyl acetate.
2-Phenylpropyl butyrate.
3-Phenylpropyl cinnamate.
3-Phenylpropyl formate.

3-Phenylpropyl hexanoate.
2-Phenylpropyl isobutyrate.
3-Phenylpropyl isobutyrate.
3-Phenylpropyl isovalerate.
3-Phenylpropyl propionate.
2-(3-Phenylpropyl)-tetrahydrofuran.
α-Pinene; 2-pinene.
β-Pinene; 2(10)-pinene.
Pine tar oil.
Pinocarveol; 2(10)-pinen-3-ol.
Piperidine.
Piperine.
d-Piperitone; p-menth-1-en-3-one.
Piperitenone; p-mentha-1,4(8)-dien-3-one.
Piperitenone oxide; 1,2-epoxy-p-menth-4(8)-en-3-one.
Piperonyl acetate; heliotropyl acetate.
Piperonyl isobutyrate.
Polylimonene.
Polysorbate 20; polyoxyethylene (20) sorbitan monolaurate.
Polysorbate 60; polyoxyethylene (20) sorbitan monostearate.
Polysorbate 80; polyoxyethylene (20) sorbitan monooleate.
Potassium acetate.
Propenylguaethol; 6-ethoxy-m-anol.
Propionaldehyde.
Propyl acetate.
Propyl alcohol; 1-propanol.
p-Propyl anisole; dihydroanethole.
Propyl benzoate.
Propyl butyrate.
Propyl cinnamate.
Propyl disulfide.
Propyl formate.
Propyl 2-furanacrylate.
Propyl heptanoate.
Propyl hexanoate.
Propyl p-hydroxybenzoate; propylparaben.
3-Propylidenephthalide.
Propyl isobutyrate.
Propyl isovalerate.
Propyl mercaptan.
α-Propylphenethyl alcohol.
Propyl phenylacetate.
Propyl propionate.
Pulegone; p-menth-4(8)-en-3-one.

Pyridine.
Pyroligneous acid extract.
Pyruvaldehyde.
Pyruvic acid.
Rhodinol; 3,7-dimethyl-7-octen-1-ol; citronellol.
Rhodinyl acetate.
Rhodinyl butyrate.
Rhodinyl formate.
Rhodinyl isobutyrate.
Rhodinyl isovalerate.
Rhodinyl phenylacetate.
Rhodinyl propionate.
Rum ether; ethyl oxyhydrate.
Salicylaldehyde.
Santalol, α and β.
Santalyl acetate.
Santalyl phenylacetate.
Skatole.
Sorbitan monostearate.
Styrene.
Sucrose octaacetate.
α-Terpinene.
γ-Terpinene.
α-Terpineol; *p*-menth-1-en-8-ol.
β-Terpineol.
Terpinolene; *p*-menth-1,4(8)-diene.
Terpinyl acetate.
Terpinyl anthranilate.
Terpinyl butyrate.
Terpinyl cinnamate.
Terpinyl formate.
Terpinyl isobutyrate.
Terpinyl isovalerate.
Terpinyl propionate.
Tetrahydrofurfuryl acetate.
Tetrahydrofurfuryl alcohol.
Tetrahydrofurfuryl butyrate.
Tetrahydrofurfuryl propionate.
Tetrahydro-pseudo-ionone; 6,10-dimethyl-9 undecen-2-one.

Tetrahydrolinalool; 3,7-dimethyloctan-3-ol.
Tetramethyl ethylcyclohexenone; mixture of 5-ethyl-2,3,4,5-tetramethyl-2-cyclohexen-1-one and 5-ethyl-3,4,5,6-tetramethyl-2-cyclohexen-1-one.
2-Thienyl mercaptan; 2-thienylthiol.
Thymol.
Tolualdehyde glyceryl acetal, mixed *o, m, p.*
Tolualdehydes, mixed *o, m, p.*
p-Tolylacetaldehyde.
o-Tolyl acetate; *o*-cresyl acetate.
p-Tolyl acetate; *p*-cresyl acetate.
4-(*p*-Tolyl)-2-butanone; *p*-methylbenzylacetone.
p-Tolyl isobutyrate.
p-Tolyl laurate.
p-Tolyl phenylacetate.
2-(*p*-Tolyl)-propionaldehyde; *p*-methylhydratropic aldehyde.
Tributyl acetylcitrate.
2-Tridecenal.
2,3-Undecadione; acetyl nonyryl.
γ-Undecalactone; 4-hydroxyundecanoic acid γ-lactone; peach aldehyde; aldehyde C−14.
Undecenal.
2-Undecanone; methyl nonyl ketone.
9-Undecanal; undecenoic aldehyde.
10-Undecenal.
Undecen-1-ol; undecylenic alcohol.
10-Undecen-1-yl acetate.
Undecyl alcohol.
Valeraldehyde; pentanal.
Valeric acid; pentanoic acid.
Vanillin acetate; acetyl vanillin.
Veratraldehyde.
Verbenol; 2-pinen-4-ol.
Zingerone; 4-(4-hydroxy-3-methoxyphenyl)-2-butanone.

VI

Some Commonly Used Chemical Preservatives

Ascorbic acid
Ascorbyl palmitate
Benzoic acid
Butylated hydroxyanisole
Butylated hydroxytoluene
Calcium ascorbate
Calcium propionate
Calcium sorbate
Caprylic acid
Dilauryl thiodipropionate
Erythorbic acid
Gum guaiac
Methylparaben (methyl-p-hydroxy-benzoate)
Potassium bisulfite
Potassium metabisulfite
Potassium sorbate
Propionic acid
Propyl gallate
Propylparaben (propyl-p-hydroxy-benzoate)
Sodium ascorbate
Sodium benzoate
Sodium bisulfite
Sodium metabisulfite
Sodium nitrate
Sodium nitrite
Sodium propionate
Sodium sorbate
Sodium sulfite
Sorbic acid
Stannous chloride
Sulfur dioxide
Thiodipropionic acid
Tocopherols

Limitations on the Use of BHA and BHT

BHA. The food additive BHA (butylated hydroxyanisole) alone or in combination with other antioxidants permitted in food for human consumption in subpart D of *Code of Federal Regulations* may be safely used in or on specified foods, as follows: (a) If the BHA meets the specification assay (total BHA), 98.5 percent minimum; melting point 48°C minimum. (b) If the BHA is used alone (as indicated by asterisk) or in combination with BHT, as an antioxidant in foods, as follows:

Food	Limitation (Total BHA and BHT) ppm
Dehydrated potato shreds	50
Active dry yeast*	1000
Beverages and desserts prepared from dry mixes*	2
Dry breakfast cereals	50
Dry diced glacé fruit*	32
Dry mixes for beverages and desserts*	90
Emulsion stabilizers for shortenings	200
Potato flakes	50
Potato granules	10
Sweet potato flakes	50

(c) To assure safe use of the additive: (1) The label of any market package of the additive shall bear, in addition to the other information required by the act, the name of the additive. (2) When the additive is marketed in a suitable carrier, in addition to meeting the requirement of (1), the label shall declare the percentage of the additive in the mixture. (3) The label or labeling of dry mixes for beverages and desserts shall bear adequate directions for use to provide that beverages and desserts prepared from the dry mixes contain no more than 2 ppm BHA.

BHT. The food additive BHT (butylated hydroxytoluene), alone or in combination with other antioxidants permitted in subpart D of Code of Federal Regulations, may be safely used in or on specified foods, as follows: (a) If the BHT meets the specification assay (total BHT) 99 percent minimum. (b) If the BHT is used alone or in combination with BHA, as an antioxidant in foods, as follows:

Food	Limitation (Total BHA and BHT) ppm
Dehydrated potato shreds	50
Dry breakfast cereals	50
Emulsion stabilizers for shortenings	200
Potato flakes	50
Potato granules	10
Sweet potato flakes	50

(c) To assure safe use of the additive: (1) The label of any market package of the additive shall bear, in addition to the other information required by the act, the name of the additive. (2) When the additive is marketed in a suitable carrier, in addition to meeting the requirement of (1), the label shall declare the percentage of the additive in the mixture.

Source: *Code of Federal Regulations*. Revised, April 1, 1974. pts. 10-129, pp. 414-415.

VII

Cyclamates: A Chronology
of Confusion

The timetable of events which led eventually to the banning of cyclamates has now been pieced together. It is an intriguing history replete with serendipity, opportunism, claim, counterclaim, confusion, and more confusion. It pits a billion-dollar business against even bigger business; it pits business against government, scientist against scientist, fact against fact. But if you have ever wondered how a chemical later suspected of being a carcinogen could ever have been placed on the GRAS list, that most elite of additive listings, the cyclamate story provides some interesting insights. In chronological order of events as they apparently occurred, here is that history.

1937. In a University of Illinois laboratory, graduate student Michael Sveda, seeking to find new compounds to reduce fever, tastes sweetness on his cigarette, checks the chemicals with which he is working, and discovers the forerunner to sucaryl sodium, a cyclamate product. (Curiously enough, the sweetener aspartame was discovered in 1965 in much the same way, also in a laboratory engaged in a search for new drugs, this time for the treatment of ulcers. James Schlatter, before reaching for a thin piece of laboratory paper, wet a finger to provide some friction. Since he was working with peptides (small units of amino acids) not generally recognized for their pleasant flavor, he was surprised at the sweet flavor he tasted. His curiosity aroused, Schlatter sought the chemical from among the compounds he was studying. Eventually, aspartame came to light.) Before this discovery the only other nonnutritive sweetener in use is saccharin, ten times as sweet as this newly discovered chemical, but yielding a strong, bitter aftertaste.

1950s. Early in the fifties Abbott Laboratories submits to the FDA data supporting a claim that cyclamate is safe for use in foods for diabetics and others who need to reduce their intake of sugar. This petition requests approval for sucaryl (cyclamate) as an ingredient in foods for special dietary, though not general, use.

With a touch of bureaucratic arrogance, Abbott's petition is dismissed, the experimental work labeled "an illustration of how an experiment should not be conducted." It should be stressed that if toxicity testing is to this day in its infancy, in the early fifties it was at best only the crudest of sciences.

The FDA decides to run its own tests, conducts a two-year animal feeding study, and finds cyclamate safe for its intended use. With appropriate data secured, permission for use is granted, the only restriction being a label notation that the product is intended for persons who must limit their intake of ordinary sweets.

Given opportunity to evaluate the scanty backlog of data, the National Research Council Food Protection Committee decides that based upon evidence then available cyclamates "need not be classified as unsafe." A casual warning appends the study, to the effect that intakes of five grams or more of cyclamate have been found to be mildly laxative. There is also what has become a traditional scientific assertion (expressing a valid need but serving equally as hedge against an error in judgment), i.e., that more work is needed.

Limited tests (using the crude methods of the time) on the product's potential to cause cancer prove negative.

Cyclamate begins a steady, though far from spectacular, growth in usage for "special dietary needs."

1958. Congress passes the Food Additives Amendment; GRAS category is established. At this time, there is no evidence to prove cyclamate a health risk; there is at worst a lack of information. When the GRAS list is circulated to the scientific community for evaluation, one scientist out of 190 respondents suggests that data are inadequate to include cyclamate in a listing of chemicals "generally recognized . . . to be safe under the conditions of . . . intended use. . . ." By default as much as anything else, cyclamate is given GRAS status. The door is opened for wide-scale exploitation of the compound in consumer food products generally, and at almost any use level. (Concurrently, and with implications which will shortly become apparent, the FDA moves to restrict the use of the compound cyclohexylamine, unaware of the relationship between this chemical, also an additive, and cyclamate.

The reason: cyclohexylamine is found to cause dermatitis and convulsions when applied to the skin or inhaled.)

1960. Seeds of a diet craze take root; use of cyclamate in diet beverages shows a marked rise; the sugar industry gears up for a major industrial battle with Abbott Laboratories.

1962. NAS/NRC Food and Nutrition Board revises earlier thinking and questions use of artificial sweeteners in weight-reducing diets; questions whether data are adequate to assure safety under such expanded usage.

1963-1967. Amid growing battle between Abbott Laboratories and the sugar industry, cyclamate consumption jumps threefold, from 6 to 18 million pounds annually. During this period Abbott Laboratories, defending the safety of cyclamates, issues what proves to be an erroneous statement that cyclamate cannot be considered harmful because it is not metabolized in the body; rather, it is simply excreted in its original chemistry.

1964. The sugar industry organizes the promotion agency Sugar Information, Inc.

1965. The FDA reviews scientific data on cyclamate, finds no basis for assuming the chemical to pose any risk to human health.

1966. Two Japanese investigators find cyclohexylamine excreted from humans and dogs ingesting cyclamate. Cyclohexylamine, a metabolite (a breakdown product) of cyclamate, has previously come under suspicion and been restricted in use. The Japanese work stimulates much new research, but scientists differ in interpretation of results.

1968. The FDA calls on the NAS/NRC to judge the data and to appraise the FDA about use of both cyclamate and saccharin; report due in late 1968. At about this time another research finding is made: cyclamate interferes with the absorption of the antibiotic lincomycin hydrochloride.

1968 (November). The NAS/NRC reaffirms its original position: cyclamates considered safe for use levels up to 5 grams/day, cautions against unrestricted use, indicates more work needed; the FDA staff summarizes existing information.

1968 (December). The FDA cautions the public to limit cyclamate usage to no more than 3.5 grams/165 pounds of body weight.

1969 (April). The FDA proposes label requirements for cyclamates, the label to caution against intake of over 3.5 grams for adults, 1.2 grams for children; also, cyclohexylamine content of food grade cyclamate restricted to 25 ppm.

Also in 1969, or thereabouts, an FDA research scientist, using a new, highly sensitive, untested research method finds deformities in chick embryos hatching from eggs injected with cyclamate. Other work shows cyclohexylamine causes breakage in test animal chromosomes.

April proposal is not finalized; more drastic action is contemplated.

1969 (June). In research on cyclamate using white Swiss mice as test animals, University of Wisconsin scientists find a significant incidence of bladder tumors; Abbott Laboratories is informed, in turn Abbott informs the FDA and the National Cancer Institute (NCI). Neither the FDA nor the NCI feels the test results are relevant to humans; however, attention is now focused on the bladder as a probable site for adverse reactions.

1969 (October). The FDA lab finds bladder lesions in rats fed cyclamate and saccharin in a ratio of 10:1 (this is the ratio used in many soft drinks); one-half of the bladder tumors are cancerous; many rats shown to convert cyclamates to cyclohexylamine.

The Industrial Test Laboratories of Illinois, in a two-year rat-feeding trial, report a finding of one bladder tumor in one rat.

Latest findings are evaluated by Abbott, the FDA, NCI, and HEW (FDA's administrative parent).

1969 (October 17) A.M. The NAS/NRC is informed of new information and collective evaluation by the bodies above.

P.M. The NAS/NRC recommends the removal of cyclamate from GRAS list.

1969 (October 18). HEW Secretary Robert Finch orders NAS/NRC recommendation into effect. (But what this means in reality is that cyclamates are still on the market, but restricted to usage by persons whose health needs demand it.)

1970 (June). Congressman L. H. Fountain of North Carolina, chairman of the House Intergovernmental Operations Subcommittee, convenes a hearing to investigate the cyclamate issue; FDA fails to convince the subcommittee of the product's safety; Congressman Fountain, in writing, requests the new HEW secretary Elliot Richardson to consider rescinding decision to allow continued cyclamate use.

1970 (August). FDA's medical advisory group considers the evidence and supports the request above.

1970 (September). The FDA orders ban on all cyclamate-containing products; food industry caught unaware, warehouse stocks of processed cyclamate-containing foods to be discarded; industry eventually to suffer the financial loss.

Abbott Laboratories begins a major review, hoping to compile evidence that will ultimately convince the FDA of the safety of cyclamate, get cyclamate officially reinstated as an additive, and enable the company to recoup some of its billion-dollar-a-year business.

1973 (November). Abbott Laboratories petitions the FDA for approval of cyclamate use for special dietary and technological purposes; in all, over 300 individual reports on toxicology are submitted as evidence; several post-1970, long-term animal feeding research reports are included.

1974 (February). Abbott petition made available for public scrutiny.

1974 (September). The FDA finds Abbott data inconclusive, requests the company to withdraw its petition; the cancer issue is deemed unresolved.

Here the cyclamate issue now rests. In the near future the FDA will again consider further evidence, but only evidence of the kind the agency feels will resolve the question of safety. Even then, one wonders, Is safety a resolvable issue? (Source: W. L. Pines. 1974. The cyclamate story. FDA Consumer 8(10):19. Adapted from the honors thesis of B. J. Boneparth, under the direction of J. H. Young, Emory University, Atlanta, Georgia.)

Glossary of Terms

Additive: A substance, either synthetic or derived from naturally occurring foods, permitted (by regulation) as a food ingredient.

Antinutritional factors: Components of food, often of natural origin, known to hinder nutritional responses.

Antioxidant: A preservative compound whose primary function is the prevention of off-flavors caused by oxidation of food fats and oils.

Calorie: Term used to express the energy value of a food. It is generally more useful to express energy value as kilocalories, defined as the amount of heat necessary to raise one kilogram (1000 grams) of water from 15 to 16°C. It is in fact the kilocalorie that is most commonly used to express food energy value. However, a trend is now underway to replace the calorie with another common unit of energy expression, the joule.

Carbohydrate: A class of chemically defined nutrients whose main function is to supply energy. Sugars and starches are carbohydrates in nature.

Cholesterol: A fatlike substance (sterol) important in the synthesis of certain hormones and an essential constituent of many cells.

Color, certified color: Synthetic dyes (synthetic color) certified by the FDA for safety and purity for use in food.

Color, lakes: Oil-soluble alumina or calcium salts of synthetic colors.

Color (natural): Food color derived from sources found in nature.

Common or usual name: Any food name, possibly coined, that accurately and directly identifies or describes the basic nature of a food or its characterizing properties or ingredients.

Emulsifier: Any substance whose main function is to prevent separation of oil and water ingredients in food.

Enriched: Containing added nutrient(s) found naturally associated with a given food.

Fat: A class of nutrients consisting of glycerol and three fatty acids i.e. triglycerides.

Fat, saturated: A fat consisting of fatty acids with no double bonds, no bonds capable of accepting hydrogen. Such fats are usually solid (hard) rather than liquid.

Fat, unsaturated: A fat containing fatty acids with one or more unsaturated (capable of accepting hydrogen) bonds, technically called double bonds. Unsaturated fats are soft fats.

Filled milk products: Dairy products in which part or all of the milk fat has been removed and replaced by other animal or vegetable fat or oil.

Flavor (artificial): Flavor not found to occur in nature.

Flavor enhancer (potentiator): A substance that brings out the flavor in food without itself contributing significantly (if at all) to that flavor.

Flavor (natural): Flavor derived from specified animal and plant products.

Formulated meal replacements: Foods designed to provide all the nutrients required of a single meal.

Fortified: Containing added nutrient(s) usually not found associated with the unfortified food.

Gram: A unit of weight in the metric system, one gram being equal to approximately 1/28 of an ounce.

GRAS: Acronym for Generally Recognized as Safe, a special category of food additives adapted from language of the Food Additives Amendment of 1958, which defines certain additives as "any ... substance ... generally recognized, among experts qualified by scientific training to evaluate its safety, as having been adequately shown ... to be safe under the conditions of its intended use."

Heath food: A misnomer commonly applied to natural/organic-type foods. However, any product whose main function in the diet is nutritional could properly be termed a "health" food.

Heavy metals: Highly toxic metallic elements (lead, cadmium, mercury, etc.); termed "heavy" because of their relatively dense nature.

Imitation foods: Foods processed to appear and taste similar to conventional foods. If such foods are of "equivalent" nutritive value, they need not be labeled "imitation."

International Unit (I.U.): A measure of the biological value (vitamin activity) of a food or vitamin preparation based upon comparison with an internationally recognized reference standard, usually a given quantity of purified (crystalline) vitamin. One I.U. of vitamin A equals 0.6 mg of beta carotene, 0.3 mg retinol; one I.U. of vitamin D equals .025 mg calciferol; one I.U. of vitamin E equals 1 mg of dl-alpha-tocopheryl acetate. Activity of other vitamins is usually expressed as weight (milligrams or micrograms) of the chemically pure compound.

Invert sugar: The product of acid or enzyme treatment of cane sugar (sucrose). The process, called inversion, gives rise to the term *invert.*

Joule (kilojoule): The international unit for expressing energy value of a food. One joule equals 10^7 ergs, or the energy expended when 1 kilogram is moved 1 meter by 1 newton. One kilocalorie equals exactly 4.184 kilojoules.

Kilogram: 1000 grams, about 2.2 pounds.

Microgram: One-millionth of a gram.

Milligram: One-thousandth of a gram.

Natural food: Unrefined, wild-growing (theoretically) foods, untreated with manmade chemicals of any kind. This definition cannot be assumed to apply to all (or even many) foods labeled as "natural."

Nonstandardized food: A processed food that is not regulated by a standard of identity.

Nutrient: A food component that provides a nutritional function or functions. Such functions include energy production, tissue building (growth) and repair, and specialized regulatory action of various life processes.

Nutrition Information Panel: That section on the food package (usually the right side panel) where nutrition information is located.

Nutritional Quality Guideline: A regulatory guideline for establishing the nutrient content of certain classes of food.

Organic food (or organically grown food): Food grown (theoretically) on soil treated only with organic fertilizer (manure, garbage, compost, etc.).

Preservative: Natural or synthetic compounds with the ability to retard or prevent microbial or chemical spoilage.

Principal Display Panel: The regulatory term used to define that area of the food package on which the food name appears.

Protein: Nutrients consisting of combinations of components called amino acids. Protein is the nutrient primarily responsible for growth and maintenance of body tissue.

Protein Efficiency Ratio (PER): A method of expressing protein quality and defined as weight gained divided by protein consumed for each rat after four weeks of feeding trials with a test protein and a casein-based diet.

Protein extenders and fillers: Protein isolated and concentrated from plant sources and processed to the texture and flavor of animal products (meat, fish, etc.) and used as an ingredient in animal products.

Protein, textured vegetable: Protein isolated from plant sources and processed to achieve the texture and flavor of animal products.

Recommended Daily Allowance (RDA): The term used to replace Minimum Daily Requirement (MDR) and intended as a goal or guide to nutrient intake, i.e., good nutrition.

Restoration: The addition to a food of nutrients lost during processing.

Restructured foods: Traditional foods structured from bits and fragments into common shapes and sizes of the original food, i.e., onion rings formed from onion fragments, shrimp-shaped product formed from shrimp fragments.

Safe and suitable: A regulatory term indicating a food ingredient that (1) performs an appropriate function in a food, (2) is used at a level no higher than that required to achieve its intended purpose, and (3) is not a food additive as defined by the federal Food, Drug and Cosmetic Act, unless used in conformity with appropriate regulations covering these additives.

Special dietary use food: A "medical" food designed to fulfill those special needs of persons with diet-related diseases.

Spice: Any substance in whole, broken, or ground form (except for products such as onion, garlic, or celery, which are traditionally regarded as food) whose main function in food is seasoning rather than nutritional.

Stabilizer: Any substance that binds and smooths food ingredients through a thickening action.

Standard of identity: A regulation specifying the kind and amount of ingredients and processing requirements of a given food—a processed food "recipe."

Statement of ingredients: A statement on the food package of ingredients in descending order by quantity.

Toxicity: The property of a substance to cause any kind of harmful effect on life.

USP (United States Pharmacopoeia): A measure of the vitamin activity of a food or vitamin preparation. This standard has been replaced by and is now designated as the International Unit. For all intents and purposes USP and I.U. are identical, the USP having originated from the scientific body

(U.S. Pharmacopical Convention) formerly charged with the responsibility for setting reference standards. Now that most vitamins can be isolated or synthesized in pure form, a unit of weight of the vitamin can serve much the same purpose.

Vitamins: Complex organic compounds needed by the body as governors and regulators of life-sustaining processes.

Vitamin supplement: A vitamin or combination of vitamins (and/or minerals) ordinarily consumed as assurance against inadequate nutrient intake due either to improper dietary habits or to special dietary needs.

Metric Equivalents

Unit	Metric Equivalent
ounce	28 grams
3½ ounces	100 grams
8 ounces (½ pound)	227 grams
1 pound	454 grams
2.2 pounds	1 kilogram
1 gallon	3.79 liters
1 quart	0.95 liters (950 milliliters)
1 pint	0.48 liters (480 milliliters)
1 cup (8 fluid ounces)	0.24 liters (240 milliliters)
1 tablespoon	15 milliliters
1 teaspoon	5 milliliters

Select Bibliography

Select Bibliography

Chapter I. Food Definitions and Standards

Code of Federal Regulations. Revised January 1, 1974. Title 21, pts. 1-119, pp. 119-417.
Federal Register. 1972. 37(137):12934-12936.
Federal Register. 1973. 38(195):27924-27929.
Federal Register. 1974. 39(89):15993-15995.
Federal Register. 1974. 39(95):17304-17305.
Federal Register. 1974. 39(116):20879-20882.
Federal Register. 1974. 39(116):20882-20884.
Federal Register. 1974. 39(217):39554-39556.
Federal Register. 1975. 40(121):26266-26267.

Chapter II. Food Names

Federal Register. 1973. 38(86):11091-11094.
Federal Register. 1973. 38(49):6950-6975.
Federal Register. 1973. 38(178):20702-20750.
Federal Register. 1974. 39(116):20878-20908.
Spaeth, R. S. 1974. Part IV: Innovative processed soy foods find markets in affluent and poor societies. Food Product Development, 8(7):92-93.

Chapter III. Food Additives

Federal Register. 1974. 39(185):34172-34219.
Hall, Richard L. 1973. Food additives. Nutrition Today, 8(4):20-28.
McNamara, Stephen H. 1975. Some legal aspects of providing foods for hungry populations. Food Product Development, 9(2):54-60.
Manufacturing Chemists' Association. 1963. *Agricultural Chemicals, What They Are, How They Are Used.* Washington, D.C.
NAS/NRC Report. 1973. *A Comprehensive Survey of Industry in the Use of Food Chemicals Generally Recognized as Safe (GRAS).* Washington, D.C.: U.S. Department of Commerce, National Technical Information Service.

National Academy of Sciences, Subcommittee on Food Technology of the Food Protection Committee. 1973. The use of chemicals in food production, processing, storage and distribution. Nutrition Reviews, 30(6):191-198.

Wodicka, Virgil O. 1971. FDA review of the GRAS list. Presentation made before the 75th Annual AFDOUS Conference, Columbus, Ohio, June 23, 1971.

Chapter IV. Food Flavor

Anonymous abstracts. 1970. Monosodium glutamate: Blood levels and absorption after oral and intravenous administration. Nutrition Reviews, 28:158-162.

Arthur, D. Little report to International Minerals and Chemical Corporation. March 15, 1953.

Arthur, D. Little report no. C58049 to International Minerals and Chemical Corporation. March 1953.

Code of Federal Regulations. Revised April 1, 1974. Pts. 120-129, pp. 450-459.

CRC handbook of food additives. 1972. Cleveland: CRC Press.

Eiserle, R. J., and W. J. Downey. 1971. A review of the literature concerned with flavor research as it applies to problems of the flavor industry. CRC Critical Reviews in Food Technology.

Federal Register. 1973. 38(13):2139-2141.

Federal Register. 1973. 38(148):20718-20723.

Fenaroli's Handbook of Flavor Ingredients. 1971. Cleveland: CRC Press.

Flavor thoughts. Modern Flavor Analysis. No. 5. Naarden, Holland.

Geremia, Ramon J. 1970. HEW News. News release of the Public Health Service, Food and Drug Administration. Washington, D.C.

Hara, S., T. Shibuya, R. Nakakawaji, M. Kyy, Y. Nakamura, H. Hoshikawa, T. Takeuchi, T. Iwao, and H. Ino. 1962. Observations of pharmacological actions and toxicity of sodium glutamate with comparisons between natural and synthetic products. J. Tokyo Medical College, 20:69.

Hazelton Laboratories Report. 1966. Project no. 466-103 to International Minerals and Chemical Corporation. November 3, 1966.

MacGregor, D. R., H. Sugisawa, and J. S. Mathews. 1974. Apple juice volatiles. J. Food Sci., 29:448.

National Academy of Sciences. 1973. Evaluating the safety of food chemicals. Appendix: Guidelines for estimating toxicologically insignificant levels of chemicals in food. Federal Register, 38(143):20051.

Oser, Bernard L., and Richard A. Ford. 1974. Recent progress in the consideration of flavoring ingredients under the food additives amendment. 8. GRAS substances. Food Technology, 28(8):76-80.

Radolph, T. G., and J. P. Rollins. 1950. Beet sensitivity: Allergic reactions from the ingestion of beet sugar (sucrose) and monosodium glutamate of beet origin. J. Lab. Clin. Med., 36:407.

Schultz, T. H., R. A. Flath, D. R. Black, D. G. Guadagni, W. G. Schultz, and R. Teranishi. 1967. Volatiles from Delicious apple essence—Extraction methods. Journal of Food Science, 32(3):279-283.

Sjostrom, L. B., S. E. Cairncross, and J. F. Caul. 1955. Effect of glutamate on the flavor and odor of foods. Monosodium glutamate (a second symposium), pp. 31-38.

Chapter V. Food Color

Allied Chemical, National Aniline Division. 1963. National certified food and drug and cosmetic colors. Technical Bulletin, March 15, 1963, pp. 1-18.

Anonymous. 1971. Natural foodstuff colours. The Flavour Industry, pp. 217-221.
Anonymous. 1973. Red food coloring–how safe is it? Consumer Reports, 38:130-133.
Code of Federal Regulations. 1975. Title 21, Food and Drugs, pts. 1-9.
FDA Advisory Committee on Protocols for Safety Evaluations. 1970. Toxicology and Applied Pharmacology, 16:264.
Food Chemical News. 1974. 16(12):36-38.
Food Chemical News. 1975. 16(52):27-28.
Food Chemical News Guide to the Current Status of Food Additives and Color Additives, The. Washington, D.C.: Food Chemical News, Inc.
Goldberg, L. 1971. Trace chemical contaminants in food: Potential for harm. Food and Cosmetic Toxicology, 9:65-80.
National Academy of Sciences. 1971. *Food Colors.* Washington, D.C.: Printing and Publishing Office, National Academy of Sciences.
Shnackenberg, Robert C. 1973. Caffein as a substitute for Schedule II stimulants in hyperkinetic children. Amer. J. of Psychiatry, 130:796.

Chapter VI. Food Preservatives

Cassens, R. G., J. G. Sebranek, G. Kubberod, and G. Woolford. 1974. Where does the nitrite go? Food Product Development, 8(10):50-56.
Epley, Richard J., P. B. Addis, C. E. Allen, and J. J. Wartheson. 1974. Nitrite in meat. Animal Science Fact Sheet no. 28. St. Paul: University of Minnesota Agricultural Extension Service.
Greenberg, Richard A. 1972. Nitrite in the control of *Clostridium botulinum.* Proc. Meat Ind. Res. Conf., pp. 25-33.
Ingredients Newsnotes. 1972. Food Technology, 28(5):25.
Johnson, F. C. 1971. A critical review of the safety of phenolic antioxidants in foods. CRC Critical Reviews in Food Technology, 2(3):267-304.
Marth, E. H. 1966. Antibiotics in foods–naturally occurring, developed, and added. Residue Reviews, 12:66-161.
Rhoades, J. W., and D. E. Johnson. 1972. N-dimethylnitrosamine in tobacco smoke condensate. Nature, 236:307-308.

Chapter VII. Emulsifiers and Stabilizers (Thickeners and Binders)

Code of Federal Regulations. 1974. Title 21, pts. 10-129, subpart D.
CRC Handbook of Food Additives. 1972. Cleveland: CRC Press.
Federal Register. 1973. 38(13):2137.
Institute of Food Technologists' Expert Panel on Food Safety and Nutrition. 1973. *Carrageenan, a scientific status summary by the Institute of Food Technologists' Expert Panel on Food Safety and Nutrition.* Chicago: Institute of Food Technologists.
Thrust. 1973. No. 6-M., February 9. Washington, D.C.: Milk Industry Foundation.

Chapter VIII. Toxic Metals in Food

Anonymous. 1971. Pica and lead poisoning. Nutrition Reviews, 29(12):267.
Anonymous. 1973. Mercury toxicity reduced by selenium. Nutrition Reviews, 31(1):25.
Bingham, F. T., and C. Nelson. 1972. Cadmium adsorption and growth of various plant species as influenced by solution concentration. Jour. Envir. Qual., 1:288.
Coon, J. M. 1969. Naturally occurring toxicants in foods. Food Technology, 23:55.
Dairy Council Digest. 1973. The role of essential trace minerals in nutrition. 44(4).
Federal Register. 1974. 39(236):42738-42740.
Federal Register. 1974. 39(236):42740-42743.

Ganther, H. E., et al. 1972. Selenium: relation to decrease of toxicity of methylmercury added to diets containing tuna. Science, 175:1122.

Ganther, H. E., et al. 1972. *Proceedings Sixth Annual Conference on Trace Substances in Environmental Health.* University of Missouri, Columbia. P. 45.

Institute of Food Technologists' Expert Panel on Food Safety and Nutrition. 1973. Mercury in food. Jour. Food Science, 38:1.

John, M. K. 1973. Cadmium uptake by eight food crops as influenced by various soil levels of cadmium. Envir. Pollution, 4:7-15.

Lagerwerff, J. V. 1971. Uptake of cadmium, lead, and zinc by radish from soil and air. Soil Science, 111:129.

Lunde, G. 1970. Analyses of arsenic and selenium in marine raw materials. Jour. of the Sci. of Food and Agri., 21(5):242-247.

Page, A. L., and F. T. Bingham. 1973. Cadmium residues in the environment. Residue Reviews, 48:1-44.

Saha, J. G. 1972. Significance of mercury in the environment. Residue Reviews, 43:103-163.

Chapter IX. Poisons and Antinutritional Factors in "Natural" Foods

Anonymous. 1969. Cottonseed meal and tumors. Nutrition Reviews, 27(10):292-294.

Arkcoll, D. B. 1969. Preservation of leaf protein preparations by air-drying. Jour. Sci. Food Agric., 20:600.

Coon, J. M. 1969. Naturally occurring toxicants in foods. Food Technology, 23:55-59.

Fassett, D. W. 1966. *Toxicants Occurring Naturally in Foods.* Publication 1354:257-266. Washington, D.C.: NAS/NRC.

Food and Nutrition Board, National Academy of Sciences, National Research Council. 1975. Hazards of overuse of vitamin D. Nutrition Reviews, 33(2):61-62.

Gerarde, H. W. 1972. Survey update: Determining safety for flavor chemicals. Presentation made before the Flavor and Extract Manufacturers' Association of the United States. Fall Symposium. November 16, 1972. Washington, D.C.

Hall, R. L. 1970. Toxic substances naturally present in food. Presentation made at Food Update Conference. March 23, 1970. Chicago, Illinois.

Joint Committee Statement of American Academy of Pediatrics Committees on Drugs and on Nutrition. 1974. Nutrition Reviews, Special Supplement. July 1974.

Liener, I. E. 1974. Other naturally occurring toxicants in foods. Presentation made on short course program, Topics in Food Chemistry, University of Minnesota, St. Paul. February 13, 1974.

Rackis, Joseph J. 1973. Biological and physiological factors in soybeans. Proceedings World Soybean Conference. Munich, Germany. November 11-14, 1973.

Sever, J. L. 1973. Potatoes and birth defects: Summary. Teratology, 8(3):319-320.

U.S. Department of Health, Education, and Welfare, Public Health Service. 1974. *The Toxic Substances List.* HEW publication no. (NIDSH) 74-134.

Wu, M. T., and D. K. Salunkhe. 1972. Research note. Control of chlorophyll and solanine syntheses and sprouting of potato tubers by hot paraffin wax. Journal of Food Sciences, 37:629.

Chapter X. Natural/Organic Foods

Anonymous. 1960. Chemicals in food. NAC News and Pesticide Review, October, pp. 3-10.

Anonymous. 1973. Food faddism. Dairy Council Digest, 44(1).

Darling, M. 1973. *Natural, Organic and Health Foods.* Extension Folder 280. University of Minnesota, St. Paul.

Erhard, D. 1974. The new vegetarians, part two. Nutrition Today, 9(1):20-27.

Institute of Food Technologists Expert Panel on Food Safety and Nutrition and the Committee on Public Information. 1974. Organic foods. Food Technology, 28(1).

Kilgore, Wendell W., and Ming-Tu Li. 1973. The carcinogenicity of pesticides. Residue Reviews, 48:41-161.

Lachance, Paul A. 1974. Nebulous health food terms: We need relevant legal definitions. Food Product Development, 8(5):48-50.

Chapter XI. Nutrients, Nutrient Sources, and Label Information

Anonymous. 1972. Functions and interrelations of vitamins. Dairy Council Digest. 43(5):25-30.

Anonymous. 1973. The role of essential trace elements in nutrition. Dairy Council Digest, 44(4):19-22.

Anonymous. 1975. The role of fiber in the diet. Dairy Council Digest, 46(1):1-4.

Federal Register. 1973. 38(13):2124-2164.

Federal Register. 1974. 39(145):27317-17320.

Feeley, R. M., P. E. Criner, and B. K. Watt. 1972. Cholesterol content of foods. J. Am. Diet Assoc. 61:134.

FDA. 1974. *Nutritional Labeling–Terms You Should Know.* Department of Health, Education, and Welfare publication no. (FDA) 74-2010.

Food and Nutrition Board, National Research Council. 1974. *Recommended Dietary Allowances.* 8th ed. Washington, D.C.: National Academy of Sciences.

Food and Nutrition Board, National Academy of Sciences, National Research Council. 1975. Hazards of overuse of vitamin D. Nutrition Reviews, 33(2):61-62.

Harper, A. E. 1974. The new recommended dietary allowances. Nutrition News, 37(2): 5.

Hegarty, P. V. J. 1975. Some biological considerations in the nutritional evaluation of foods. Food Tech., 9(4):52.

Hegsted, D. M. 1972. Problems in the use and interpretation of the recommended dietary allowances. Ecology of Food and Nutrition, 1:255-265.

Klevay, L. M. Quoted in Minneapolis Tribune, April 7, 1975.

Lachance, Paul A. 1973. Carbohydrates as nutrients. Food Product Development, 7(5): 29-37.

Miller, G. A., and P. A. Lachance. 1973. Protein: chemistry and nutrition. Food Product Development, 7(10):23-33.

National Dairy Council. 1974. Focus, 15(18).

National Dairy Council. 1974. Focus, 15(19).

National Dairy Council. 1974. Focus, 15(20).

Oster, Kurt A., J. B. Oster, and D. J. Ross. 1974. Immune responses to bovine xanthine oxidase in atherosclerotic patients. American Laboratory, 6(8):41-47.

Packard, V. S. 1974. *Nutritional Labeling of Food Products.* Special Report 49. St. Paul: Agricultural Extension Service.

Rosenfield, D. 1973. Utilizable protein: quality and quantity concepts in assessing foods. Food Product Development, 7(3):57-62.

U.S. Department of Agriculture. 1974. *Fats in Food and Diet.* Agricultural Information Bulletin No.361. Washington, D.C.: Agricultural Research Service.

Wilk, I. J. Quoted in Minneapolis Tribune, April 8, 1975.

Chapter XII. Food and Food Supplements, Their Labels and Use

Anonymous. 1963. The nonfat composition of milk. Dairy Council Digest, 34(3):1-3.

Cereal Institute, Inc. 1974. *The Role of Presweetened Breakfast Cereals in the American Diet.* Chicago: Cereal Institute, Inc.

Federal Register. 1973. 38(13):2143-2150.

Federal Register. 1973. 38(148):20730-20740.

Federal Register. 1975. 40(103):23244-23250.

Hackler, L. R. 1973. Evaluating protein quality in breakfast foods. New York's Food and Life Sciences, 6(2):407.

Harper, Alfred E. 1974. Statement submitted by Alfred E. Harper to the Subcommittee on Health of the Senate Committee on Labor and Public Welfare in connection with hearings on Senate Bill S-2801. August 15, 1974.

Heenan, Jane. 1974. Myths of vitamins. FDA Consumer, 8(2):4-9.

Packard, V. S. 1974. *Nutritional Labeling of Food Products.* Special Report 49. St. Paul: University of Minnesota Agricultural Extension Service.

Saperstein, S., G. A. Spilles, and R. J. Amen. 1974. Nutritional problems and the use of special dietary foods. Food Product Development, 8(3):58-64.

Chapter XIII. Nutritional Quality Guidelines

Federal Register. 1973. 38(49):6969-6973.

Federal Register. 1973. 38(148):20742-20744.

Federal Register. 1974. 39(116):20892-20908.

Index

Index

Acetone, 47

Acetosulfam, 192

Acidophilus milk, and disease prevention, 104-105

Additives, food: classification of, 50-52; consumption of, annual per capita, 55-57; contaminants, chance, compared to, 46; cyclamates, 48-49, 51; definition of, FDA, 43; functions of, 54-55, 57; generally recognized as safe (GRAS) ingredients, 50-51, 52-53, 55; incidental, 44, 45, 46; intentional, 45; and pesticides, 46; prohibited, 48-50; regulated, 50, 51-52; toxicity of, 47-53; uses of, appropriate and inappropriate, 58

Adulterated food, FDA definition of, 145. *See also* Antinutritional factors, naturally occurring

Agar, 122

Alcohol, 148

Alcohol, wood, 47

Alfalfa, 151, 174

Algin, 122

Allergens. *See* Allergy, food

Allergy, food, 14, 40, 41, 95-96, 149, 162, 233

Allyl isothiocyanate, 149

Amines, 149

Amino acid metabolism, diseases of, 232

Amygdalin, 156

Anencephaly spina bifida, 154

Annatto, 84

Antibiotics. *See* Preservatives

Antinutritional factors, naturally occurring: alcohol, 148; and alfalfa, 151; allergens, 149; allyl isothiocyanate, 149; amines, 149; antivitamin factors, 149-151; capsaicin, 151; carbohydrates, 151; carcinogens, 151-152; cholesterol, 152-153; definition of, 145-146; diseases caused by, table of, 165-168; enzyme inhibitors, 153-155; estrogens, 155; of Faba bean, 155; fatty acids, 155; in flavor components, 155; and food industry, 146-147; in foods, various, 147, 165-168; glycosides, 156-157; goitrogens, 157; gossypol, 158; hemagglutinins, 158-159; hydroxyphenylisatin, 159; lathyrogens, 159; and life processes, 148; lycopene, 159-160; menthol, 161; metal binding factors, 160-161; and mushrooms, 161; myristicin, 161; nitrates and nitrites, 161; in protein, 161-162; purines and pyrimidines, 162; salt, 162; shellfish poisons, 162; in soybean meal, 150; sterculic acid, 158; stimulants, 162-163; toxicity of, summarized, 164-168; vitamins, 163

Antioxidants: naturally occurring, 105-106; synthetic, 106-111

353

CPSIA information can be obtained
at www.ICGtesting.com
Printed in the USA
LVHW081112210919

63182ILV00028B/465/P

9 780816 658435